Die Grundlehren der mathematischen Wissenschaften

in Einzeldarstellungen
mit besonderer Berücksichtigung
der Anwendungsgebiete

Band 127

Herausgegeben von

J. L. Doob · E. Heinz · F. Hirzebruch · E. Hopf · H. Hopf
W. Maak · S. MacLane · W. Magnus
M. M. Postnikov · F. K. Schmidt · D. S. Scott · K. Stein

Geschäftsführende Herausgeber

B. Eckmann und B. L. van der Waerden

Hans Hermes

Enumerability · Decidability Computability

An Introduction to the Theory of Recursive Functions

Translated by

G. T. Hermann and O. Plassmann

Second revised Edition

Springer-Verlag New York Inc. 1969

Prof. Dr. H. Hermes
Mathematisches Institut
der Albert-Ludwigs-Universität, Freiburg i. Br.

Translation of
Aufzählbarkeit, Entscheidbarkeit, Berechenbarkeit, 1965
(Grundlehren der mathematischen Wissenschaften, Band 109)

 Library of Congress Catalog Card Number 76-89552. Printed in Germany
Title No. 5110

PREFACE TO THE ORIGINAL EDITION

The task of developing algorithms to solve problems has always been considered by mathematicians to be an especially interesting and important one. Normally an algorithm is applicable only to a narrowly limited group of problems. Such is for instance the Euclidean algorithm, which determines the greatest common divisor of two numbers, or the well-known procedure which is used to obtain the square root of a natural number in decimal notation. The more important these *special* algorithms are, all the more desirable it seems to have algorithms of a greater range of applicability at one's disposal. Throughout the centuries, attempts to provide algorithms applicable as widely as possible were rather unsuccessful. It was only in the second half of the last century that the first appreciable advance took place. Namely, an important group of the inferences of the logic of predicates was given in the form of a calculus. (Here the Boolean algebra played an essential pioneer role.) One could now perhaps have conjectured that *all* mathematical problems are solvable by algorithms. However, well-known, yet unsolved problems (problems like the word problem of group theory or Hilbert's tenth problem, which considers the question of solvability of Diophantine equations) were warnings to be careful. Nevertheless, the impulse had been given to search for the essence of algorithms. *Leibniz* already had inquired into this problem, but without success. The mathematicians of our century however, experienced in dealing with abstract problems and especially in operating with formal languages, were successful. About 1936 several suggestions to make precise the concept of algorithm and related concepts were made at almost the same time (Church's thesis). Although these suggestions (the number of which has been increased since) often originated from widely different initial considerations, they have been proved to be equivalent. The motivations for these precise replacements, the fact of their equivalence, and the experimental fact that all algorithms which have occurred in mathematics so far are cases of these precise concepts (at least if we concentrate on their essential nucleus) have convinced nearly all research workers of this field that these precise replacements are adequate interpretations of the at first intuitively given concept of algorithm.

Once we have accepted a precise replacement of the concept of algorithm, it becomes possible to attempt the problem whether there exist well-defined collections of problems which cannot be handled by algorithms, and if that is the case, to give concrete cases of this kind. Many such investigations were carried out during the last few decades. The undecidability of arithmetic and other mathematical theories was shown, further the unsolvability of the word problem of group theory. Many mathematicians consider these results and the theory on which they are based to be the most characteristic achievements of mathematics in the first half of the twentieth century.

If we grant the legitimacy of the suggested precise replacements of the concept of algorithm and related concepts, then we can say that the mathematicians have shown by strictly mathematical methods that there exist mathematical problems which cannot be dealt with by the methods of calculating mathematics. In view of the important role which mathematics plays today in our conception of the world this fact is of great philosophical interest. *Post* speaks of a natural law about the "limitations of the mathematicizing power of Homo Sapiens". Here we also find a starting point for the discussion of the question, what the actual creative activity of the mathematician consists in.

In this book we shall give an introduction to the theory of algorithms. First of all we shall try to convince the reader that the given precise replacements represent the intuitive concepts adequately. The best way to do this is to use one of these precise replacements, namely that of the Turing machine, as a starting point. We shall deal with the most important constructive concepts, like the concepts of computable function, of decidable property, and of set generated by a system of rules, using Turing machines as a basis. We shall discuss several other precise replacements of the concept of algorithm (e.g. μ-recursiveness, recursiveness) and prove their equivalence. As applications we shall, among others, prove the undecidability of the predicate calculus and the incompleteness of arithmetic; further, the most important preliminary step for the proof of the unsolvability of the word problem of group theory, i.e. the proof of the unsolvability for the word problem for Thue-systems, will be shown.

The theory will be developed from the point of view of the classical logic. This will be especially noticeable by the application of the classical existence operator, e.g. in the definition of computability. We call a function computable if *there exists* an algorithm to find the values for arbitrarily given arguments. — However, this will be made especially clear in all cases where proofs are carried out constructively.

In contrast to many publications on the subject we shall be careful to distinguish between the formulae of a symbolic language and the things denoted by them, in any case in the basic definitions.

At the end of several paragraphs there are a few, mainly easy, exercises, which the reader should attempt to solve.

It corresponds to the introductory character of this book that not all results of the subject will be discussed. However, the references given at the end of most paragraphs will inform the reader of the latest developments in this theory. Besides, we would like, once and for all, to refer the reader to a basic work, namely the "Introduction to Metamathematics" by *S. C. Kleene* (Amsterdam 1952), and also to the papers published in *The Journal of Symbolic Logic* (1936–). Another book in which the theory is based on Turing machines is *M. Davis*: "Computability and Unsolvability" (New York 1958).

The present book is based upon the lectures which the author has given on this subject regularly since 1949. In 1955 a manuscript of the lecture course was published by Verlag Aschendorff (Münster) under the title "Entscheidungsprobleme in Mathematik und Logik".

I would like to express my gratitude to Dr. H. Kiesow and Dr. W. Oberschelp for the valuable assistance in preparing the manuscript. In this respect I am also grateful to Miss T. Hessling, Miss E. Herting, Mr. D. Titgemeyer and Mr. K. Hornung.

Münster i. W., Spring 1960 H. HERMES

PREFACE TO THE ENGLISH TRANSLATION

In the translated edition the text follows the original very closely. A few alterations were made and some errors corrected.

The translation was taken care of by Messrs. Gabor T. Herman, B. Sc., and O. Plassmann. I would like to express my gratitude to the translators, especially to Mr. Herman, who critically examined the whole text and suggested corrections and improvements at several places.

Münster i. W., Spring 1964 H. HERMES

PREFACE TO THE SECOND EDITION

Some errors have been corrected. For valuable hints I am grateful to Professor H. B. Curry.

The section concerning the minimal logic of Fitch has been rewritten. It seems that the proof of the main theorem becomes more transparent by introducing primitive recursive functions whose arguments and values are expressions. For this idea I am indebted to Dr. F.-K. Mahn.

For recent literature I want to refer the reader to the Bibliography of the book *H. Rogers, Jr.:* "Theory of Recursive Functions and Effective Computability" (New York 1967).

Freiburg i. Br., Spring 1969 H. HERMES

CONTENTS

Chapter 1. *Introductory Reflections on Algorithms* 1

§ 1. The Concept of Algorithm 1
§ 2. The Fundamental Concepts of the Theory of Constructivity 9
§ 3. The Concept of Turing Machine as an Exact Mathematical Substitute for the Concept of Algorithm 17
§ 4. Historical Remarks 26

Chapter 2. *Turing Machines* 31

§ 5. Definition of Turing Machines 31
§ 6. Precise Definition of Constructive Concepts by means of Turing Machines 35
§ 7. Combination of Turing Machines 44
§ 8. Special Turing Machines. 48
§ 9. Examples of Turing-Computability and Turing-Decidability 55

Chapter 3. *μ-Recursive Functions* 59

§ 10. Primitive Recursive Functions 59
§ 11. Primitive Recursive Predicates 66
§ 12. The μ-Operator 74
§ 13. Example of a Computable Function which is not Primitive Recursive 82
§ 14. μ-Recursive Functions and Predicates. 88

Chapter 4. *The Equivalence of Turing-Computability and μ-Recursiveness* . . 93

§ 15. Survey. Standard Turing-Computability 94
§ 16. The Turing-Computability of μ-Recursive Functions 98
§ 17. Gödel Numbering of Turing Machines 103
§ 18. The μ-Recursiveness of Turing-Computable Functions. Kleene's Normal Form 108

Chapter 5. *Recursive Functions* 113

§ 19. Definition of Recursive Functions 113
§ 20. The Recursiveness of μ-Recursive Functions 118
§ 21. The μ-Recursiveness of Recursive Functions 130

Chapter 6. *Undecidable Predicates* 141

§ 22. Simple Undecidable Predicates 141
§ 23. The Unsolvability of the Word Problem for Semi-Thue Systems and Thue Systems 145
§ 24. The Predicate Calculus 155
§ 25. The Undecidability of the Predicate Calculus 163
§ 26. The Incompleteness of the Predicate Calculus of the Second Order . 171
§ 27. The Undecidability and Incompleteness of Arithmetic 175

Chapter 7. *Miscellaneous* 187

§ 28. Enumerable Predicates 187
§ 29. Arithmetical Predicates 192
§ 30. Universal Turing Machines 203
§ 31. **λ**-*K*-Definability 206
§ 32. The Minimal Logic of Fitch 219
§ 33. Further Precise Mathematical Replacements of the Concept of Algorithm 231
§ 34. Recursive Analysis 234

Author and Subject Index 241

LIST OF SYMBOLS

Symbols of the propositional calculus $\neg, \wedge, \vee, \rightarrow, \leftrightarrow$ 67

Symbols of the predicate calculus $\bigwedge, \bigvee, \bigwedge_{x=0}^{n}, \bigvee_{x=0}^{n}$ 67

Operator $\boldsymbol{\mu}$ 74; $\underset{x=0}{\overset{n}{\boldsymbol{\mu}}}$ 75

Empty word $\square$ 146

Turing machines r 39; l 39; a_j 40; $*$, I 42;
R, L 49; $\boldsymbol{\rho}, \boldsymbol{\lambda}$ 49; S 49; $\mathfrak{R}, \mathfrak{L}$ 49;
T 52; $\boldsymbol{\sigma}$ 52; C 52; K 53; K_n 54;
$\overset{n}{\rightarrow}, \overset{\pm n}{\rightarrow}, \rightarrow$ 45

Tape inscriptions m, $\sim$, $*$, $*\cdots*$, $*\cdots$, W, X 48

Special functions χ_M 14; $S(x)$, x', U_n^i, C_0^0 60; C_n^k 63;
$V(x)$, $x \dot{-} y$, $|x-y|$, $sg(x)$, $\overline{sg}(x)$ 64;
$\varepsilon(x, y)$, $\delta(x, y)$ 65;
$p(n)$, p_n, $\exp(n, x)$, $l(x)$ 77;
$\sigma_n(x_1, \ldots, x_n)$, $\sigma_{nj}(x)$ 78;
$U(u)$ 111

Special predicates O_n, A_n 69; E, D, Pr 72; T_n 110

CHAPTER 1

INTRODUCTORY REFLECTIONS ON ALGORITHMS

§ 1. The Concept of Algorithm

The concept of algorithm, i. e. of a "general procedure", is more or less known to all mathematicians. In this introductory paragraph we want to make this concept more precise. In doing this we want to stress what is to be considered essential.

1. Algorithms as general procedures. The specific way of mathematicians to draw up and to enlarge theories has various aspects. Here we want to single out and discuss more precisely an aspect characteristic of many developments. Whenever mathematicians are occupied with a group of problems it is at first mostly isolated facts that captivate their interests. Soon however they will proceed to finding a connection between these facts. They will try to systematize the research more and more with the aim of attaining a comprehensive view and an eventual complete mastery of the field in question. Frequently the method of attaining such mastery consists in separating special classes of questions such that each class can be dealt with by the help of an algorithm. An algorithm is a general procedure such that for any appropriate question the answer can be obtained by the use of a simple computation according to a specified method.

Examples of general procedures can be found in every mathematical discipline. We only need to think of the division procedure for the natural numbers given in the decimal notation, of the algorithm for the computation of approximating decimal expressions of the square root of a natural number, or of the method of the decomposition into partial fractions for the computation of integrals with rational functions as integrands.

In this book we shall understand by a *general procedure* a process the execution of which is clearly specified to the smallest details. Among other things this means that we must be able to express the instructions for the execution of the process in a *finitely long* text.[1]

[1] One cannot produce an infinitely long instruction. We can however imagine the construction of one which is potentially infinitely long. This can be obtained by

There is no room left for the practice of the creative imagination of the executer. He has to work slavishly according to the instructions given to him, which determine everything to the smallest detail.[1]

The requirements for a process to be a general procedure are very strict. It must be clear that the ways and means which a mathematician is used to of describing a general procedure are in general too vague to come up really to the required standard of exactness. This applies for instance to the usual description of methods for the solution of a linear equation system. Among other things it is left open in this description in which way the necessary additions and multiplications should be executed. It is however clear to every mathematician that in this case and in cases of the same sort the instruction can be supplemented to make a complete instruction which does not leave anything open. — The instructions according to which the not specially mathematically trained assistants work in a calculating pool come relatively near to the ideal we have fixed our eyes upon.

There is a case, which we feel is worth mentioning here, in which a mathematician is used to speaking of a general procedure by which he does not intend to characterize an unambiguous way of proceeding. We are thinking of calculi with several rules such that it is not determined in which sequence the rules should be applied. But these calculi are closely connected with the completely unambiguously described procedures. We shall deal with them in Section 6 of this paragraph. In this book we want to adopt the convention of *calling procedures general procedures only if the way of proceeding is completely unambiguous.*

There are *terminating algorithms,* whereas other algorithms can be continued as long as we like. The Euclidean algorithm for the determination of the greatest common divisor of two numbers terminates; after a finite number of steps in the computation we obtain the answer, and the procedure is at an end. The well-known algorithm of the computation of the square root of a natural number given in decimal notation does *not,* in general, terminate. We can continue with the algorithm as long as we like, and we obtain further and further decimal fractions as closer approximations to the root.

---

first giving a finite beginning of the instruction, and then giving a finitely long set of rules which determines exactly how in every case the already existing part of our instruction is to be extended. But then we can say that the finite beginning together with the finitely long set of rules is the actual (finite) instruction.

[1] Obviously the schematical execution of a given general procedure is (after a few tries) of no special interest to a mathematician. Thus we can state the remarkable fact that by the specifically mathematical achievement of developing a general method a creative mathematician, so to speak, mathematically depreciates the field he becomes master of by this very method.

2. *Realization of algorithms.* A general procedure, as it is meant here, means in any case primarily an operation (action) with concrete things. The separation of these things from each other must be sufficiently clear. They can be pebbles[1] (counters, small wood beads) as e.g. on the classical *abacus* or on the Japanese *soroban*, they can be symbols as in mathematical usage (e.g. 2, x, $+$, $($, $\int$), but they can also be the cog-wheels of a small calculating-machine, or electrical impulses as it is usual in big computers. The operation consists in bringing spatially and temporally ordered things into new configurations.

For the practice of applied mathematics it is absolutely essential which *material* is used to execute a procedure. However, we want to deal with the algorithms from the theoretical point of view. In this case the material is irrelevant. If a procedure is known to work with a certain material then this procedure can also be transferred (more or less successfully) to another material. Thus, the addition in the domain of natural numbers can be realized by the attachment of strokes to a line of strokes, by the adding or taking away of beads on an abacus, or by the turning of wheels in a calculating-machine.

Since we are only interested in such questions in the domain of general procedures which are independent of the material realization of these procedures, we can take as a basis of our considerations a realization which is mathematically especially easy to deal with. It is therefore preferred in the mathematical theory of algorithms to consider such algorithms which take effect in altering a *line of signs.* A line of signs is a finite linear sequence of *symbols* (*single signs, letters*). It will be taken for granted that for each algorithm there is a finite number of letters (at least one) the collection of which forms the *alphabet* which is the basis of the algorithm. The finite lines of signs, which can be composed from the alphabet, are called *words.* It is sometimes convenient to allow the *empty word* $\square$, which contains no letters. — If $\mathfrak{A}$ is an alphabet and W a word which is composed only of letters of $\mathfrak{A}$, we call W a *word over* $\mathfrak{A}$.

The letters of an alphabet $\mathfrak{A}$ which is the basis of an algorithm are in a certain sense nonessential. Namely, if we alter the letters of $\mathfrak{A}$ and so obtain a corresponding new alphabet $\mathfrak{A}'$, then we can, without difficulty, give an account of an algorithm for $\mathfrak{A}'$ which is "isomorphic" to the original algorithm, and which functions, fundamentally, in the same way.

---

[1] Algorithms (or in any case the procedures discussed in Section 4) are frequently called *calculi.* This name originates from the *calculi* (small pieces of limestone) which the Romans used for calculations. — It should be noted that the word "calculus" is sometimes used to signify a concept not identical with that of algorithm.

3. Gödel numbering.[1] We can, in principle, make do with an alphabet which contains only a single letter, e.g. the letter |.[2] The words of this alphabet are (apart from the empty word): |, ||, |||, etc. These words can in a trivial way be identified with the natural numbers 0, 1, 2, ... Such an extreme standardization of the "material" is advisable for some considerations. On the other hand it is often convenient to have at our disposal the diversity of an alphabet consisting of several elements, e.g. when we use auxiliary letters (cf. however § 15.2). Normally we want to take an alphabet with more than one element as the basis of our considerations.

The use of an alphabet consisting of *one element* only does not imply an essential limitation. We can, as a matter of fact, associate the words W over an alphabet $\mathfrak{A}$ consisting of N elements with natural numbers $G(W)$ (in such a way that each natural number is associated with at most one word), i.e. with words of an alphabet consisting of *one* element. Such a representation G is called a *Gödel numbering,* and $G(W)$ the *Gödel number* (with respect to G) of the word W. GÖDEL, in his article cited at the end of this paragraph, was the first to use such a representation. The following are the requirements for an arithmetization G.

1) If $W_1 \neq W_2$, then $G(W_1) \neq G(W_2)$ (one-one mapping).

2) There exists an algorithm such that for any given word W the corresponding natural number $G(W)$ can be computed in a finite number of steps by the help of this algorithm.

3) For any natural number n it can be decided in a finite number of steps, whether n is the Gödel number of a word W over $\mathfrak{A}$.

4) There exists an algorithm such that if n is the Gödel number of a word W over $\mathfrak{A}$, then this word W (which according to 1) must be unique) can be constructed in a finite number of steps by the help of this algorithm.

We want to give here a simple arithmetization for words over the alphabet $\mathfrak{A} = \{a_1, \ldots, a_N\}$. We perceive a_j as a "digit" which represents the natural number j. Then we can consider every non-empty word W as a representation of a number in a $(N + 1)$-al number-system. In addition we define $G(\square) = 0$. It is easily seen that the requirements 1) to 4) are satisfied. — Later on we shall study various other Gödel numberings (among others in §§ 17, 18, 21, 32).

---

[1] Also often called *arithmetization.*

[2] In the interest of the clarity of the printing the inverted commas will frequently be omitted. Thus, we shall write | instead of "|" (or actually "|" instead of ""|"").

4. Remarks on the empty word. For some reflections it is convenient to allow the empty word. There are, however, considerations which are carried out more simply if the empty word is excluded. Nothing is lost fundamentally by the exclusion of the empty word. As a matter of fact we can map, unambiguously and in a constructive manner, the words W over an alphabet $\mathfrak{A} = \{a_1, \ldots, a_N\}$ onto the non-empty words W' over $\mathfrak{A}$. One such mapping φ, for example, can be given as follows: $\varphi(\square) = a_1$; $\varphi(W) = a_1 W$, if the word W contains only symbols which coincide with a_1; $\varphi(W) = W$ for all other words.

In considering functions and predicates with regard to their constructive properties we want to assume generally that the occurring arguments and values are non-empty words.

The simplest procedure of representing the natural numbers by words consists in using an alphabet $\mathfrak{A} = \{|\}$ of one element only. Then the number n will be represented by the word consisting of n strokes. Thus, the number 0 will be represented by the empty word. However, if we want to use non-empty words only, then by the help of the above mentioned mapping φ we can represent the number n by $n + 1$ strokes. This *number representation* (which has already been mentioned in Section 2) *will be made use of later on* (cf. § 6.2).

5. Idealization of algorithms. At this stage we want to refer to an extrapolation which is in general done in the theories which will be dealt with in this book. Everyone knows that there exists a general procedure by the help of which for any two natural numbers n and m (given in decimal notation) the power n^m can be calculated (also in decimal notation). For *small* numbers (e.g. $n < 100$, $m < 10$) this calculation can actually be carried out. With larger numbers this becomes doubtful. With quite large numbers (somewhere in the region of the iterated power $1000^{1000^{1000}}$) it is possible that the existence of an *actual* calculation is contradictory to the laws of nature (e.g. because there is not enough material in the world for writing down the result in decimal notation, or because mankind does not exist long enough for the effective carrying out of such a calculation). It would certainly be interesting to investigate such limitations which are possibly due to the laws of nature, and especially to the size of human memory.[1] Such investigations are however very difficult, and hardly any work has been done in their direction. We want

[1] To this class of problems belongs for example the demonstration of the fact that it is convenient for man to choose a number not very far from 10 as basis b of the number-system employed. With respect to this we have to consider that b may not be essentially smaller than 10. Otherwise the numbers of everyday life would reach a length inconvenient for the human memory. On the other hand b may not be essentially greater than 10 since in this case the multiplication table would become too heavy a burden for the memory.

here, once and for all, to *abstract* from these limitations, and assume the ideal situation that there is no shortage of time, space and material for the carrying out of a general procedure. Especially, we allow arbitrarily long (finite) lines of signs.[1]

6. Deductions. In Section 1 we strongly emphasized the fact that an algorithm operates in a completely unambiguous manner. But we have already mentioned a case in which the mathematician speaks of a "general procedure" where no such unambiguity exists. We want to analyse this case and to investigate how it is related to the concept which we understand by algorithm. We start with

Example 1. The smallest set of real numbers which contains the numbers 9 and 12 and which is closed under substraction and multiplication by $\sqrt{2}$ can be obtained by the following algorithm.

(a) 9

(b) 12

(c) $x - y$

(d) $x\sqrt{2}$,

where (a) and (b) mean that 9 and 12 respectively can be written down as numbers of the set, (c) means that for any two numbers x and y which have already been obtained as numbers of the set the difference is also a member, (d) means that the multiple by $\sqrt{2}$ of an already obtained number x also belongs to the set.

It is evidently clear that a kind of "procedure" is described here by the help of which we can obtain the elements of the number module generated by 3 and $3\sqrt{2}$. Here we have described this procedure in a manner which is in accordance with mathematical practice. However, we have to realize that this description is full of all kinds of imperfections. First of all we should not speak of numbers, but rather of *number representations* (*numerals*) $\pm a \pm b\sqrt{2}$, where a and b are natural numbers given in their decimal notation. Then, for example, (a) would be more exactly $+ 9 + 0\sqrt{2}$. In the case of (c) and (d) it is obviously meant that the known rules of calculation are to be applied to obtain the new number in the standard form. However, we shall refrain from giving complete formulations of these rules.

---

[1] The limitations which we have mentioned with regard to human beings have analogues in the case of computing machines, e.g. limited storing capacity. If we want to get a picture of the idealization undertaken here we must assume that the storing equipment can be enlarged arbitrarily according to demand without changing the basic construction of the machine. It is especially easy to imagine this with regard to the ordinary table calculating machine.

One can form derivations by the help of the rules (a), ..., (d). For instance the following derivation of seven steps.

(i)	12	(by (b)).
(ii)	9	(by (a)).
(iii)	3	(by (c), using lines (i) and (ii)).
(iv)	$3\sqrt{2}$	(by (d), using line (iii)).
(v)	$9\sqrt{2}$	(by (d), using line (ii)).
(vi)	$6\sqrt{2}$	(by (c), using lines (v) and (iv)).
(vii)	$3-6\sqrt{2}$	(by (c), using lines (iii) and (vi)).

Thus a *deduction* is a finite sequence of words which is produced according to the given rules. The number of words is called the *length* of the deduction. Our example above is of length 7.

Naturally a finite set of rules of this kind (usually called a *rule system*) does not in general provide us with an algorithm in the strict sense of Section 1, since it is not determined in which sequence the separate rules should be applied. From any given rule system we can normally produce arbitrarily many different deductions. What is now the connection between a "procedure" given by a rule system and an algorithm proper? We can at first confirm (in Example 1 and in any other example) that every single rule of a rule system describes an actual algorithm, in fact a terminating algorithm.[1] Furthermore, for any *actual* deduction we can easily give an account of a general procedure in the sense of Section 1 according to which this deduction can be performed. This algorithm consists of the rules of the rule system together with a complementary instruction which states the sequence in which the rules are to be applied. (One and the same rule may be applied more than once.) In this sense, a rule system can be conceived as an, in general, unlimited supply of algorithms.

We should keep in mind that we cannot always extend a previously given deduction by the help of a previously given rule by one more step, and that we cannot begin a deduction with an arbitrary rule. Thus, in

---

[1] "Rules" occasionally occur which cannot be conceived directly as terminating algorithms. Such a rule could be something like the following. *Write down a multiple of an (already given) number n*. Obviously there is no algorithm here in the strict sense of the word, for it is not said with which number k we should multiply, and so the procedure is ambiguous. We can however replace this "rule" by rules which are (terminating) algorithms, and whereby we have indeed to make appropriate use of a superimposed rule system (cf. the following parts of the main text). That is to say, we have first of all to produce a number k by the help of an additional rule system and then to form the product nk of n and k according to a further additional rule.

Example 1 we cannot make use of either rule (c) or rule (d) in the first step.

Example 1 has to do with an especially simple case of rule system. In general the situation is more complicated since several rule systems $R_{11}, \ldots, R_{1m_1}$; $R_{21}, \ldots, R_{2m_2}$; $R_{k1}, \ldots, R_{km_k}$ are superimposed on each other (cf. Example 2). It is generally supposed in the application of a rule of a later system that a word or several words have already been derived before in earlier systems. A word is considered to be *deductable* in the joint system if it is obtained by the use of a rule of the *last* system $R_{k1}, \ldots, R_{km_k}$.

Instead of giving a rather complicated general definition we confine ourselves to an example of the *superimposed rule systems*. In the rule system of the example some (not all) tautologies (i. e. identically true formulae) of the propositional calculus can be deduced.

Example 2. The joint system consists of the collection of three systems. The *first system* is used to produce the *propositional variables.* These are special words over the alphabet {O, I}.

R_{11}: Write down O.

R_{12}: Attach the letter I.

Propositional variables are e.g. O, OI, OII.

The *second system* provides the *formulae* (*expressions*) of the propositional calculus. These are special words over the alphabet {O, I, →, (,)}.

R_{21}: Write down a word which was obtained by the use of the first system.

R_{22}: If words W_1, W_2 have already been obtained, write down $(W_1 \to W_2)$.

Formulae are e. g. OII, ((OII → OI) → OII).

The *third* (*last*) *system* helps us to obtain *tautologies*. Tautologies are special formulae.

R_{31}: If words W_1, W_2 have already been obtained in the second system, write down $(W_1 \to (W_2 \to W_1))$.

R_{32}: If words W_1, W_2 have already been obtained in the second system, write down $((W_1 \to (W_1 \to W_2)) \to (W_1 \to W_2))$.

R_{33}: If words W_1, W_2, W_3 have already been obtained in the second system, write down $((W_1 \to W_2) \to ((W_2 \to W_3) \to (W_1 \to W_3)))$.

R_{34}: If words $(W_1 \to W_2)$ and W_1 have already been obtained in the third system, write down the word W_2 (*modus ponens*).

References

For the concept of calculus, compare

LORENZEN, P.: Einführung in die operative Logik und Mathematik. Berlin-Göttingen-Heidelberg: Springer 1955.

CURRY, H. B.: Calculuses and Formal Systems. Logica, Studia Paul Bernays dedicata, pp. 45—69. Neuchâtel: Éditions du Griffon 1959. See also: Dialectica **12**, 249—273 (1958).

Gödel numbering was used for the first time in

GÖDEL, K.: Über formal unentscheidbare Sätze der Principia Mathematica und verwandter Systeme I. Mh. Math. Phys. **38**, 173—198 (1931).

For the realization of algorithms, see

MENNINGER, K.: Zahlwort und Ziffer I, II. Göttingen: Vandenhoeck & Ruprecht 1957, 1958.

§ 2. The Fundamental Concepts of the Theory of Constructivity

From the concept of algorithm we can derive a whole line of further important concepts. Some of these are computability, enumerability, decidability and generability. In this paragraph we shall define these concepts, and establish simple relations between them.

1. Computable functions. In mathematics, functions occur quite frequently which are such that there exists a terminating algorithm which provides, for any given argument, the value of the function. Such functions are called *computable functions.* Computable are for instance the functions $x + y$, xy, x^y, where any natural numbers x and y can be taken as values of the arguments. When saying (here and in what follows) that the arguments are natural numbers, we have to realize that it would be more correct in this context to say that the arguments (if we for instance prefer the decimal notation) are words over the alphabet $\{0, 1, 2, \ldots, 9\}$. Only those functions belong to the domain of our considerations whose arguments and values are words, or in any case can be characterized unambiguously by words over an alphabet. To say for instance that the sumfunction is computable means that there exists a general procedure by the help of which we can for any two numbers given in the decimal notation obtain their sum in decimal notation in a purely schematic manner.

The concept of computability has meaning not only in the case when the arguments and values of a function are *natural* numbers, but also when they are *integers* or *rational* numbers. We can for instance represent integers and rational numbers by finite words (e.g. $-34\div49$) Thus e.g. the sum function $x + y$ is also computable if we allow rational arguments.

On the other hand the situation becomes quite different in the case of *real* numbers. From the point of view of classical mathematics there exist non-denumerably many real numbers. Thus, we cannot represent every real number by a word of a previously given finite alphabet any more, since there are only denumerably many such words (every word is a finite line of signs over a finite supply of letters). We shall deal with this situation in § 33.

From now on we shall assume that for a *computable function* f the domain of arguments of f consists of all words W over a finite alphabet $\mathfrak{A}$, whereas the values of f are words over a finite alphabet $\mathfrak{B}$.[1] (Functions of this type are especially those whose domain of arguments is the domain of natural numbers, but whose values need not necessarily be natural numbers.) That such a function is computable means that there exists a general procedure which can be described in finitely many sentences, and by the help of which we can obtain effectively the value of the function $f(W)$ for every possible argument W.

The generalization for functions of several arguments can be carried out without difficulty.

According to elementary theorems of the classical theory of sets (and accepting the classical concept of function originating from Dirichlet) there exist non-denumerably many functions which are defined for all natural numbers, and the values of which are natural numbers. On the other hand there are certainly only denumerably many of these functions which are computable. This is so, since for every computable function there exists a computing instruction of finite length (cf. § 3), and there can only exist denumerably many such instructions. It is so to speak an exception if a function is computable.

Later on we shall give examples of actual functions which are not computable (cf. § 22).

Let us end this section with a fundamental remark. There is an existential quantifier in the definition of the concept of computable function: a function f is computable, if *there exists* an instruction, such that .... This existential quantifier will in this book always be understood in the sense of *classical logic*. It appears that we would come up against difficulties if we tried to interprete the quantifier as an existential quantifier of a *constructive logic*.[2] In spite of the classical interpretation of the existential quantifier the computability of actual functions will, of course, in general be proved constructively by giving a computing procedure explicitly.

---

[1] If we do not assume that *every* word over $\mathfrak{A}$ is an argument of f we arrive at the concept of the *partially computable function*. With this, however, we shall not concern ourselves in this book.

[2] Cf. references on Church's thesis at the end of § 3.

The remark on the classical interpretation of the existential quantifier applies mutatis mutandis also to the concepts of enumerability, decidability and generability.

2. *Enumerable sets and relations.* Let f be a computable function whose domain of arguments is the domain of natural numbers. Then, for the domain M of the values of f a natural sequence $f(0), f(1), f(2), \ldots$ is determined which will run through M, possibly with repetitions. We can say that the function f *enumerates* the elements of M in the sequence $f(0), f(1), f(2), \ldots$. In general we shall call a set M of words over an alphabet an *enumerable set* if there exists a computable function whose domain of values coincides with M. Furthermore we shall also call the *empty set of words* enumerable.

We must, of course, differentiate strictly between the concept of enumerable set and the concept of *denumerable* set. Every enumerable set is naturally at most denumerable (i.e. finite or denumerable) in the sense of the theory of sets. However, the impossibility of all denumerable sets being enumerable can easily be made plausible. There exist namely at most denumerably many enumerable sets over a fixed finite alphabet, since (apart from the empty set) for any enumerable set there exists a computable function which enumerates it, and since, as we have seen in Section 1, there are only denumerably many computable functions. On the other hand there exist non-denumerably many sets of words over every alphabet which contains at least one letter. — For actual examples of non-enumerable sets cf. § 28.

Just as we can speak of enumerable sets we can also speak of *enumerable relations (predicates)*. Let us consider a binary relation R (i.e. a relation of two arguments) between words over an alphabet $\mathfrak{A}$. That R is enumerable means that there exist two computable singulary functions (i. e. functions of one variable) f and g such that the sequence of ordered pairs $(f(0), g(0))$, $(f(1), g(1))$, $(f(2), g(2)), \ldots$ runs through (perhaps with repetitions) the set of all ordered pairs which stand in the relation R. Furthermore we shall call the *empty relation* enumerable. — Accordingly, we can introduce for each n the concept of enumerable relation of n arguments.

3. *Decidable sets and relations.* Now we shall deal with an expression which is often used in connection with terminating algorithms. We start with some examples.

(1) It is *decidable* whether or not a natural number is a prime.

(2) It is *decidable* whether a linear equation system is solvable.

(3) Is is *not decidable* whether or not a formula of the predicate calculus is valid.

If we try to find a common structure in these examples, we see that in each instance two sets[1] M_1 and M_2 are put in a relation, namely in (1) the set M_1 of prime numbers with the set M_2 of natural numbers, in (2) the set M_1 of solvable linear equation systems with the set M_2 of all linear equation systems, in (3) the set M_1 of the identical formulae of the predicate calculus with the set M_2 of all formulae of the predicate calculus. M_1 is always a subset of M_2.

In all cases M_1 and M_2 are sets of *words* (or can in any case be understood as such). We shall confine ourselves to illustrating this more in detail using Example (2) only. It is sufficient to realize that every linear equation system can be written down as a word. In showing this we shall confine ourselves to linear equations whose coefficients are given as integers in decimal notation.[2] Then we can represent such a linear equation system by a word over the alphabet $\mathfrak{A} = \{0, 1, 2, 3, 4, 5, 6, 7, 8, 9, +, -, =, x, ;\}$, e.g. the system

$$\begin{aligned} 3x_1 - 4x_2 &= 5 \\ 2x_1 + x_2 &= -6 \end{aligned}$$

by the word

$$+3x1 - 4x2 = +5; +2x1 + 1x2 = -6.$$

Let M_1 and M_2 be sets of words over a fixed alphabet $\mathfrak{A}$, and let $M_1 \subset M_2$. In this case we shall say that *M_1 is decidable relative to M_2* if there exists a terminating algorithm[3] by the help of which we can determine effectively for every word of M_2 whether or not it belongs to M_1. Such an algorithm is called a *decision procedure.*

Besides the concept of *relative* decidability considered just now we can also introduce a concept of *absolute* decidability. A set M_1 of words over an alphabet $\mathfrak{A}$ is called plainly *decidable,* if M_1 is decidable relative to the set M_2 of all words over $\mathfrak{A}$.[4]

If, in example (1), we give the natural numbers in decimal notation by the help of the symbols 0, 1, ..., 9, then each word formed from these symbols designates a natural number. We can therefore say in the terminology introduced just now that the set of prime numbers is decidable.

---

[1] or properties (we can identify a property with the set of things which have this property).

[2] In the linear algebra assertion (2) is often stated without giving the details how the coefficients are given. It is however easily explained by an example that it is necessary to postulate the way the coefficients are given. Let $a = 0$ or 1, according to whether or not Fermat's conjecture concerning his "last problem" is true. Then we do not know today whether or not the equation $ax = 1$ is solvable.

[3] This existential quantifier is also classical; cf. the end of Section 1.

[4] We should keep in mind that with both the relative and the absolute decidability we refer to the underlying alphabet $\mathfrak{A}$.

An n-term relation can be thought of as a *set* of n-tuples. Therefore, the concept of decidability, just like that of enumerability, can be carried over to relations.

Every *finite* set (and accordingly every *finite* relation) is decidable. A decision procedure consists in writing down all the elements of the set in a list and in checking for any given word whether or not it appears in the list. Thus, especially, every set is decidable which consists of one single element only.[1]

Now we shall discuss a few simple relations between absolute and relative decidability and enumerability. The considerations which will lead to a verification (just like a few more which we shall deal with in the following sections) cannot, by the nature of this matter, be exact proofs. Together however they contribute to the clarification of these concepts given intuitively. Later on, when we replace the intuitive concepts discussed here by exact ones we shall be in the position to give a strict proof of the analogous relations.

(a) Let M_0, M_1, M_2 be sets of words over an alphabet $\mathfrak{A}$. *If M_1 is decidable relative to M_2, and M_0 is contained in M_1, then M_0 is decidable relative to M_1 if and only if M_0 is decidable relative to M_2 (transitivity of the relative decidability).* If namely M_0 is decidable relative to M_2, then a fortiori it is decidable relative to M_1. If conversely M_0 is decidable relative to M_1, then it is also decidable relative to M_2. Let W be an arbitrary word of M_2. First we determine whether or not W belongs to M_1. If W does not belong to M_1 then W definitely does not belong

---

[1] This is the place to point out that we must positively keep in mind the fact that in the everyday language of mathematicians we may come across phrases in which the word "decidable" occurs and is used in a different way to that of examples (1), (2) and (3) above. Let us consider as example a possible assertion α of the following kind. "Fermat's last problem is decidable." If we assumed that by decidability absolute decidability is meant here, the assertion α would trivially be true, since every set of one element only (here: the set which contains α as its only element) is decidable. However, somebody who states the assertion α will not, in general, agree with this interpretation. It is possible that he rather means the following. "Either there exists a counter example of Fermat's assertion (which will be found when sought long enough) or there exists a mathematical proof of it."

If we now assume that all methods of mathematical proofs allowed today and in future define together a (perhaps very complicated) algorithm Γ (this assumption is indeed more than questionable), then the following could be said, from our point of view, about the assertion α understood in this way:

(1) It could be that the motive for the assertion α is the HILBERTian "in der Mathematik gibt es kein Ignorabimus". This motivation must be rejected according to our present day knowledge, for it contradicts the undecidability of arithmetic established by GÖDEL.

(2) The answer of the question whether Fermat's assertion is provable (in case it is not refutable by a counter example) depends on Γ, and Γ is not known.

to M_0. But if W belongs to M_1 then we can decide according to the hypothesis whether or not W is an element of M_0.

If M_2 is the set of all words over $\mathfrak{A}$, then, according to definition, decidability relative to M_2 can be replaced by absolute decidability. We have according to this as an application of (a):

(b) Let M_0 and M_1 be sets of words over an alphabet $\mathfrak{A}$. *If M_1 is decidable, and M_0 is contained in M_1, then M_0 is decidable relative to M_1 if and only if M_0 is decidable.*

In many important examples it is clear which alphabet $\mathfrak{A}$ provides the basis, and whether M_2 is absolutely decidable (cf. e.g. (1) and (2) further above). In these cases the relation of decidability of M_1 relative to M_2 can be replaced by the property of absolute decidability of M_1.

(c) Let M be a set of words over an alphabet $\mathfrak{A}$. We assign to this set a function $\chi_M(W)$ which is defined for all word W over $\mathfrak{A}$ and for which

$$\chi_M(W) = \begin{cases} 0, \text{ if } W \in M \\ 1, \text{ if } W \notin M \end{cases}$$

is valid.

χ_M is called *the characteristic function* of M. We have that *M is decidable if and only if the characteristic function χ_M is computable.* Namely, if M is decidable, then we must determine for any given W whether $W \in M$ or $W \notin M$. According to which we must write down a 0 or 1 resp. Thus, we obtain the value of the function $\chi_M(W)$. This is a terminating algorithm for the computation of χ_M. — Conversely, if χ_M is computable we can decide by computation of $\chi_M(W)$ whether $W \in M$ or $W \notin M$.

4. Generable sets and relations. In § 1 Section 6 we introduced the concept of rule system and that of a deduction produced according to this rule system. We say that a word W is *deducible* by the help of the rules of a given rule system if it is possible to obtain a deduction by the help of the rule of the system the last word of which coincides with W.[1] A set M of words over an alphabet $\mathfrak{A}$ is called *generable* if there exists a rule system, such that a word W is deducible by the help of the rules of the system if and only if it belongs to M.

Just as in the cases of enumerability and decidability we can speak of *generable relations.*

The examples (1) and (2) of § 1.6 show the generability of the module generated by 3 and $3\sqrt{2}$, and that of a certain set of tautologies.

---

[1] In case of a *superimposed* rule system (cf. § 1.6) the last word of the deduction must be formed by the help of a rule of the last system.

Now we shall deal with a few relations between generability and the concepts introduced earlier on. First of all we have

(d) *A set M of words over an alphabet $\mathfrak{A}$ is generable*[1] *if and only if M is enumerable.* First let M be enumerable. Then, either M is empty (and so generated by the help of a rule which is not applicable), or M is the domain of values of a computable function f. Let R be an algorithm by the help of which we can calculate for any n the value of the function $f(n)$. Now we consider a superimposed rule system.

First system. This is used to generate the natural numbers n.

Second (last) system. This consists of *one* rule, which says: For any word n which can be obtained by the help of the first system produce $f(n)$ by means of the algorithm R.

Obviously, exactly those words are generated in the superimposed rule system which belong to M.

Conversely, let M be a generable set. Let M be generated by an (in general superimposed) rule system R with finitely many rules. We have to show that M is enumerable. For any number k there can be finitely many deductions of length k (perhaps even none).[2] Now, if we first examine all deductions of length 1, then we see that we can produce these and place them in a standard system (e.g. lexicographically or with reference to the given rules). Let $W_{01}, \ldots, W_{l_1 1}$ be the last words of these proofs in the standard sequence ($W_{01}, \ldots, W_{l_1 1}$ can partly coincide with each other). Then we define $f(0) = W_{01}, f(1) = W_{11}, \ldots, f(l_1) = W_{l_1 1}$. Now we examine all deductions of length 2. These can also be produced effectively. Let $W_{12}, \ldots, W_{l_2 2}$ be the last words of these deductions ordered by the same method as above. Then we define $f(l_1 + 1) = W_{12}, \ldots, f(l_1 + l_2) = W_{l_2 2}$. We carry on in this way. If for a k *no* deduction of length k exists we go at once to $k + 1$. Now there are two possibilities: (1) There exists no deduction of length k for any k whatsoever. Then M is empty and so enumerable by definition. (2) There is a number k for which there exists a deduction of length k. Then deductions exist also of length $2k$, $3k$, etc., which can be obtained in

---

[1] In the literature "*generable sets*" are often referred to as "*generated sets*".

[2] This applies even to the generalized deductions for which we do not demand that the last word should be formed by the help of a rule of the *last* system (cf. § 1.6). For these generalized deductions the assertion follows easily by induction on the length. For this we have to take notice of the facts that a generalized deduction of length $k + 1$ becomes a generalized deduction of length k if we take away the last word, and that from a given generalized deduction of length k we can form only finitely many generalized deductions of length $k + 1$. There are namely only finitely many rules at our disposal, and every rule is only applicable to finitely many possible combinations of lines of the generalized deduction of length k. In every single case the result is unambiguously determined since we made the condition that every rule must be an algorithm in the strict sense.

a trivial way by repetitions of the deduction of length k. This will make certain that the above defined function $f(n)$ is defined for all n. Because of the definition of f it is clear that $f(n)$ can be computed effectively for all n (the method given above describes the general procedure). Thus, f is a computable function. The domain of values of f obviously coincides with M.

(e) Let E be the set of *all* words over an alphabet $\mathfrak{A} = \{a_1, \ldots, a_N\}$. E is generated by the help of the following rules.

R_0: Write down the empty word.

R_1: Attach the letter a_1.

$\vdots$ $\quad\quad\vdots$

R_N: Attach the letter a_N.

(f) Let M be a set of words over $\mathfrak{A}$, and let $\overline{M}$ be the complement of M relative to E. Then we have that *M is decidable if and only if both M and $\overline{M}$ are generable.* Instead of *generable* we could (by (d)) also say *enumerable.* To demonstrate this assertion first we assume that M is decidable. Now, we consider the following superimposed rule system.

First system. The system of (e) for the generation of all words of E.

Second (last) system. This contains only the following rule. Check for a word W generated by the help of the first system whether or not it belongs to M. If $W \in M$, write down W. If $W \notin M$, then consider the rule inapplicable.

It is clear that with the help of this system we can generate exactly the set M. If we change the last system in the obvious way, we see that $\overline{M}$ is also generable.

Now let us assume that both M and $\overline{M}$ are generable. If M or $\overline{M}$ is empty, then M is decidable in a trivial way. Thus we shall take for granted that neither M nor $\overline{M}$ is empty. Then there exist according to (d) computable functions f and g, such that the domain of values of f coincides with M, and that of g with $\overline{M}$. To decide whether or not an arbitrarily given word W of E belongs to M we can now use the following algorithm. Compute $f(0)$, $g(0)$, $f(1)$, $g(1)$, $f(2)$, $g(2)$, ... one by one and check after each computation of an $f(n)$ or a $g(n)$ whether or not the value of the function obtained coincides with W. If it does not coincide with W, continue with the computation. If it does coincide with W, then W is equal to either an $f(n)$ or a $g(n)$. In the first case it is an element of M, in the second it is an element of $\overline{M}$, and thus it is not an element of M. At this stage the decision has been reached, and the procedure terminates. We should note that the procedure terminates in *every* case, for every word is either an element of M or $\overline{M}$ and thus lies in the domain of values of f or g.

5. The invariance of the constructive concepts under Gödel numbering. Let G be a Gödel numbering of the words over an alphabet $\mathfrak{A}$ of N elements. Let M be a set of words over $\mathfrak{A}$. We assign to M the set $\tilde{M}$ of the Gödel numbers of the elements of M. Then we have the

Theorem. M is enumerable (resp. decidable) if and only if $\tilde{M}$ is enumerable (resp. decidable).

Proof.

(a) Let M be enumerable. Say it is enumerated by the function f. Then, obviously, the function f_1, which is defined by $f_1(n) = G(f(n))$, is an enumeration of $\tilde{M}$. — Conversely, if $\tilde{M}$ is enumerable and if it is enumerated by the function h, then M is enumerated by the function h_1 which is defined by $h_1(n) = G^{-1}(h(n))$.

(b) Let $\tilde{M}$ be decidable. Then M is also decidable. In order to establish whether a word W belongs to M we must compute the number $G(W)$ and establish whether $G(W)$ belongs to $\tilde{M}$. If M is decidable, then $\tilde{M}$ is also decidable. In order to determine whether a number n belongs to $\tilde{M}$ we must determine first whether n is the Gödel number of a word over $\mathfrak{A}$ at all (cf. § 1.3). If this is not the case, then certainly $n \notin \tilde{M}$. If on the other hand n is the Gödel number of a word W over $\mathfrak{A}$, then we must construct W and determine whether W belongs to M.

Note. This theorem mutatis mutandis is also valid for relations instead of sets.

§ 3. The Concept of Turing Machine as an Exact Mathematical Substitute for the Concept of Algorithm

To be able to give proofs of undecidability or of other impossibilities in the domain of the constructive, it is not sufficient to have only a perception (however clear it may be) of the concept of algorithm. We must rather take an exact definition of what we call a general procedure as the basis of such considerations. In this paragraph we shall deal with considerations which lead to the replacement of the concept of algorithm by the exact mathematical concept of the so-called Turing machines.

The first to suggest the identification of the originally intuitively given concept of algorithm with a certain, exactly defined concept was A. Church in 1936 (cf. § 4.3). The so-called *Church's thesis* is admitted today by most mathematical logicians. A short survey of arguments for it is given in the preface. *Criticisms* of this view were given by L. Kalmár (possibly not every in the intuitive sense computable function is computable in Church's sense), and by R. PÉTER (possibly not every in Church's sense computable function is computable in the intuitive sense). The cited references should be consulted on this subject.

1. Preliminary remarks. Besides the concept of Turing machine discussed here there are other suggestions on the replacement of the concept of general procedure by an exact mathematical concept. We shall discuss a few of these suggestions later on. We choose the Turing machines as our starting point, since this way seems to be the most natural and easiest approach. We start with a preliminary remark on our method. If we want to demonstrate that the exact concept of the Turing machine is an adequate version of the intuitive concept of algorithm, then by the nature of the matter we can only give arguments for the plausibility of our assumption. Such a plausibility consideration cannot be more than an appeal to the more or less great experience with algorithms which the mathematician has gained in his life.

It will immediately be clear, once they are defined, that every Turing machine describes a general procedure. The only question is, whether every general procedure can be carried out by a suitable Turing machine. If we have any argument by which we intend to show that in principle *every* algorithm can be reproduced by a suitable Turing machine (or by any other suggested precise replacement), then we shall probably always find examples of algorithms which do not come directly under our consideration. Then we must extend our argumentation accordingly, and so forth. When we shall have worked through sufficiently many considerations and additional considerations of this kind we shall finally arrive at the conviction (and that is the experience of the last decades) that it is not worth-while undertaking experiments with new notions again and again. The assertion that *all* algorithms can be comprehended by Turing machines will finally be considered verified in the same way as the assertion in physics on the impossibility of a perpetuum mobile. This assertion is also such that people have refused for some time now to make any further experiments on it.

The fact is very remarkable that *it can be proved purely mathematically that the suggested replacements of the concept of algorithm by various exact mathematical concepts (originating from very different starting points) are all equivalent.*[1] This is at least an indication to the fact that we are dealing with a fundamental concept.

2. Algorithms and machines. We demonstrated in § 1 that a method which is to be a general procedure in our sense has to be prescribed to the smallest details so that no creative imagination is needed to carry it

[1] Cf. § 15, § 19, § 30 and § 31.

out. But if everything is thus determined in detail, then it is obviously possible to leave the carrying out of the method to a machine, and what is more, to a fully automatic machine.

Machines can have very complicated structures. The aim of the following considerations is to give an account of a relatively simple type of machine which is mathematically easy to deal with. These are the Turing machines of which we can expect that every algorithm can be executed by a suitable machine of this type. We shall give the reasons for the fact that *all* algorithms can (after suitable adjustments) be carried out by such machines, following TURING's method, i.e. by analyzing the behaviour of a calculator (i.e. human computer) working according to a prescribed method.

3. The computing material. To fix our ideas we shall start from the assumption that the task of a calculator is to compute the value of a function for a given argument according to a given instruction which contains all details. The calculator uses for his calculation a sheet of paper (or, according to demand, more sheets of paper). We assume that such a sheet is divided into squares. The calculator is not supposed to write down more than one symbol in one square. He may use all symbols of a finite alphabet $\mathfrak{A} = \{a_1, \ldots, a_N\}$.[1] The argument is written down on this sheet in these symbols at the beginning of the calculation.

For some algorithms it is without doubt convenient to have a two-dimensional calculating surface at our disposal. We can e.g. think of the usual division algorithm for natural numbers. But hardly anybody would doubt that in principle it is not necessary to use a two-dimensional calculating surface; we can always get by with a one-dimensional *computing tape* (in short *tape*). This is divided into a linear sequence of *squares*. In the course of the calculation sufficiently many squares must be available. They are available since we take the computing tape to be continued infinitely in both directions.[2] Thus the tape looks like this.

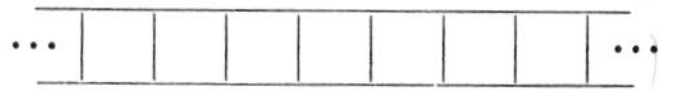

[1] In the mathematical practice of today, where different alphabets (in the everyday sense of the word) and other signs are used, the number of symbols actually used is in the region of 200.

[2] We could qualify the phrase which makes use of the actual infinite by assuming a computing tape of finite length and demanding only that the tape is extendable on either side *according to demand*. We must however realize that even such an assumption already represents a mathematical *idealization* of the situation existing in reality, in any case then, if we concede that the world contains only a finite number of material particles.

In principle it would be sufficient to require the extension of the computing tape into the infinite on one side only (cf. § 15). This would however bring complications caused by the existence of a square (the first square) which has only one other square as neighbour.

We imagine that the tape is provided with a *direction*, and we speak in this sense of *left* and *right* (beginning and end resp.). The squares of the tape are *empty* except for finitely many squares which are *marked*, i.e. have a letter printed in them. It is often convenient to include with the *actual symbols* $a_1, \ldots, a_N$ the *empty symbol* as an *ideal symbol*. We represent the empty symbol by a_0 or $*$. Thus a square is empty if and only if the symbol $*$ is printed in it.

We call the inscription on the tape the *tape expression*.

4. The computing steps. Now we shall turn our attention to the calculation itself. The calculation runs according to a finite instruction. We shall try to bring such instructions into standardized form. First of all it is clear that every calculation can be thought of as a process which can be divided up into *single steps*. Such a step could for instance consist in printing a letter into a certain empty square. The printing into a square would in reality be a continuous process. The details of this process are however of no interest for our present considerations. The only important parts of the process are the beginning (the empty square) and the end (the marked square), and therefore we are justified to speak of a discontinuous *step*.[1] We shall attempt to divide up the whole computing procedure into as elementary computing steps as possible.

A computing step leads from a certain *starting situation* or *starting configuration* to a new *situation* or *configuration*, which itself is again the starting configuration for the next computing step. But there is also the possibility that the calculator stops operating *(halts)* when a certain configuration is reached. Then the computation of the value of the function is at an end.

Let us consider first what can happen in a single computing step.

First there is the possibility that the tape expression is changed. Since we want to consider the simplest computing steps possible we shall assume that in such a step the symbol is changed in a single square only. Every change of the tape expression will consist in a change in finitely many squares, and can so be divided up into single elementary steps of this kind. A single step can consist of printing a symbol into a previously empty square, or of removing a symbol from a square (this procedure can be called *erasure*). By erasure and subsequent printing we can change the symbol of a square into an arbitrary symbol. But we shall, since it is very convenient, explicitly allow the immediate change of an arbitrary symbol a_j in a square into a new symbol a_k $(j, k = 0, \ldots, N)$ in *one* single step. We shall characterize this procedure by saying

[1] We should keep in mind that every process can be thought of as continuous or discontinuous according to the way we consider it. The *digital computers* for instance are regarded as machines working by steps (discontinuously) although the procedure itself runs continuously. On the other hand, the *integraphs* are thought of as continuously operating machines.

that the letter a_k is written down (*printed*) into the square in question. This printing is in general a printing *over* by which the previously existing letter disappears. It remains only, if $j = k$.

It is necessary in the process considered that at each instance the calculator pays special attention to one certain square of the computing tape into which (perhaps) he is going to print. We call this square the *scanned square*. In general it is necessary to change the scanned square in the course of the computation. The passage from one scanned square to an other which lies at some distance from it need not be accomplished in *one* step. We can, more conveniently, break it up into simple steps which consist in changing the scanned square for a new one which lies immediately on the right or on the left of the old one. Such a simple step which only consists in changing the scanned square we shall characterize in short by saying that *we go* (on the tape one square) *to the left* (resp. *to the right*).

Now we have already enumerated the different kinds of computing steps which we have to consider. To summarize, the possible computing steps are

a_k: The printing of a symbol a_k ($k = 0, \ldots, N$) into the scanned square.

r: The moving to the right.

l: The moving to the left.

h: The halting.

We shall refer to these steps in short by the respective symbols introduced on the left.

5. *The influence of the computing tape on a computing step.* The answer to the question *which* of the possible computing steps is carried out at a certain stage of the calculation depends on the configuration at the time. A configuration is given by three components, namely (1) the computing instruction, (2) what is already printed on the computing tape at the moment in question, and (3) the square scanned at the time.

We shall here look more closely at (2). We shall ask ourselves to what extent the tape expression[1] which is found on the tape at a certain moment influences the next step. The weakest dependance requirement which we can state here is that the symbol in one single square is decisive for the next step, while the symbols in the other squares play no part as far as the next step is concerned. We could say that the calculator has to observe the square in question and that the result of the observation,

[1] We could think the computing tape with its tape expression as an extension of the human memory. But we ought to bear in mind that a man who is calculating on a sheet of paper uses his own memory as well as that of the paper. If we want to reproduce a human calculation by a Turing machine we must transfer onto the tape the actual human memory relevant to the calculation.

i.e. the symbol in this square, is essential for the next step. Therefore, we shall (temporarily) call this critical square the *observed square*.

Now we shall give a plausibility argument to show that this weakest dependance requirement is sufficient. In other words, it is not necessary to admit that a computing step is influenced by a greater part of the tape expression or by the whole tape expression. In every such case we can as a matter of fact succeed, by changing the computing instruction, in getting along with computing steps each one of which depends on the observation of a *single* square only. Let us consider a characteristic example of this. If we want to add two numbers, each one of which is given by *one* digit in decimal notation, something like $3 + 4$ or $3 + 2$ or $6 + 2$, then first of all we can formulate the respective computing instructions in the following manner.[1]

If a 4 is observed in the last square and a 3 in the last square but two, print 7 as the result.

If a 2 is observed in the last square and a 3 in the last square but two, print 5 as the result.

If a 2 is observed in the last square and a 6 in the last square but two, print 8 as the result, etc.

Instead of this instruction let us now take the following.

I. Observe the last square. According to whether 0, 1, ... or 9 is seen, continue according to $\mathrm{II}_0, \mathrm{II}_1, \ldots, \mathrm{II}_9$ resp.[2]

. . . .

II_2. Observe the last square but two.

. . . .

If a 3 is seen, print the result 5.

. . . .

If a 6 is seen, print the result 8.

. . . .

. . . .

We see that this instruction leads to the same result as the instruction considered first and that every single step depends on the observation of *one* square only. It can be shown by the help of the same method that we can forgo the consideration of computing steps which depend on the simultaneous observation of k squares in an unbroken succession, where k is a fixed natural number.

We could have an instruction in which the computing step is determined by the inscriptions in k squares which are not immediate neigh-

[1] We confine ourselves in this case to the essential and forgo a complete version of the instruction.

[2] In future we shall express this by saying "jump $\mathrm{II}_0, \mathrm{II}_1, \ldots, \mathrm{II}_9$ resp." (i.e. we shall write "jump" instead of "continue according to").

bours. Let us, for the sake of simplicity, take $k = 2$. If these two squares are very far from each other, it is impossible for the calculator to observe these squares directly in one single act. The only way he can proceed is to observe one of the squares and then to search for the other. In case of larger distances the search must be done according to an instruction which can only be carried out by steps.[1] In this way we see that computing steps which depend on the observation of several squares which can lie arbitrarily far from each other need not appear as elementary computing steps.

A similar consideration shows that we need not consider any computing steps which depend on the observation of the *whole* computing tape.

Thus, we only need to allow such computing steps which depend on the symbol in a single square, the *observed square*. Now we have for every computing step *two* squares which are essential (for this step), namely the scanned square (Section 4) and the observed square. A consideration analogous to the one carried out just now to show that is it possible to do with *one* observed square, shows that it must also be possible to *identify* the observed square with the scanned square. We shall do this from now on. Thus we shall suppose that *at any instant the only relevant part of the tape expression as far as the next computing step is concerned is the symbol in the scanned square.*

From this simplification follows another which we have already made tacitly in Section 4. If the observed square could be different from the scanned square, then we must increase the number of possible computing steps by two, which make it possible for us to change the observed square in the course of the computation.

6. *The computing instruction.* The number of computing steps necessary for the determination of a value $f(n)$ of a computable function f will in general increase as n gets larger. But the instruction for the computation of f is of finite length. Therefore, in the computation of $f(n)$, for sufficiently large n, one and the same part of the instruction will have to be applied *several times* according to necessity. Thus, the instruction will be broken up into certain *instruction parts,* which if necessary will have to be used repeatedly in the course of the calculation (how often they will have to be used depends on the given argument n).

A simple example may illustrate the concept of instruction part. It deals with the calculation of an approximating value for $\sqrt{2}$. We take 1.4 as initial approximation x_0. We want to find an approximation x_n with

[1] Let us make this clear by an example. It could perhaps be that the second effective square is the first square on the left of the square observed first which is immediately left of a square containing the symbol |. This second effective square can then be found in principle only by starting from the square observed first and by going left step by step until a square of the given kind is come upon. This whole exercise can be solved by the coupling of several simple computing steps.

an absolute error less than $2 \cdot 10^{-10}$. This can be done by the help of a computing instruction which consists of the following instruction parts, starting with the

0th instruction part. Write down 1.4 as an approximation. Afterwards jump 1st instruction part.

1st instruction part. Starting from an approximation x compute a new approximation $y = x + (2 - x^2)/2x$ in decimal notation to 12 places. Afterwards jump 2nd instruction part.

2nd instruction part. The computation is at an end if $|y - x| < 10^{-10}$. Otherwise jump 1st instruction part, starting from the new approximation y.

We must not confuse such a system of instruction parts with a generating process (§ 1.6). There it was left open in which sequence the rules have to be applied, while here the sequence is prescribed unambiguously.

The instruction parts of this example contain in general many single computing steps. Now we shall imagine that an instruction is broken up into instruction parts which are so small that every one of them, together with the tape expression and the square scanned at the time, determines *one computing step only*. In this case this instruction part need only tell which computing step is to be carried out and according to which instruction part we have to proceed afterwards, all in consideration of the contents of the scanned square.

We imagine the instruction parts numbered successively, where the 0th instruction part is the one that determines the beginning of the computation. Let the symbol b mark the behaviour (thus b stands for l, r, h, or $a_0, \ldots, a_N$ (cf. Section 4)). Then we have

the kth instruction part

If the scanned square is marked by the symbol	a_0,	carry out	b_0	and then jump instruction part	k_0.
. . . .	a_1,	. . . .	b_1	. . . .	k_1.
. . . .	. . . .	. . . .	. . . .	. . . .	. . . .
. . . .	a_N,	. . . .	b_N	. . . .	k_N.

Instead of the kth instruction part we shall also speak in short of the *k^{th} state.*

All lines have the same schema. Thus, we can confine ourselves to reproducing the instruction part by a table, namely by

$$
\begin{array}{l}
k \; a_0 \; b_0 \; k_0 \\
k \; a_1 \; b_1 \; k_1 \\
\cdots\cdots\cdots \\
k \; a_N \; b_N \; k_N
\end{array}
$$

The *complete computing instruction* is obtained in standardized form by placing the tables for the single instruction parts underneath each other. The 0th instruction, which provides the beginning of the computation, is placed appropriately at the top. An instruction standardized in this manner is called a *Turing table* or *machine table*. It is evident by the previous considerations that every computing instruction which has to do with a calculation on a computing tape can be reproduced in the form of a Turing table.

For us a *Turing machine* is nothing else than a Turing table. As far as the principal considerations (which will be carried out in this book) are concerned, it is as a matter of fact of no importance in which way the calculation according to a Turing table is *actually* carried out (cf. § 1.2). However, it is recommendable in our contemplations to imagine a Turing machine as an actually working apparatus.[1]

The symbols which appear in the second column of a Turing table are called *the symbols of this machine*.

Naturally, the way in which a Turing machine operates depends on the tape expression at the beginning of the calculation and on the square taken as the scanned square at the start.

At the beginning of the next chapter we shall again go through the definition of the Turing machine without the heuristic considerations

---

[1] With the technical possibilities of today every Turing machine can be realized in different ways as an actual machine, through which a computing tape (e.g. in form of a punched tape or a magnetic tape) is running. In this realization a certain stage of the calculation will correspond (if we think of a mechanical machine) to a certain arrangement of the parts of the machine. This explains the term *internal configuration* used by Turing to denote the states. Turing identified the internal configurations with possible *states of mind* of the calculator. Many authors followed him in this respect. But it seems that this identification is problematic. For if we assume that each "state of mind" of a calculator has a characteristic physical constituent (by which it can be distinguished from the others), then the conclusion that for every single person there is an upper *bound* M for the number of his states of mind seems to be evident. (By the way, Turing himself undertook similar considerations, e. g. concerning the number of the directly comprehensible squares of the computing tape.) But then a Turing machine which has more than M states would not (or in any case not directly) fall under these considerations.

The fact that a calculator can work according to arbitrarily long instructions (programs) does not support the view that there exist arbitrarily many possible "states of mind". For a calculator need not learn the program by heart. Usually the instruction exists outside the calculator, e.g. on a special computing tape, the program tape, which is finite, but may be arbitrarily long. The calculator must be able to "read" such a program tape, i.e. to carry out the instructions contained in it. If only *one* computing tape is available, then the program can be written down on this tape, and the calculation itself can also be carried out on the free part. A person's ability to carry out in the above sense *arbitrary* programs shows that he corresponds in this respect to a "universal Turing machine" (cf. § 30).

we have made in this chapter. Examples of Turing machines will also be given there.

References

POST, E. L.: Finite Combinatory Processes — Formulation 1. J. symbolic Logic **1**, 103—105 (1936).

TURING, A. M.: On Computable Numbers, with an Application to the Entscheidungsproblem. Proc. London math. Soc. (2) **42**, 230—265 (1937).

— On Computable Numbers, with an Application to the Entscheidungsproblem. A Correction. Proc. London. math. Soc. (2) **43**, 544—546 (1937).

WANG, H.: A Variant to Turing's Theory of Computing Machines. J. Assoc. Computing Mach. **4**, 63—92 (1957).

For criticisms on *Church's* thesis, compare

KALMÁR, L.: An Argument against the Plausibility of Church's Thesis. Constructivity in Mathematics, ed. by A. Heyting, pp. 72—80. Amsterdam: North-Holland Publishing Company 1959.

PÉTER, R.: Rekursivität und Konstruktivität. Ibid., pp. 226—233.

The problem, how far a *human being* can, with regard to his intellectual abilities, be considered as a machine, is outside the frame of this book. Among others the following articles may be consulted on this subject.

TURING, A. M.: Computing Machinery and Intelligence. Mind **59**, 433—460 (1950).

KEMENY, J. G.: Man Viewed as a Machine. Sci. Amer. **192**, 58—67 (1955).

§ 4. Historical Remarks

In the course of the centuries mathematicians have had great success in discovering or inventing new algorithms. It was therefore quite natural for the question to arise, whether we can, with sufficient effort, reach a stage in which eventually every group of problems can be attacked by a general procedure. The more or less clear idea that the whole of mathematics can be comprehended by algorithmic methods has at different times strongly influenced the development of this science. We shall confine ourselves to a few remarks on these questions.

1. Ars magna. Algorithmic methods to deal with algebraic problems were developed by the Arabs under the influence of the Indians. The name algorithm, by which a general procedure is commonly called today, originates from the name of the Arab mathematician AL CHWARIZMI (about 800). The mathematical methods introduced by the Arabs inspired the Spaniard RAYMUNDUS LULLUS (about 1300). His *Ars magna* was supposed to be a general procedure on a combinatorical basis to find out all "truths". Regarded from a sensible point of view the procedures given by Lullus are not very valuable. The important thing is that here an idea was conceived, and a splendid one at that. This is what MORITZ

CANTOR says about the "art of Lullus": "... ein Gemenge von Logik, kabbalistischer und eigener Tollheit, unter welches, man weiss nicht wie, einige Körner gesunden Menschenverstandes geraten sind"[1]. Lullus had a great influence, especially on the mathematical posterity. Even two-hundred years later CARDANO (1545) perceives the algebraic algorithms published by him in the spirit of the art of Lullus. This is shown by the title of his work: *Artis magnae seu de regulis algebraicis liber unus.* The way in which algebra developed in the 16th century seemed to be an indication that all questions, in any case in the domain of this discipline, can be dealt with by the help of general procedures. But things were indeed different at first in geometry, the second mathematical science known at that time. It is worth mentioning that DESCARTES (1598—1650) was of the opinion that the vital part of his analytic geometry was the fact that by its help all geometrical problems could be translated into algebraic problems and thus made susceptible to the algorithms developed in algebra.[2] After Descartes had given an account of his method he was of the opinion that there were no interesting problems left for the creative mathematician at all. "Aussy que ie n'y remarque rien de si difficile, que ceux qui seront vn peu versés en la Geometrie commune, & en l'Algebre, & qui prendront garde a tout ce qui est en ce traité, ne puissent trouuer."[3] — Descartes was indeed wrong (as we know today) when he assumed that all algebraic problems can be solved by general methods.

LEIBNIZ (1646—1716) strived hard to develop algorithms and to gain an insight into the essence of algorithms. He stated expressly that he was influenced by Lullus. He saw that Lullus's concept of *ars magna* comprehended several concepts and that it would be better if these were seperated from each other. He distinguished (not always clearly) an *ars inveniendi* from an *ars iudicandi.* Some of his remarks indicate that by an *ars inveniendi* a generating procedure and by an *ars iudicandi* a decision procedure is to be understood. He saw clearly that it must be possible to leave a general procedure to the charge of a machine. In this

---

[1] "... a mixture of logic, cabbalistic madness and madness of his own, into which fell, one cannot imagine how, some grains of healthy common sense". M. Cantor, Vorlesungen über Geschichte der Mathematik, vol. II, p. 104. Leipzig: Teubner 1892.

[2] Thus, for him the decisive factor was *not* the application of coordinates as such (coordinates had already been used in antiquity) but the application of coordinates with the aim of making it possible to deal with geometry by algorithms.

[3] "Because, I find nothing here so difficult that it cannot be worked out by any one at all familiar with ordinary geometry and with algebra, who will consider carefully all that is set forth in this treatise."

Œuvres, ed. Adam-Tannery, vol. VI, p. 374. Paris 1902. Translation quoted from: The Geometry of René Descartes. New York 1954.

connection let us remember that Leibniz was one of the first to build a computing machine. In spite of intensive efforts however he did not succeed in realizing his above mentioned projects (some of his remarks in this respect were not generally known until 1901 when published by COUTURAT).

2. *Modern logic.* If we want to create algorithms which are as universally applicable as possible, the forming of general procedures in the domain of *logic* suggests itself. As a matter of fact, logic is applicable in many fields, and especially in theories with axiomatic foundation. The syllogistic originating from ARISTOTLE was a first attempt in this direction. But it certainly comprehends only a very modest fraction of what a mathematician actually uses as logical conclusions. Modern logic came into existence in the last century. Its foremost pioneer was G. BOOLE (1815—1869). His book "The Mathematical Analysis of Logic, being an Essay towards a Calculus of Deductive Reasoning" appeared in 1847. Boole's computing methods are formed extensively according to the usual methods of algebra, so that in the last part of the last century people spoke of an *algebra of logic*. In the further development however the narrow analogy to algebra was dropped. G. PEANO (1858 — 1932) chose a symbolism which is connected in its construction with the structure of the natural languages. G. FREGE (1848—1925) especially strived to give an exact form to the logical rules. These endeavours found a temporary termination in the monumental work "Principia Mathematica" (3 vols., 1910—1913) by A. N. WHITEHEAD and B. RUSSELL. In this work it was established that a large part of mathematics can be deduced by the help of a logical calculus. The crowning conclusion of the development started by Boole was the so-called GÖDEL's *completeness theorem* (1930). This says that the given logical rules are sufficient to draw all inferences from an arbitrary system of axioms, provided we confine ourselves to the language of the predicate calculus.[1] Gödel's result can be expressed also in the terminology introduced in § 2.4 by saying that if we restrict ourselves to the language of the predicate calculus, then the set of inferences from a system of axioms is generable (enumerable).

3. *Impossibility proofs.* Attempts to give an account of a system of rules for the logic of higher order (cf. § 26), such that all inferences from an arbitrary system of axioms can be obtained by the help of this rule system, have not been successful. Furthermore, people tried in vain to find a procedure by the help of which we can decide in a finite number of steps for an arbitrary finite system of axioms and an arbitrary formula

[1] The decisive restriction of this language consists of the fact that the operations *for all* and *there exists* may only be applied to individual variables and not to predicate variables. Cf. also § 24 and § 26.

of the language of the predicate calculus whether the formula follows from the system of axioms *(decision problem for the predicate calculus)*. We can therefore make the conjecture that no such algorithm exists. An assertion of the kind that a well defined group of problems can *not* be mastered by any algorithm is a statement about *all* imaginable algorithms in general. To prove such a statement we must have more than the natural feeling for the concept of general procedure possessed by every mathematician. We must rather give an exact definition which comprehends all algorithms (in the intuitive sense), but which however may be too wide under certain circumstances. For, if it could be proved that a given group of problems cannot be solved by an algorithm in the sense of the definition, then that would a fortiori show that no algorithm in the intuitive sense produces a solution.[1]

GÖDEL was the first to prove (1931) that in the domain of the second order logic it is impossible to define a system of algorithms of a certain kind which would serve to obtain all inferences (so-called *incompleteness* of the second order logic).[2]

Today it is generally believed that *every* system of algorithms can be defined by recursive functions. From the definition of recursive functions it follows that GÖDEL's result holds for arbitrary systems of algorithms. This gives a deeper meaning to it. Compare what is said further down when dealing with CHURCH. The concept of μ-recursive function and the related concept of recursive function was very intensively studied by KLEENE.

The line of thought of Gödel's proof follows the classical antinomy of the liar. We can namely specify a formula F depending on the arbitrarily given system of rules in the language of the second order logic such that the contents of F can be interpreted by saying that F is not deducible in the rule system. F contains no free variables. Thus, either F is valid or the negation $\neg F$ is valid.[3] If F is valid, then F is not deducible and therefore the presumed system of rules is incomplete. But if $\neg F$ is valid and if the rule system is complete, then $\neg F$ must be deducible. This implies that F is not deducible (if we assume the consistency of the rule system), and this means by the definition of F that F is valid. Thus this case cannot arise.[4]

In 1936 A. CHURCH expressed his opinion that the concepts of λ-definable function (§ 30) and of recursive function (§ 19) are both

[1] This remark is important with regard to the fact that the opinion is sometimes encountered that the exact replacements of the concept of algorithm which are generally used today are too wide.

[2] For the incompleteness theorem cf. § 26, for the concept of μ-recursive function cf. § 19. We can consider FINSLER (1926) to be a forerunner of GÖDEL.

[3] On this terminology cf. § 26.

[4] In § 26 we shall prove the incompleteness of the second order logic in a different way by reducing it to the undecidability of the first order logic.

identifiable with the concept of computable function *(Church's thesis)*. Some time later Church showed that the decision problem for the predicate calculus is unsolvable. The same was proved by A. M. TURING at about the same time. Turing introduced the concept of Turing machine for his proof.[1]

During the last years the undecidability has also been proved for different theories which are given by systems of axioms in the language of the predicate calculus (or in languages which are unessential extensions of it). The first such proof was given by CHURCH in 1936 for Peano's arithmetic. Among the many results gained in this field let us mention the undecidability of the elementary group theory proved by TARSKI in 1946.

A well known mathematical problem, the unsolvability of which has already been established, is the *word problem for groups*. The task of this problem is to find an algorithm by the help of which we can decide in finitely many steps (for any group which is given by finitely many generators and by finitely many defining relations between these generators) for arbitrary words W_1 and W_2 (composed of these generators) whether or not W_1 and W_2 represent the same element of the group. The existing proofs of the unsolvability of the word problem for groups are still very complicated. We shall content ourselves in this book to show in § 23 the unsolvability of the word problem for semi-Thue systems and for Thue systems.

A problem about the solvability or unsolvability of which nothing is known yet is *Hilbert's tenth problem* (1900) (cf. § 29.3). The task of this problem is to find a method by the help of which we can decide for an arbitrarily given diophantine equation whether or not it is solvable.

References

FINSLER, P.: Formale Beweise und die Entscheidbarkeit. Math. Z. **25**, 676—682 (1926).

GÖDEL, K.: Die Vollständigkeit der Axiome des logischen Funktionenkalküls. Mh. Math. Phys. **37**, 349—360 (1930).

— Über formal unentscheidbare Sätze der Principia Mathematica und verwandter Systeme I. Mh. Math. Phys. **38**, 173—198 (1931). (Incompleteness theorem.)

CHURCH, A.: An Unsolvable Problem of Elementary Number Theory. Amer. J. Math. **58**, 345—363 (1936). (Church's thesis on p. 346.)

KLEENE, S. C.: General Recursive Functions of Natural Numbers. Math. Ann. **112**, 727—742 (1936).

On Turing and Post cf. the references given in § 3. Special questions raised in this paragraph will be discussed in detail later on. References will be given in the pertinent paragraphs.

---

[1] A concept of such a machine was developed by POST simultaneously with, but independently of Turing. Turing and Post identified the concept of computable function with the concept of function computable by such a machine.

CHAPTER 2

TURING MACHINES

In this chapter the concept of Turing machine will be introduced without the heuristic considerations of the last chapter. Moreover, the most important constructive concepts, which we have already come across in Chapter 1, will be defined by Turing machines. We should convince ourselves that the suggested definitions of Turing-decidability, Turing-computability and Turing-enumerability are precise replacements of the corresponding intuitive concepts. These definitions can be considered to be really obvious if we agree that the Turing machines represent a legitimate precise substitute for the concept of algorithm. In the end we shall give an account of a few simple examples of Turing machines. The machines a_j, r and l introduced in § 6.5 are fundamentally important.

The concepts defined in this chapter can be considered to be concepts of pure mathematics. It is however very suggestive to choose a technico-physical terminology suggested by the mental image of a machine.

§ 5. Definition of Turing Machines

1. Definitions. Let an alphabet $\mathfrak{A} = \{a_1, \ldots, a_N\}$, $N \geqq 1$, be given. Let a_0 denote the empty (ideal) symbol. We assume, once and for all, that none of the symbols r, l, h belongs to $\mathfrak{A}$.

A *Turing machine* M *over* $\mathfrak{A}$ is given by an $M(N+1) \times 4$ matrix (a table with $M(N+1)$ rows and 4 columns) ($M \geqq 1$) of the form

c_1	a_0	b_1	c'_1
...	...	...	...
c_1	a_N	b_{N+1}	c'_{N+1}
c_2	a_0	b_{N+2}	c'_{N+2}
...	...	...	...
c_2	a_N	b_{2N+2}	c'_{2N+2}
...	...	...	...
...	...	...	...
c_M	a_0	$b_{(M-1)N+M}$	$c'_{(M-1)N+M}$
...	...	...	...
c_M	a_N	b_{MN+M}	c'_{MN+M}

where $c_1, \ldots, c_M$ are different natural numbers $\geqq 0$[1] and $c'_j \in \{c_1, \ldots, c_M\}$ for $j = 1, \ldots, MN + N$. Furthermore, each b_j is an element of $\{a_0, \ldots, a_N, r, l, h\}$.

[1] In § 3 we took $c_1 = 0$, $c_2 = 1$ ..., $c_M = M - 1$. On the generalization made here cf. also Section 6.

We can identify a Turing machine with its table.

We should note that for each pair c_j, a_i there exists exactly one line in **M** which begins with $c_j a_i$. The c_j's are called *states*, c_1 is called the *initial state*. We shall also denote the initial state of **M** by $c_{\mathsf{M}} \cdot c_{\mathsf{M}}$ is the first state mentioned by name in the table. If a line begins with $c_j a_k h$, then c_j is called a *terminal state*. $\mathfrak{A}$ is called the alphabet of **M**.

2. The computing tape. In this book we shall frequently consider *functions B* which are defined for all *integers x*, and the values $B(x)$ of which are elements of $\{a_0, \ldots, a_N\}$. We shall regard the arguments x as numbers of *squares* of a *computing tape* arranged in such a way that the square with the number n (or for short the square n) lies immediately *on the left* of the square $n + 1$. We assume that the square n has the symbol $B(n)$ *written* (*printed*) in it. A square which has the symbol a_0 printed in it is called *empty*. Thus, we can consider the function B to be a *tape expression* of the computing tape. We consider only *such functions for which* $B(x) = a_0$ *for almost all* x; thus we assume that only finitely many squares have actual symbols printed in them, e.g.

...			a_4		a_4	a_3		a_6			...
Square No.		−4	−3	−2	−1	0	1	2	3	4	

3. Configurations[1] *and computing steps.* By a configuration K of a Turing machine **M** we understand any ordered triple (A, B, C) where A is a square (given by its number), B is a tape expression (thus a function) and C is a state of **M**.

A configuration (A, B, C) is called an *initial configuration* if $C = c_{\mathsf{M}}$.

Every configuration $K = (A, B, C)$ corresponds unambiguously to the line of the table **M** which begins with $C\,B(A)$. This we shall call *the line of the configuration K. K* is called a *terminal configuration* if the line of the configuration K begins with $C\,B(A)\,h$.

Let $K = (A, B, C)$ be a configuration which is *not a terminal configuration*. Let $C\,B(A)\,b\,c'$ be the line of the configuration K (thus $b \neq h$). We shall associate (unambiguously) with K a *consecutive configuration* $F(K) = (A', B', C')$, in which we have[2]

$$A' = \begin{cases} A, & \text{if } b \neq r \text{ and } b \neq l \\ A + 1, & \text{if } b = r \\ A - 1, & \text{if } b = l \end{cases}$$

[1] The concept which we call configuration corresponds to what is called *complete configuration* in Turing's original paper.

[2] We should convince ourselves that this definition is in accordance with the contents of § 3.4.

$$B'(x) = \begin{cases} B(x), & \text{if } x \neq A \\ B(x), & \text{if } x = A \text{ and } (b = r \text{ or } b = l) \\ b, & \text{if } x = A \text{ and } b \neq r \text{ and } b \neq l \end{cases}$$

$$C' = c'.$$

If we start a computation with a certain initial configuration, then we call this configuration *the* 0th *configuration.* Furthermore, if the nth configuration in this computation has a consecutive configuration, then we call the latter *the* $(n+1)^{th}$ *configuration.*

We introduce some more phrases which have to do with the concept of computing step or, in short, step. We imagine that the nth step ($n \geqq 1$) leads from the $(n-1)^{th}$ configuration to the nth. In this sense we shall also call the nth and $(n-1)^{th}$ configurations *the configurations after and before the* n^{th} *step respectively.* The 0th configuration we call also *the configuration after the* 0th *step.* Finally, we shall say that a machine *stops operating after the* n^{th} *step* if the nth configuration is a terminal configuration.

4. Definitions. We shall introduce here a few phrases by giving exact definitions of them. These phrases are drawn from the conception of Turing machines as actual machines. We shall express the fact that in a certain connection we choose a machine M, a number A, and a function B, and thus a certain initial configuration $K_0 = (A, B, c_{\mathsf{M}})$, by saying that *we place* M *on the tape expression* B *over the square* A. If K_0 is not a terminal configuration, then there exists for K_0 a unique consecutive configuration $K_1 = F(K_0)$. We say in this case that M *changes in the first step from* K_0 *into* K_1. If K_1 is also not a terminal configuration, then there exists a unique $K_2 = F(K_1)$, and we say that M changes in the second step from K_1 into K_2, etc. Now we can imagine two cases. Either none of the configurations $K_0, K_1, K_2, \ldots$ is a terminal configuration; then K_n is defined for each n and M never stops operating. Or there exists a first n such that K_n is a terminal configuration. Then K_{n+1} is no longer defined and we shall say that M *stops operating (halts)* after the nth step (we can have $n = 0$), and what is more, we say (if $K_n = (A_n, B_n, C_n)$) *that* M *stops operating on the tape expression* B_n *and over the square* A_n. The last tape expression B_n and the last square A_n determine especially the letter $B_n(A_n)$ which is printed in A_n at the end. It is also convenient to say that M placed on B over A *stops operating over the letter* a_j. It can happen that $a_j \neq a_0$. Then there exists a greatest continuous part of the tape which contains A_n and which is such that only actual letters are printed in its squares. These form, in the sequence in which they are printed on the tape, a word W. We shall say in this case that M *stops operating over the word* W. If K_j and K_{j+1} are defined and if the line of the configuration K_j is $C_j A_j bc$, then we shall say that *in the* $(j+1)^{th}$ *step the machine moves to the right or to the left, in case* $b = r$ *or* $b = l$ *respectively, or that the symbol* a *is printed into* A_j, *in case* $b = a$.

5. *Shifts*. If m is an integer and if $B(x) = \tilde{B}(x - m)$ for each x, then we can say that $\tilde{B}$ arises from B due to a *shift* by m places. If we place a machine M on B over A, then we shall obtain the configurations $K_0, K_1, K_2, \ldots$, where $K_n = (A_n, B_n, C_n)$. If we place M on $\tilde{B}$ over $\tilde{A} = A + m$, then we shall obtain the configurations $\tilde{K}_0, \tilde{K}_1, \tilde{K}_2, \ldots$, where $\tilde{K}_n = (\tilde{A}_n, \tilde{B}_n, \tilde{C}_n)$. We can convince ourselves without too much trouble that K_n and $\tilde{K}_n$ are defined for the same arguments n, and that for each such n we have[1]

$$\tilde{A}_n = A_n + m, \quad \tilde{B}_n(x) = B_n(x - m), \quad \tilde{C}_n = C_n.$$

Thus every scanned square $\tilde{A}_n$ and every tape expression $\tilde{B}_n$ arise from the corresponding scanned square A_n and tape expression B_n respectively due to a shift by m places. If furthermore K_n is a terminal configuration, then so is $\tilde{K}_n$, and vice versa.

The facts discussed just now permit in certain cases the introduction of a terminology which is abstracted from the numbering of the squares of the tape. Let W be a word which is a non-empty sequence of letters of the alphabet $\mathfrak{A} = \{a_1, \ldots, a_N\}$. The statement that *we place* M *behind* W will mean that we choose a tape expression B which contains in successive squares the letters of the word W in the order in which they appear in W and the remaining squares of which are empty, and that we take as initial square A the first (empty) square on the right of W. Here B and A are only determined up to a shift by m places. According to the above remarks this ambiguity plays no part in determing whether or not M stops operating. If M stops operating, then it will do so in each case over the same letter. Thus it is convenient to say that M *placed behind a word* W (*or placed behind the empty word*) *stops operating over the letter* a_j *or behind a word* W'.

6. *Equivalence of Turing machines*. We say that a Turing machine M_1 is *equivalent* to a Turing machine M_2, if there exists a one-one mapping φ of the states of M_1 onto the states of M_2 which is such that

(1) each line $c\,a\,b\,c'$ of M_1 is transferred into a line $\varphi(c)\,a\,b\,\varphi(c')$ of M_2, and

(2) $\varphi(c_{\mathsf{M}_1}) = c_{\mathsf{M}_2}$.

Given any machine M_1 we can always find an equivalent machine M_2 where the states are numbered $0, 1, 2, \ldots$, successively, and where 0 is

[1] The following equations can also be interpreted by saying that the translations of the tape are not essential.

the initial state. (In the last paragraph we numbered the states in this standard way.) It is however convenient, especially in view of the combinations of machines discussed in the next paragraphs, to have at our disposal states not given in this standard way.

Only the sequences A_n and $B_n(x)$ are essential for the use to which we shall put Turing machines. Let us suppose that the Turing machine M_1 (placed on B over A) will yield the configurations $(A_n, B_n(x), C_n)$. Let us further suppose that M_2 is equivalent to M_1 by virtue of the mapping φ of the states of M_1 onto the states of M_2. Then we see at once that the sequence of the configurations of M_2 (if placed on the same tape expression B over the same square A) is $(A_n, B_n(x), \varphi(C_n))$. *Thus, equivalent Turing machines will yield the same sequence $(A_n, B_n(x))$ (if placed on the same tape expression B over the same square A)*. Thus, they can be replaced by each other in all definitions which make use of these sequences only. This applies especially to the definitions of the following paragraph.

7. *Equivalence of Turing machines in a wider sense.* In § 1.2 we hinted at the fact that the choice of the single signs in algorithms is in principle unimportant. This leads to an enlarged concept of the equivalence of Turing machines. We say that a Turing machine M_1 over an alphabet $\mathfrak{A}_1$ is *equivalent in the wider sense* to a Turing machine M_2 over an alphabet $\mathfrak{A}_2$ if there exist one-one mappings φ, ψ, such that

1. ψ provides a mapping of $\mathfrak{A}_1$ onto $\mathfrak{A}_2$, and in addition we have that $\psi(l) = l$, $\psi(r) = r$, $\psi(h) = h$.
2. φ is a mapping of the states of M_1 onto the states of M_2.
3. Every line

   $c \quad a \quad b \quad c'$ of M_1 is transferred into a line

   $\varphi(c) \; \psi(a) \; \psi(b) \; \varphi(c')$ of M_2.
4. $\varphi(c_{\mathsf{M}_1}) = c_{\mathsf{M}_2}$.

The remarks in Section 6 about equivalent Turing machines apply mutatis mutandis also to Turing machines which are quivalent in the wider sense.

References

Cf. references in § 3, and the *Appendix* of Post's paper cited in § 23.

§ 6. Precise Definition of Constructive Concepts by means of Turing Machines

In the following sections we shall give exact definitions for the concepts of computability, decidability, enumerability. We shall call these exact concepts Turing-computability, Turing-decidability, Turing-enumerability. The reader should convince himself that these exact concepts are natural replacements of the intuitive concepts described in § 2 (assuming that Turing machines correspond to algorithms).

In the following considerations we shall take the alphabet $\mathfrak{A}_0 = \{a_1, \ldots, a_N\}$ as fixed. We shall only consider *non-empty words* over this alphabet.[1] On the terminology used in the following sections cf. also § 5.4.

1. Turing-computability. First we give a definition in the special case of singulary functions.

Let f be a singulary function which is defined for all words, and the values of which are words. We say that f is *Turing-computable* if there exists a Turing machine **M** over an alphabet $\mathfrak{A}$ with $\mathfrak{A}_0 \subset \mathfrak{A}$ such that, if we print an arbitrary word W (over $\mathfrak{A}_0$) onto the otherwise empty computing tape and if we place **M** over an arbitrary square of the tape[2], then **M** will stop operating after finitely many steps behind a word which represents the value $f(W)$ of the function.[3]

Now let $f(W_1, \ldots, W_k)$ be a k-ary function the arguments of which are arbitrary words over $\mathfrak{A}_0$ and the values of which are words over $\mathfrak{A}_0$. Then we say that f is Turing-computable if there exists a Turing machine **M** over an alphabet $\mathfrak{A}$ with $\mathfrak{A}_0 \subset \mathfrak{A}$ such that if we print onto the otherwise empty computing tape arbitrary arguments $W_1, \ldots, W_k$ in this sequence leaving a square empty each time between two words (i.e. the *ideal word* $W_1 a_0 W_2 a_0 \ldots a_0 W_k$), and we place **M** over an arbitrary square of the computing tape, then **M** will stop operating after finitely many steps and it will do so behind the word $f(W_1, \ldots, W_k)$.

---

[1] That this is not an essential limitation was shown in § 1.4.

[2] We could surmise that too much is demanded if we make it a requirement for a function to be computable that there exists a machine which carries out the computation *independently* of the square of the tape over which it is placed. We would perhaps expect that the class of computable functions defined this way is smaller than it would be if e.g. we only required the machine to carry out the computation if placed over or behind the last symbol of the argument. That this is however not the case is shown in a theorem in § 9.1. In general it will naturally be easier to give an account of a machine which carries out the computation of the value of the function if it (at the beginning of the computation) is placed over a *particular* square. However, the choice of such a square must be arbitrary to a certain extent. The definition given in the text saves us from such arbitrariness.

[3] Contrary to the beginning of the computation where we allowed an arbitrary square it does not seem plausible (if we choose to represent the value as we have done) to permit an arbitrary scanned square for the end as well. For, in general, there will be still other words on the tape besides the value of the function (e.g. arguments or secondary calculations). The machine has (by its final position) to point out the value of the function. The convention which we have given here for this purpose proves to be very convenient in many cases. But we could also adopt another convention without difficulty and demand e.g. that M stops operating over an arbitrary symbol of the value of the function, or something similar. (The equivalence of such requirements can easily be established: § 7, Exercise 2.)

It is advisable for systematical reasons (cf. e.g. § 10) to introduce also functions of zero arguments. The value of a function of *two* arguments can be determined if we are given both arguments. The value of a function of *one* argument can be determined if we are given this argument. Similarly the value of a function of *zero* arguments can be determined if we are given no argument. Thus a function of zero arguments has *only one value*. This can be an arbitrary word W. We shall denote the function of zero arguments which has the value W by C_0^W, or also in short by W if no misunderstanding may result from it.

We shall say, enlarging the above given definition of Turing-computability of a function in the natural way, that a function of zero arguments is Turing-computable if there exists a Turing machine which, if placed on an *empty* tape, will stop operating behind the value of the function after finitely many steps. Here it is immediately obvious that *every* function of zero arguments is Turing-computable in a trivial way. In order to compute C_0^W we only need to take a Turing machine which will print the word W onto the empty tape and stop operating behind this word (Exercise 2).

2. Turing-enumerability. We use here a standardized representation of the natural numbers. For this we agree upon the representation of the natural number n by $n+1$ strokes (cf. § 1.4), thus e.g. we represent the number three by IIII. In doing this we shall identify the stroke I with a_1. We say that a set M of words is *Turing-enumerable*, if M is empty or if M coincides with the domain of values of a Turing-computable function whose domain of arguments is the set of the natural numbers. — Since according to § 2.4 generability is equivalent to enumerability we can save ourselves the trouble of giving a special definition for Turing-generability.

3. Turing-decidability. Let M be a *set* of words over $\mathfrak{A}_0$. We say that M is *Turing-decidable* if there exists a Turing machine **M** over an alphabet $\mathfrak{A}$ (with $\mathfrak{A}_0 < \mathfrak{A}$) and two different (actual or ideal) symbols a_i, a_j of $\mathfrak{A}$ such that if we print an arbitrary word W over $\mathfrak{A}_0$ onto the otherwise empty computing tape and if we place **M** over an arbitrary square of the computing tape, then **M** will stop operating after finitely many steps over a_i or a_j according to whether $W \in M$ or $W \notin M$ respectively.

Let M_1 and M_2 be sets of words over $\mathfrak{A}_0$. Let $M_2 < M_1$. Then we say that M_2 *is Turing-decidable relative to* M_1 if there exists a Turing machine **M** over an alphabet $\mathfrak{A}$ (with $\mathfrak{A}_0 < \mathfrak{A}$) and two (actual or ideal) symbols a_i, a_j of $\mathfrak{A}$ such that if we print an arbitrary word $W \in M_1$ onto the the otherwise empty computing tape, and if we place **M** over an arbitrary square of the computing tape, then **M** will stop operating after a finite number of steps over a_i or a_j according to whether $W \in M_2$ or $W \notin M_2$ respectively.

We say that a *property* (a *predicate*) of words is *Turing-decidable* if the set of words with this property is Turing-decidable.

We define the *Turing-decidability of n-ary relations,* i.e. of n-ary properties ($n \geqq 2$), in a similar way (we print the n-tuple of words onto the computing tape as described in Section 1).

We can show both in the case of absolute decidability and that of relative decidability that the choice of the letters a_i, a_j is unessential. We confine ourselves here to the

Theorem. Let M be a set of words over $\mathfrak{A}_0$. Let M be a Turing machine over an alphabet $\mathfrak{A} = \{a_1, \ldots, a_L\}$ (with $\mathfrak{A}_0 < \mathfrak{A}$) and a_i, a_j different (actual or ideal) symbols of $\mathfrak{A}$. Let M be such that if it is placed on the computing tape which has nothing but an arbitrary word W over $\mathfrak{A}_0$ printed on it, then it will stop operating after finitely many steps over a_i or a_j according to whether $W \in M$ or $W \notin M$ respectively (from now on we shall express this in short by saying that M decides M by the help of a_i, a_j). Further let a_k, a_l be two arbitrary different (actual or ideal) symbols of $\mathfrak{A}$. Then, we can give an account (using M) of a Turing machine M' over $\mathfrak{A}$ which decides M by the help of a_k, a_l.

Proof. We obtain the table for such a machine M' from the table of M as follows. Let $\bar{c}$ be a new state which does not appear in the table of M. We change every line of M which is of the form $c a_i h c'$ into $c a_i a_k \bar{c}$, every line of the form $c a_j h c'$ into $c a_j a_l \bar{c}$, and finally attach a further $L + 1$ lines of the form $\bar{c} a_p h \bar{c}$ ($p = 0, \ldots, L$). We see at once that in every case in which M stops operating over a_i or a_j the machine M' makes a further step and then stops operating over a_k or a_l respectively.[1]

4. Remark. In all definitions of this paragraph we permitted that M is a Turing machine over an alphabet $\mathfrak{A}$ which may contain more letters than the alphabet $\mathfrak{A}_0$. The symbols of $\mathfrak{A}$ which do not belong to $\mathfrak{A}_0$ are used as "auxiliary letters". The question arises whether we can do *without* auxiliary letters in every case, i.e. whether we may demand that M is a machine over $\mathfrak{A}_0$. We can make intuitively clear that this is indeed possible. Let, for instance, $\mathfrak{A}_0 = \{a_1, \ldots, a_N\}$, $\mathfrak{A} = \{a_1, \ldots, a_L\}$, $L > N$. Suppose we have a problem which has to do with words over $\mathfrak{A}_0$ only, and an algorithm which solves this problem using the symbols of $\mathfrak{A}$. We can "translate" every *symbol* a_j of $\mathfrak{A}$ into a sequence of j symbols a_1, e.g. a_3 into $a_1 a_1 a_1$. We can translate a *finite tape expression,* e.g. $a_4 a_0 a_2 a_3$, using the translation of the individual symbols into $a_1 a_1 a_1 a_1 a_0 a_0 a_0 a_1 a_1 a_0 a_1 a_1 a_1$.

---

[1] We can write this, using the notation introduced in the next paragraph, by

$$\mathsf{M}' = \mathsf{M} \begin{matrix} \overset{i}{\nearrow} \mathsf{a}_k \\ \underset{j}{\searrow} \mathsf{a}_l \end{matrix}$$

First we translate the problem in this way. Then we apply to it a sort of "translated algorithm" which will finally give the translated solution which we shall have to translate back again. It is plausible that in this way we obtain an algorithm which uses only the symbols of $\mathfrak{A}_0$. — We shall nevertheless retain the general definition. However, we shall show in an important special case that the alphabet $\mathfrak{A}_0$ is sufficient for our purpose (cf. § 15).

5. *The elementary machines* r, l, a_j. Now we shall introduce a few especially simple Turing machines, from which more complicated machines will be constructed in the next paragraph. For every alphabet there are machines of these kinds, *and these depend on the alphabet chosen.* In this section we shall denote the underlying alphabet by $\{a_1, \ldots, a_N\}$.

Definition 1. The *right machine* r is given by the following

Table for the right machine r

$$\begin{array}{cccc} 0 & a_0 & r & 1 \\ \cdots & \cdots & \cdots & \cdots \\ 0 & a_N & r & 1 \\ 1 & a_0 & h & 1 \\ \cdots & \cdots & \cdots & \cdots \\ 1 & a_N & h & 1 \end{array}$$

The right machine, if placed on an arbitrary tape expression over an arbitrary square A, will stop operating after one step over the square which is immediately on the right of A. The original tape expression will not be altered. The machine will move to the right by one square.

Definition 2. The *left machine* l is given by the following

Table for the left machine l

$$\begin{array}{cccc} 0 & a_0 & l & 1 \\ \cdots & \cdots & \cdots & \cdots \\ 0 & a_N & l & 1 \\ 1 & a_0 & h & 1 \\ \cdots & \cdots & \cdots & \cdots \\ 1 & a_N & h & 1 \end{array}$$

The left machine, if placed on an arbitrary tape expression over an arbitrary square A, will stop operating after one step over the square which is immediately on the left of A. The original tape expression will not be altered. The machine will move to the left by one square.

Definition 3. The *machine* a_j ($j = 0, \ldots, N$) is given by the following

Table for the machine a_j

$$\begin{array}{cccc} 0 & a_0 & a_j & 0 \\ 0 & a_1 & a_j & 0 \\ \cdots & \cdots & \cdots & \cdots \\ 0 & a_j & h & 0 \\ \cdots & \cdots & \cdots & \cdots \\ 0 & a_N & a_j & 0 \end{array}$$

In all lines we have the symbol a_j in the third place except for the line $0 a_j h 0$. Let the machine a_j be placed on the tape expression B over the square A. If $B(A) = a_j$, then the machine a_j will stop operating after zero steps over the square A without changing the tape expression. If $B(A) \neq a_j$, the machine a_j will stop operating after one step over the square A. The tape expression B_1 will differ from B inasmuch that the letter a_j will be printed in the square A. Thus we have in *either case* that a_j stops operating over A. The last tape expression coincides with the original everywhere with the possible exception of the symbol printed in A which is a_j at the end of the computation. Thus the machine a_j prints the symbol a_j into the scanned square A.

We shall often represent a_0 by $*$ and a_1 by I. We shall use for typographical reasons the same symbols $*$ and I respectively for the corresponding Turing machines.

6. Examples. All machines given in this section are machines over an alphabet $\{\mathsf{I}\}$ of one element only. This alphabet is sufficient for the representation of the *natural numbers* (cf. § 1.4).

Table 1	*Table 2*	*Table 3*
$0 * \mathsf{I}\ 0$	$0 * \mathsf{I}\ 2$	$0 * \mathsf{I}\ 0$
$0\ \mathsf{I}\ r\ 1$	$0\ \mathsf{I}\ l\ 1$	$0\ \mathsf{I}\ r\ 1$
$1 * h\ 1$	$1 * h\ 1$	$1 * r\ 0$
$1\ \mathsf{I}\ h\ 1$	$1\ \mathsf{I}\ l\ 0$	$1\ \mathsf{I}\ h\ 1$
	$2 * h\ 2$	
	$2\ \mathsf{I}\ h\ 2$	

We shall examine how these machines work in special cases.

(1) If the machine M given in *Table 1* is placed immediately behind the last stroke of a sequence of strokes which are printed on the otherwise empty tape, then M will attach a further stroke to the sequence and will stop operating on the right of this stroke. (The lines 1, 2, 3 are decisive here; line 4 is unessential and could be stated differently.) Thus, M com-

putes the successor function, *provided* that it is placed on the tape bearing the argument in the manner described. In other cases this does not apply. We cannot prove just by giving an account of **M** that the successor function is Turing computable in the sense of Section 1. For that we must give an account of a machine which performs the computation independently of the square over which it is placed at the beginning of the computation (see § 9).

(2) If we place the machine **M** given in *Table 2* over the last stroke of a word W, then **M** will stop operating after finitely many steps over the symbol $*$ or | according to whether W represents an even or an odd natural number respectively. (Similarly to example (1) it is not yet shown by this that the set of even numbers is Turing decidable in the sense of Section 3.) To see that this is so we follow the computing process. At first line 2 is decisive. Now we must distinguish between two cases, according to whether or not W consists of only one stroke. In the first case line 3 is decisive for the next step. The machine stops operating over the symbol $*$ and thus shows that 0 (represented by *one* stroke) is an even number. In the second case line 4 is decisive for the next step, and afterwards line 1 or 2 is decisive according to whether $W =$ || or $W \neq$ ||. In the first case line 6 will be the next decisive line. The machine stops operating over | and thus shows that 1 is an odd number. In the second case we have reached a situation which corresponds to the initial situation, and so the whole process repeats itself modulo 2. The machine stops operating, if the word W is exhausted, over $*$ or | according to whether W represents an even or an odd number respectively. (Line 5 is unessential.)

(3) We require a Turing machine which will, if placed over an arbitrary square of the originally empty tape, print onto the tape (beginning at the initial scanned square) the infinite sequence |$*$|$*$|$*$ ... This is carried out by the machine given in *Table 3*.

(4) Example of a Turing machine which computes the *difference* $x \dot{-} y$. Let $x \dot{-} y = x - y$, if $x \geqq y$, and $x \dot{-} y = 0$, if $x < y$. Let $W_1 * W_2$ be printed onto the otherwise empty computing tape, where W_1 and W_2 represent numbers in the sense of Section 2. We seek a machine **M** which performs the following. If we place **M** on the empty square behind W_2, then **M** stops operating after finitely many steps and, indeed, it does so over an empty square behind a sequence W of strokes, which has an empty square in front of it and which represents (again in the sense of Section 2) the value of the function $x \dot{-} y$.

The table of **M** is produced in three stages.

(a) We describe in quite general terms the procedure which carries out the calculation.

(b) We break up the procedure into single (larger) parts, which ought to be carried out in a fixed order.

(c) By further breaking up of the procedure parts we construct a table for **M**.

To (a). By erasing alternatively one stroke on the right end of W_2 and at the left end of W_1 we get that either W_2 or W_1 is completely erased. If W_2 is erased first, then $y \leqq x$, and so $x \dot{-} y = x - y$. Then the remainder of W_1 (without the erasure of a further stroke of it) represents the required value of the function. On the other hand, if W_1 is erased first, then $x < y$, and so $x \dot{-} y = 0$. This value 0 must now be written down in the form of a stroke.

In future we shall speak of the right word and left word and we shall mean by it W_2 and W_1 respectively, or the shortened words originating from these words.

To (b). We consider the following procedure parts (together with their sequence which starts with (α)).

(α) Erase one stroke at the right end of the right word. Jump (β).

(β) Check whether the right word is completely erased. If no, jump (γ); if yes, jump (η).

(γ) Go to the left word. Jump (δ).

(δ) Erase a stroke at the left end of the left word. Jump (ε).

(ε) Check whether the left word is completely erased. If no, jump (ζ); if yes, jump (θ).

(ζ) Go to the right word. Jump again (α).

(η) Go to the square immediately behind the left word.

(θ) Go to an empty square immediately behind an isolated stroke.

The instruction given just now can easily be followed up when represented in a so-called *flow diagram*.

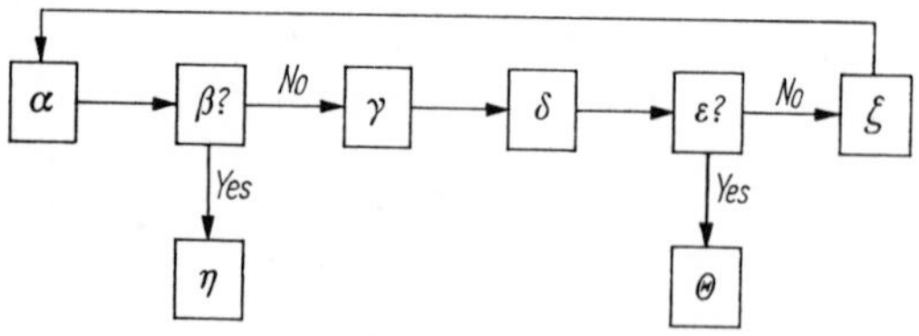

To (c). We can easily see now that the following machine table carries out the program (b). We shall show for each line to which procedure part it belongs. Some lines necessary for a complete description of a machine table have been left out, e.g. the line beginning with 1 *. These lines play no part in the computation, they can therefore be given arbitrarily. To achieve that the table corresponds entirely to the flow diagram it has not been attempted to obtain the shortest table possible. The given table can be made shorter, e.g. by replacing the two lines $3 * * 10$ and $10 * h\ 10$ by a new line $3 * h\ 3$.

$$(\alpha)\left\{\begin{array}{llll} 0 & * & l & 1\\ 0 & | & r & 0\\ 1 & | & * & 2 \end{array}\right.$$

$$(\beta)\left\{\begin{array}{llll} 2 & * & l & 3\\ 3 & * & * & 10\\ 3 & | & l & 4 \end{array}\right.$$

$$(\gamma)\left\{\begin{array}{llll} 4 & * & l & 5\\ 4 & | & l & 4 \end{array}\right.$$

$$(\delta)\left\{\begin{array}{llll} 5 & * & r & 6\\ 5 & | & l & 5\\ 6 & | & * & 7 \end{array}\right.$$

$$(\varepsilon)\left\{\begin{array}{llll} 7 & * & r & 8\\ 8 & * & * & 11\\ 8 & | & r & 9 \end{array}\right.$$

$$(\zeta)\left\{\begin{array}{llll} 9 & * & r & 0\\ 9 & | & r & 0 \end{array}\right.$$

$$(\eta)\quad 10 \; * \; h \; 10$$

$$(\theta)\left\{\begin{array}{llll} 11 & * & l & 12\\ 12 & * & | & 12\\ 12 & | & r & 13\\ 13 & * & h & 13 \end{array}\right.$$

7. Non-periodic computing procedures. One could, in the first instance, believe that if a Turing machine never stops operating if placed on the empty tape, then the procedures produced by it (if it is placed on the empty tape) must be periodic. We could for instance put forward the following considerations. If we follow a machine in its computation, then at each step certain lines of the machine table will be decisive. A machine table has only finitely many lines. Thus, there must be a line which is the first to be decisive for the second time in the process considered. Since the behaviour is determined by the lines, the computing procedure must be periodic after this stage.

This argument however is not valid because it does not pay attention to the fact that a computation is governed not only by the machine table but also by the contents of the computing tape. More exactly we could say the following. Let us assume for instance that at the k_1^{th} computing step we must proceed according to the line n I rm of the machine table and that we must do the same again in the k_2^{th} step ($k_2 > k_1$). Let us further assume that before the k_1^{th} step there is an empty square on the computing tape immediately right of the square scanned at the time. In this case we have to proceed in the $(k_1 + 1)^{th}$ step of the computation according to the line which begins with m ∗. But we only need to proceed according to the same line in the $(k_2 + 1)^{th}$ step if before the k_2^{th} step, too, there is an empty square next on the right of the scanned square. This need not be the case at all.

That no modified argument would suffice follows from the fact that we can give an example of a Turing machine which prints onto the empty tape the non-periodic sequence I ∗ II ∗ III ∗ IIII ∗ ... (cf. § 8.10).

Exercise 1. Construct a Turing machine which computes the function $f(n)$ = remainder of n when divided by 3, if it is placed behind the last square marked by n on an otherwise empty tape.

Exercise 2. Let W be an arbitrary word over an alphabet $\mathfrak{A}$. Give an account of a Turing machine which prints W onto the empty tape and stops operating over the square behind W.

§ 7. Combination of Turing Machines

It is often difficult to read off the behaviour of the machine from a large machine table. It seems recommendable to introduce operations by the help of which we can combine simple tables into more complicated ones. Our aim is to construct all machines given in this book from the machines introduced in § 6.5. The kind of combination which will be discussed here is analoguous to the "flow diagram" (or "block diagram") which is used in programming for electronic computers (cf. example (4) in § 6.6).

Let us consider here an important remark. When we combine the machines $\mathsf{M}_1, \ldots, \mathsf{M}_r$ into a machine M, then M is not determined unambiguously, but only "up to equivalence", i.e. up to a renumbering of the states. This does not hinder us in any way since we have worked out in § 5.6 that for the ends pursued here we can replace equivalent machines by each other. We could determine the machine M completely and not only up to equivalence by supplementary instructions. This would however only be possible by unnecessary, troublesome and arbitrary standardizations.

1. Diagrams. Let $\mathsf{M}_1, \ldots, \mathsf{M}_r$ be symbols for given Turing machines, all of them over a fixed alphabet $\{a_1, \ldots, a_N\}$. By the help of these symbols M_i we can produce diagrams D which satisfy the following requirements.

(1) At least one symbol M_j occurs in D. A symbol may occur more than once. Altogether only finitely many of these symbols occur.

(2) Exactly one occurrence of one of these symbols is marked as the *initial symbol* (e.g. by having a circle drawn around it).

(3) The symbols occuring in D can be connected by oriented lines (arrows). An arrow begins at one of these symbols and ends at a symbol which may coincide with the first one (returning arrow). Every arrow carries a number j ($j = 0, \ldots, N$).

(4) From any one symbol at most one arrow may leave carrying the number j ($j = 0, \ldots, N$).

It is advisable to use a few *abbreviations* (cf. Fig. 7.2 which is an abbreviation of Fig. 7.1, where $N = 2$). If a machine symbol is connected with another by *all* arrows $\xrightarrow{0}, \ldots, \xrightarrow{N}$ (in the same direction), then we use *one* arrow without a number. If only $\xrightarrow{j}$ is missing, then we write $\xrightarrow{\neq j}$. If only one machine symbol occurs which has no arrow ending at it, then this must be the initial symbol and so in this case we need not use the circle mentioned under (2). If in any other case no initial symbol is given, then the *leftmost* symbol must be the initial symbol. For $\mathsf{M} \to \mathsf{M}$ we write M^2, for $\mathsf{M}^2 \to \mathsf{M}$ M^3, etc. For $\mathsf{M}_1 \to \mathsf{M}_2$ we write $\mathsf{M}_1\mathsf{M}_2$, for $\mathsf{M}_1 \to \mathsf{M}_2 \to \mathsf{M}_3$ $\mathsf{M}_1\mathsf{M}_2\mathsf{M}_3$, etc. By M^0 we understand a machine which is given by the following table.

$$\begin{array}{cccc} 0 & a_0 & h & 0 \\ 0 & a_1 & h & 0 \\ \multicolumn{4}{c}{\cdots\cdots} \\ 0 & a_N & h & 0 \end{array}$$

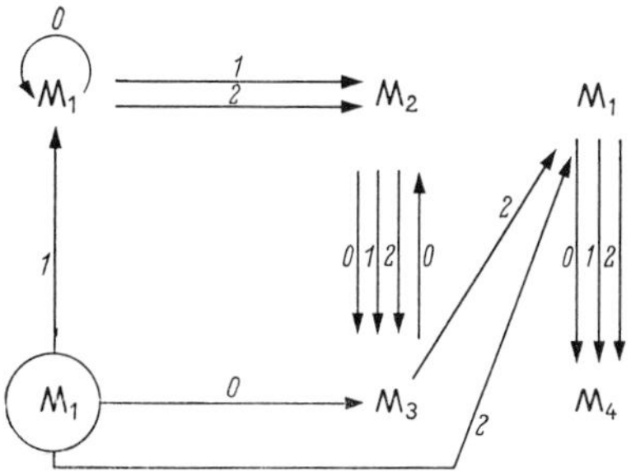

Fig. 7.1

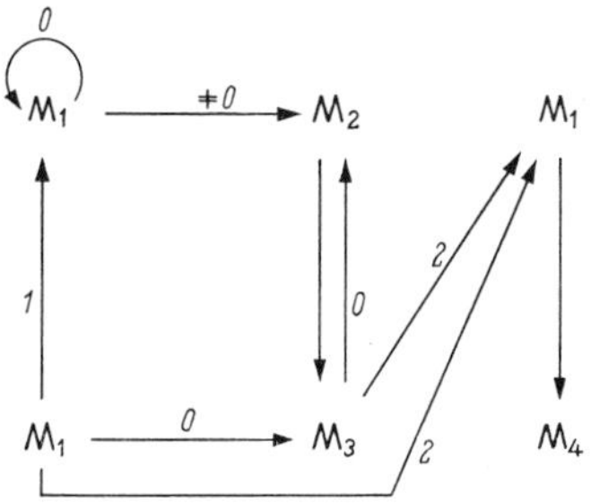

Fig. 7.2

Examples for diagrams

2. *Definition of the machine* M *represented by a diagram* D. This machine, as it has already been explained, is only determined up to equivalence. We obtain a *table for* M as follows.

(1) First we produce the tables for the machines M_j which are represented by symbols in the diagram. If a machine appears in the diagram more than once, then we produce correspondingly many tables for this machine and associate these tables with the corresponding symbols in the diagram.

The tables must be chosen so that no two separate tables contain the same *state*.

(2) After this we produce a "large" table by writing the single tables underneath each other. The sequence is arbitrary with the single exception that the table which is associated with the initial symbol should be placed at the top.[1]

(3) In order to obtain from this preliminary table the definitive table for M we carry out the following alterations. If the symbol M_i occurring at a certain place in the diagram is connected by $\xrightarrow{k}$ to the symbol M_j occurring at another place, then we change each line of the form

$$(*) \qquad c\, a_k\, h\, c'$$

which occurs in the table corresponding to the first symbol (in case such a line exists at all) into

$$(**) \qquad c\, a_k\, a_k\, c_{\mathsf{M}_j},$$

where c_{M_j} is the initial state in the table corresponding to the symbol M_j. (This instruction can be carried out unambiguously, since we have presumed that for each symbol there is at most one arrow $\xrightarrow{k}$ leaving it. It is possible in special cases that no alteration at all need be carried out according to (3).)

It follows from the definition of the table of M that this table is unambiguously determined up to equivalence.

3. *The method of operation of the machine* M *obtained from such a diagram* D can easily be followed. If we place M on a tape expression $B(x)$ over a square A, then M will at first perform the same steps as the machine M' which is denoted by the *initial symbol*, since the table of this machine is placed at the top and so provides the initial state. This will be the case until the machine M' eventually stops operating after a step

[1] This process could be carried out for the example given in Fig. 7.1 or 7.2 in the following way.

Table for machine M_1 given bottom left.
Table for M_4.
Table for machine M_1 given top left.
Table for M_2.
Table for M_3.
Table for machine M_1 given top right.

at the configuration (A_n, B_n, C_n). In this case a line of the form (*) is decisive for M'. It is possible that M' is not connected by an arrow $\xrightarrow{k}$ to another machine symbol. Then the line (*) is not altered, and M also stops operating. If however M' is connected by an arrow $\xrightarrow{k}$ to a symbol for a machine M'', then M contains instead of (*) the line (**) which is now decisive. The effect of this is that the tape expression and the square scanned at the time are not altered, but the initial state of M'' is put into operation. This means that $A_{n+1} = A_n$, $B_{n+1} = B_n$, $C_{n+1} = c_{\mathsf{M}''}$. The further configuration will now correspond to the configuration of the machine M'' placed on B_{n+1} over A_{n+1}, etc.

Thus, the configurations of M placed on B over A are at first those of M' placed on B over A; and then (if M' stops operating on the tape expression B' over A') those of M'' placed on B' over A', etc.

We can express this intuitively by saying that M carries out successively the work of the machines M', M'', ... in a sequence which is determined by the diagram (together with the original B and A).

4. Example. In § 6.1 we introduced the functions C_0^k of zero arguments which have the constant value k. We can in general consider for every n a function of n arguments which has the constant value k. (For these functions also we shall use the abbreviation k if no misunderstanding is possible.) All these functions C_n^k are Turing-computable. A computation will be carried out e.g. by the machine

$$* (\mathsf{r}\,\mathsf{I})^{k+1}\, \mathsf{r} * .$$

Using $*$ we produce an empty square which marks off the beginning of the value of the function to be computed. Then using $(\mathsf{r}\,\mathsf{I})^{k+1}$ we print the value of the function k (during this, signs already printed in the squares in question will perhaps be "overprinted"). Using $\mathsf{r} *$ we make certain that there is an empty square on the right of the value of the function.

Exercise 1. We can say that two machines M and M' are *interchangeable* if for any tape expression B and any initial square A the following applies. If M reaches (if placed on B over A) a terminal configuration (A_n, B_n, C_n), then M' also reaches (if placed on B over A) a terminal configuration (A_m, B_m, C_m) with $A_n = A_m$, $B_n = B_m$, and vice versa. Show, that for any Turing machine M over $\{a_1, \ldots, a_N\}$ it is possible to give effectively a combination of elementary machines r, l, $\mathsf{a}_0, \ldots, \mathsf{a}_N$ (§ 6.5) which is interchangeable with M.

Exercise 2. Prove that if a function f is computable according to the definition of § 6.1, then it is also computable if we change this definition inasmuch that we require the computing machine M to stop operating over an arbitrary symbol of the value of the function, and vice versa (cf. note 2 on p. 36).

§ 8. Special Turing Machines

Here we shall construct, using the processes discussed in § 7, a few Turing machines (which we shall need later on) from the elementary machines **r**, **l**, $\mathsf{a}_0, \ldots, \mathsf{a}_N$ introduced in § 6.5. Definitions and hints on the method of operation of the most important machines are summarized in the *Summary* on pp. 50/51. We shall use the following simple symbolism in the characterization of the method of operation.

m a square with an actual symbol printed in it, called in short a *marked square*,

$\sim$ a marked or empty square,

$*$ an empty square,

$* \ldots *$ a finite sequence of empty squares (at least one),

$* \ldots$ an empty part of the tape reaching to infinity on the right,

W a part of the tape on which the non-empty word W is printed and which contains no empty squares. For such a part of the tape we shall often say in short *the word* W,

X shall stand for $W_1 * W_2 * W_3 \ldots W_{n-1} * W_n$, where $W_1, \ldots, W_n$ ($n \geqq 1$) are non-empty words. In this paragraph, we shall also speak in short of *the sentence* X.

The squares which are not given an account of can have arbitrary symbols printed in them.[1] We shall characterize the scanned square at any instant by underlining it. On the left of the arrow $\Rightarrow$ the initial tape expression and the original scanned square are described, on the right the tape expression and the scanned square which arise after the machine in question has stopped operating. All machines given in our table (except possibly **ρ**, **λ** and **S**) stop operating after finitely many steps.

The machines which we are going to define are not the only ones which would solve the tasks in question. They are chosen so that their *method of operation* can easily be followed. They are therefore not always as simple as possible. Let us indicate at this point that it is not clear when to call a machine simpler than another. If we actually carried out the construction of the machines out of elementary machines (which are now considered to be constituting elements), then we could consider the number of the elementary machines needed for the construction decisive for the simplicity. On the other hand we might want to take into consideration the time (i.e. the number of steps) of the solution of the task. This would lead to more difficult considerations into which we shall not go in this book.

---

[1] Here we should keep in mind the assumption which is binding for us, that only finitely many squares may be marked.

We shall now make a few comments on the machines defined in the Summary.

1. The large right machine R (*the large left machine* L) moves from the square over which it is placed to the neighbouring square on the right (left). If this square is empty, it stops operating. On the other hand, if the square is marked, then R (L) moves over all the marked squares on the right (left) until the first empty square is reached, over which it stops operating. In any case the tape expression will not be altered (cf. the Summary on pp. 50/51. This Summary should also be consulted for the other machines of this paragraph).

2. The right search machine ρ (*left search machine* λ) is in a certain sense dual to R (L). ρ (λ) moves from the square over which it is placed one square to the right (left). If this square is marked, it stops operating. However, if this square is not marked, ρ (λ) moves to the right (left) until the first marked square is reached, over which it stops operating. The tape expression will not be altered. Thus, ρ (λ) "searches" for the first marked square on the right (left) of the original scanned square (and stops operating over it if such square exists at all).

3. The search machine S performs the following operation. If we place S over an *arbitrary square* of the tape, on which at least one square is marked, then S will stop operating after finitely many steps over a marked square. The original tape expression will coincide with the terminal tape expression (during the computation however the tape expression may be changed).

The method of the construction of S is based on the following remarks. If we could assume that there exists a marked square on the right (left) of the original scanned square, then we would reach our goal simply by using ρ (λ). Since however we may not assume this, we must search systematically alternatively on the right and on the left of the original scanned square until we finally reach a square which is marked. The limits (both on the right and on the left) of the search we have done so far must be remembered all the time. For this purpose we use a "marker" on both sides which consists of the letter a_1. This marker will be shifted step by step further outside until we finally come across a marked square. These limit markers must of course be erased at the end of the computation.

4. The right (left) end machine $\mathfrak{R}$ ($\mathfrak{L}$) moves from the square over which it is placed to the right (left) until it comes, for the first time, to the second square of two neighbouring empty squares ("double gap") (not counting the original scanned square). Then it moves one square back to the left (right). The tape expression is not altered.

Summary of important Turing machines

serial no	nomen-clature	name	illustration of the method of operation[1]	structure
1a	R	large right machine	$\simeq W * \Rightarrow \sim W \underline{*}$ $\simeq * \Rightarrow \sim \underline{*}$	$\rightarrow \mathsf{r} \circlearrowleft^{\neq 0}$
1b	L	large left machine	$* W \simeq \Rightarrow \underline{*} W \sim$ $* \simeq \Rightarrow \underline{*} \sim$	$\rightarrow \mathfrak{l} \circlearrowleft^{\neq 0}$
2a	ρ	right search machine	$\simeq * \ldots * m \Rightarrow \sim * \ldots * \underline{m}$ $\simeq m \Rightarrow \sim \underline{m}$	$\rightarrow \mathsf{r} \circlearrowleft^{0}$
2b	λ	left search machine	$m * \ldots * \simeq \Rightarrow \underline{m} * \ldots * \sim$ $m \simeq \Rightarrow \underline{m} \sim$	$\rightarrow \mathfrak{l} \circlearrowleft^{0}$
3	S	search machine	searches for a marked square	$\mathsf{r} \xrightarrow{0} \mathsf{a}_1\, \mathfrak{l} \xrightarrow{0} \mathsf{a}_1\, \rho\, \mathsf{a}_0\, \mathsf{r} \xrightarrow{0} \mathsf{a}_1\, \lambda\, \mathsf{a}_0$ (loop from a_0 back to $\mathfrak{l}$) $\mathfrak{l} \downarrow^{\neq 0} \rho\, \mathsf{a}_0\, \lambda$; $\mathsf{r} \downarrow^{\neq 0} \lambda\, \mathsf{a}_0\, \rho$
4a	$\mathfrak{R}$	right end machine	$\simeq X * * \Rightarrow \sim X \underline{*} *$ $\simeq * X * * \Rightarrow \sim * X \underline{*} *$ $\simeq * * \Rightarrow \sim \underline{*} *$	$\rightarrow \mathsf{R}\, \mathsf{r} \xrightarrow[0]{} \mathfrak{l}$ (loop $\neq 0$ from r back to R)
4b	$\mathfrak{L}$	left end machine	$* * X \simeq \Rightarrow * \underline{*} X \sim$ $* * X * \simeq \Rightarrow * \underline{*} X * \sim$ $* * \simeq \Rightarrow * \underline{*} \sim$	$\rightarrow \mathsf{L}\, \mathfrak{l} \xrightarrow[0]{} \mathsf{r}$ (loop $\neq 0$ from $\mathfrak{l}$ back to L)

5	T	left translation machine	$\simeq * W * \Rightarrow \sim W \underline{*} *$	$\mathsf{r}^2 \begin{cases} \xrightarrow{0} \iota \\ \xrightarrow{1} \mathsf{a}_0 \iota \mathsf{a}_1 \\ \cdots\cdots \\ \xrightarrow{N} \mathsf{a}_0 \iota \mathsf{a}_N \end{cases}$ (branches 1 to N loop back to r^2)
6	σ	shifting machine	$* W_1 * W_2 \underline{*} \Rightarrow * W_2 \underline{*} \ldots *$	$\mathsf{L}\,\iota \xrightarrow{\neq 0} \mathsf{a}_0 \mathsf{T}$ (loops back to L); $\downarrow 0$ T
7	C	cleaning up machine	$\sim * * X * W \underline{*} \Rightarrow \sim W \underline{*} \ldots *$	$\mathsf{L}\,\iota \xrightarrow{\neq 0} \mathsf{r}\,\mathsf{R}\,\sigma$ (loops back to L); $\downarrow 0$ $\mathsf{T}\,\mathsf{L}\,\iota\,\mathsf{T}$
8	K	copying machine	$* W \underline{*} \ldots \Rightarrow * W * W \underline{*} \ldots$	$\mathsf{L}\,\mathsf{r} \begin{cases} \xrightarrow{0} \mathsf{R} \\ \xrightarrow{1} \mathsf{a}_0 \mathsf{R}^2 \mathsf{a}_1 \mathsf{L}^2 \mathsf{a}_1 \\ \cdots\cdots \\ \xrightarrow{N} \mathsf{a}_0 \mathsf{R}^2 \mathsf{a}_N \mathsf{L}^2 \mathsf{a}_N \end{cases}$ (branches 1 to N loop back to r)
9	K_n	n-copying machine ($n \geqq 1$)	$* W_n * W_{n-1} \cdots * W_1 \underline{*} \ldots \Rightarrow * W_n * W_{n-1} \cdots * W_1 * W_n \underline{*} \ldots$	$\mathsf{L}^n \mathsf{r} \begin{cases} \xrightarrow{0} \mathsf{R}^n \\ \xrightarrow{1} \mathsf{a}_0 \mathsf{R}^{n+1} \mathsf{a}_1 \mathsf{L}^{n+1} \mathsf{a}_1 \\ \cdots\cdots \\ \xrightarrow{N} \mathsf{a}_0 \mathsf{R}^{n+1} \mathsf{a}_N \mathsf{L}^{n+1} \mathsf{a}_N \end{cases}$ (branches 1 to N loop back to r)

[1] Not all possible cases are enumerated here but only those which are interesting in connection with the applications considered. *The squares which are not specified are never scanned, and so remain unaltered.* For the notation cf. the introduction to § 8, p. 48.

5. *The left translation machine* T shifts a word one square to the left. (A right translation machine could be produced according to the same principle, but it will not be used in this book.) The moving of the word is done letter by letter starting from the left.

6. *The shifting machine* $\boldsymbol{\sigma}$ carries out the following. Given two words following each other: $* W_1 * W_2 *$, let the square behind W_2 be the scanned square. Then W_1 is erased and W_2 is shifted to the left so far that the beginning of the shifted word W_2' coincides with the original beginning of W_1. After this $\boldsymbol{\sigma}$ stops operating over the square behind the shifted word W_2'.

We shall follow up the process more exactly.

(1) Due to L ʟ the machine will after a few steps be over the second square on the left of W_2. This square could be marked (see (2)) or unmarked (see (3)). (When we reach this square for the first time we find, on account of our hypothesis, that it is marked. Thus, in this case the machine proceeds further according to (2).)

(2) If the square is marked, the machine erases the marking symbol by $\overset{\neq 0}{\rightarrow}$ a_0 and moves W_2 one square to the left by means of T. Now we are again, in principle, in the initial situation with a shortened word W_1'. We have to keep in mind however that the word W_1 may already have completely disappeared. Further computation is carried out according to a backcoupling to (1).

(3) The word W_1' has now completely disappeared, but W_2 has still to be shifted one more square to the left. This happens according to $\overset{0}{\rightarrow}$ T.

7. *The cleaning up machine* C. We shall later on need a machine C for *cleaning up computations*. The task of this machine is to erase secondary calculations and to move forward the final result into a certain position. We shall assume that the secondary calculations are printed on the tape in the form of a sentence X (i.e. in a sequence of words which are separated from each other by gaps). The result in the form of a word W is situated behind these secondary calculations with one square in between. All squares are empty on the right of the result. Left of the secondary calculations are at least two empty squares (otherwise we would not be able to recognize the "beginning" of the secondary calculations) (cf. the *Summary*). C erases the secondary calculations starting from the right and brings the result forward so far that the beginning of the shifted word W is in the first square of the two square gap mentioned just now.

The way of operation of C is followed up in detail as follows.

(1) Due to L ʟ the machine will after a few steps be over the second square on the left of W. This square could be marked (see (2)) or unmarked

(see (3)). (When we reach the square for the first time we find, on account of the hypothesis, that it is marked. Thus, in this case the machine proceeds further according to (2).)

(2) The machine shifts, by $\overset{\neq 0}{\rightarrow}$ $\mathsf{r\,R}\boldsymbol{\sigma}$, W to the beginning of the last part of sentence X. Then it is coupled back to L.

(3) The secondary calculations have disappeared completely. The task remains to move W two squares further left. This happens according to $\overset{0}{\rightarrow}$ $\mathsf{T\,L\,l\,T}$.

8. The copying machine K. The translation machine T shifts a word and so copies it in a certain way. However, if we apply this computation the original tape expression will be lost. But we often have the task to copy a word so that the original expression is preserved. For this purpose we construct a *copying machine* K. K performs the following computation. We presume that we place K over the square behind the word W. On the right of this word all squares of the tape are empty. Then K stops operating after finitely many steps. When it does so all the squares of the computing tape which were marked at the beginning of the computation are marked in the same way. In addition to this a copy of the word W is printed on the right of the original word W with a square in between. K stops operating behind the last of the squares which are marked at the end of the computation. We shall follow up the method of operation of K by considering the single parts of the procedure in the example $W = babb$. (We presume here that the alphabet $\{a_1, a_2\} = \{a, b\}$ provides the basis.)

(1) The computing tape at the beginning of the computation is

$$* \, b\,a\,b\,b \, \underline{*} \cdots .$$

(2) The machine moves back to the first square in front of the first b by the help of L:

$$\underline{*} \, b\,a\,b\,b \, * \cdots .$$

(3) The machine goes over to the neighbouring square on the right by means of r:

$$* \, \underline{b}\,a\,b\,b \, * \cdots .$$

(4) The symbol in the square scanned at the time is erased by means of a_0, by means of R the machine moves to the square on the right of W and by means of the second R moves one more square to the right, where it prints b by means of b:

$$* \, * \, a\,b\,b \, * \, \underline{b} \, * \cdots .$$

(5) By the help of L^2 the machine moves back to the place where the letter b was erased and reproduces b by means of b:

$$* \underline{b} abb * b * \cdots.$$

(6) The machine goes over to the square immediately on the right by means of r:

$$* b \underline{a} bb * b * \cdots.$$

(7) The symbol in this square is erased by means of a_0. By means of R^2 the machine moves to the square on the right of the letter b already copied and prints a by means of a:

$$* b * bb * b \underline{a} * \cdots.$$

(8) By means of L^2 the machine returns to the place where a was erased. Here we see that the importance of the erasure of the letters of the original word lies in the fact that by the help of this marker we can find the letter which is to be copied next. The machine reproduces a by means of a:

$$* b \underline{a} bb * ba * \cdots.$$

(9) We give an account of the tape expressions and scanned squares after steps which are characteristic later on:

$$
\begin{array}{l}
* ba\underline{b}b * ba * \cdots \\
* ba\underline{*}b * ba * \cdots \\
* ba*b * ba\underline{b} * \cdots \\
* ba\underline{b}b * bab * \cdots \\
* bab\underline{b} * bab * \cdots \\
* bab\underline{*} * bab * \cdots \\
* bab* * bab\underline{b} * \cdots \\
* bab\underline{b} * babb * \cdots \\
* babb \underline{*} babb * \cdots.
\end{array}
$$

We should convince ourselves that the procedure also functions in the last stage in which the last letter of W is already erased and reproduced. Now the scanned square is empty. Therefore nothing else will be copied, the machine moves instead by the help of R behind the last letter of the copied word and we finally obtain

$$* babb * babb \underline{*} \cdots.$$

9. The n-copying machine K_n. We often have the task (especially when computing functions with several arguments) of copying a word which is not placed on the very right, so that the copying procedure has to be carried out over a few words printed in between. The copying machine K_n ($n \geqq 1$) carries out this computation. *We assume for this computation that the words in question are separated from each other by gaps of one square only.* K_1 is identical to K. K_n is built according to the pattern of K with the difference that the uninteresting words lying in between are jumped over every time. We should bear in mind that n is a *fixed* number; for every n there exists a machine K_n.

10. Turing machines and periodicity. We asserted in § 6.7 that we can give an account of a Turing machine which prints onto the initially empty computing tape the unperiodic sequence

$$\mathsf{I} * \mathsf{II} * \mathsf{III} * \mathsf{IIII} * \cdots.$$

This is carried out by the machine $\overset{\downarrow\!\overline{\quad\;\;}\!\mid}{\mathsf{I}\,\mathsf{r}\,\mathsf{K}}$. The process can easily be followed. First a square is marked by I, after which the machine moves one square to the right. Now K copies this stroke. By $\mathsf{I}\,\mathsf{r}$ a further stroke is attached and the machine is moved a square further to the right. Now the last word, consisting of two strokes, is copied by K, then a new stroke is adjoined, etc.

§ 9. Examples of Turing-Computability and Turing-Decidability

We have demanded in the definition of Turing-computability of a function (§ 6.1) and of Turing-decidability of a predicate (§ 6.3) that the machine which performs the task may be placed over an *arbitrary* square of the computing tape. As we have already emphasized, these definitions have the advantage that they are free from the arbitrariness which arises from prescribing the choice of the initial square. On the other hand it is not immediately clear how this aggravating condition can be fulfilled. This is one of the reasons why so far we have only given one trivial example for Turing-computability (cf. § 7.4). By the help of the machines developed in the last paragraph, especially by the help of the search machine, we are now in a position to solve this problem.

1. Special and arbitrary initial squares in the computation of functions and in the decision of predicates. We discuss here the case of the computability of functions. The same applies mutatis mutandis also to the decidability of predicates. We start from the assumption that we know a machine M' which carries out the computation of the value of an n-ary function ($n \geqq 1$), provided that we place M' over a *certain* initial square $a_{W_1 \cdots W_n}$ defined by the arguments $W_1, \ldots, W_n$. (We can for instance take the last square which has a symbol of the arguments printed in it to be the initial square, or we can also take the empty square immediately following it; cf. the examples in § 6.6.) We also assume that we can find the square $a_{W_1 \cdots W_n}$ by the help of a given machine N in such a way that if we place N behind the last square marked by the arguments $W_1, \ldots, W_n$, then it will stop operating after finitely many steps over $a_{W_1 \cdots W_n}$. Then we can effectively describe a machine M by the help of M' and N which computes f in the sense of § 6.1; thus it may initially be placed over an arbitrary square. This we show in

Theorem 1. Let f be an n-ary function ($n \geqq 1$) which is defined for all words over an alphabet $\mathfrak{A}_0 = \{a_1, \ldots, a_{N_0}\}$ and which assumes words over this alphabet as values.[1] Let N be a machine over $\mathfrak{A}_0$ which performs the following. If we place N behind the last symbol of an arbitrary n-tuple $(W_1, \ldots, W_n)$ of words, then N will stop operating over a square which we shall call $a_{W_1 \cdots W_n}$. The original tape expression should not be altered at the end of the computation by N. Let M' be a machine such that if we print onto the otherwise empty computing tape an n-tuple $W_1, \ldots, W_n$ of arguments and if we place M' over the square $a_{W_1 \cdots W_n}$ of the tape marked in this way, then M' will stop operating after finitely many steps behind the value $f(W_1, \ldots, W_n)$ of the function. *Then f is Turing-computable. A machine M which Turing-computes f will be given by*

$$\mathsf{M} = \mathsf{S}\,\mathfrak{R}\,\mathsf{N}\,\mathsf{M}'.$$

The assertion is evident. S searches for a square which is marked by the arguments; then $\mathfrak{R}$ leads to the square behind the last square marked by the argument and N to the square $a_{W_1 \cdots W_n}$ over which M' must be placed in order to compute $f(W_1, \ldots, W_n)$.[2]

Similarly we can prove

Theorem 2. Let R be an n-ary relation ($n \geqq 1$) in the domain of words over an alphabet $\mathfrak{A}_0 = \{a_1, \ldots, a_{N_0}\}$. Let N be the same type of machine as N of the previous theorem. Let M' be a machine such that if we print onto the otherwise empty tape an n-tuple $W_1, \ldots, W_n$ of words and if we place M' on the tape marked in this way over $a_{W_1 \cdots W_n}$, then M' will stop operating after finitely many steps over the symbol a_i or a_j ($i \neq j$,

---

[1] It should be remembered that in § 1.4 we agreed to allow only non-empty words as arguments and values of a function.

[2] It can happen that M' uses auxiliary letters, i.e. it is defined over an alphabet $\mathfrak{A} = \{a_1, \ldots, a_N\}$ with $N > N_0$. In this case the machines S, $\mathfrak{R}$ and N must be replaced by machines $\overline{\mathsf{S}}$, $\overline{\mathfrak{R}}$ and $\overline{\mathsf{N}}$ resp. which are also defined over $\mathfrak{A}$. In this case the tables of $\overline{\mathsf{S}}$, $\overline{\mathfrak{R}}$ and $\overline{\mathsf{N}}$ are enlargements of the tables of S, $\mathfrak{R}$ and N resp. In the additional lines the last two columns can be chosen arbitrarily because $\overline{\mathsf{S}}$, $\overline{\mathfrak{R}}$ and $\overline{\mathsf{N}}$ will in our case only be placed on a tape expression over an alphabet $\mathfrak{A}_0$.

$1 \leqq i, j \leqq N_0$), according to whether or not $RW_1 \ldots W_n$ is valid. *Then R is Turing-decidable. A machine* M *which Turing-decides R will be given by*

$$\mathsf{M} = \mathsf{S}\,\mathfrak{R}\,\mathsf{N}\,\mathsf{M}'.$$

2. *Examples of Turing-computable functions.* Here we consider functions whose arguments and values are natural numbers. The natural numbers are represented in the manner described in § 1.4. All the following Turing machines are machines over the alphabet {|}. *We shall further presume that these machines are placed over the first square behind the given arguments.* This square is already the initial square for M′. Thus, we can let e. g. $\mathsf{N} = \mathsf{r}^0$. Therefore we can apply Theorem 1 of the previous section and by its help we can effectively find machines which compute the functions considered in the sense of the definition in § 6.1.

(1) The *successor function $S(x)$ is computed by* | r (cf. § 6.6).

(2) The *sum function* $x + y$ is computed by

$$\mathsf{S}_0 = \mathsf{L}\,|\,\mathsf{R}\,\mathfrak{l} * \mathfrak{l} *.$$

First the machine connects the arguments, which are separated by a one square gap, by filling up this gap. Then we must remove two strokes, since the number n is represented by $n + 1$ strokes. – The interpretation of the sum as the cardinal number of a union of sets is reflected in this computing process.

(3) $f(x) = 2x$ is computed by

$$\mathsf{K}\,\mathsf{S}_0.$$

First the argument is copied by K, and then the sum is computed by S_0 (cf. (2)).

(4) The *product function* $x \cdot y$ is computed by means of

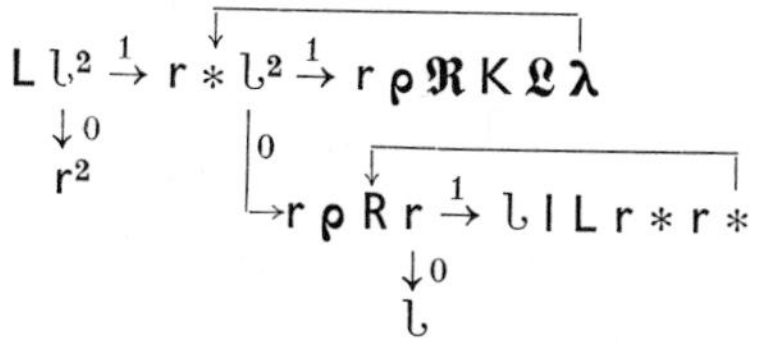

Let the method of operation be illustrated only in short by the remark that (apart from the special case when the first argument is equal to zero) the word represented by the second argument is copied by means of K as often as it is determined by the first argument. Then these copies (together with the original second argument) are joined together by filling up the gaps. In doing this we have to remove two strokes each time (cf. (2)).

(5) The function max (x, y) can be Turing-computed by means of

$$\mathsf{K}_2^2\, \overset{\curvearrowleft}{\mathfrak{l}} * \mathfrak{l} \overset{1}{\to} \mathfrak{L}\, r * r \overset{1}{\to} \mathfrak{R}$$
$$\downarrow 0$$
$$\mathsf{R}$$

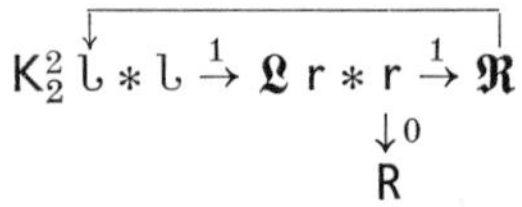

The computation is carried out essentially as follows. We reproduce both arguments by K_2^2. Then we shorten alternatively the rightmost and leftmost words by one stroke, checking each time whether these words are exhausted after the shortening. If the *rightmost* word is erased first, then the *first* argument is the required maximum. This is still preserved intact as a copy. We are just behind it and only need to stop. If on the other hand the *leftmost* word is exhausted first, then the *second* argument is equal to the maximum. This is still in its original position. We have to go behind it.

3. *Examples of Turing-decidable relations.* Similarly to Section 2 we restrict ourselves to giving an account of machines which carry out the decision (with $a_i = a_0$ and $a_j = a_1$) if placed behind the last square marked by an argument.

(1) The property of a number x that it is divisible by a fixed number n $(n \geq 1)$ is Turing-decidable. We give an account of a machine which if placed behind the last stroke of x on an otherwise empty tape stops operating after finitely many steps and more precisely over I if n does not divide x and over $*$ if n divides x.

$$\overset{\pm 0}{\underset{\downarrow 0}{\mathfrak{l}\, \overset{\curvearrowleft}{\mathfrak{l}}\, {}^n\_|}}$$
$$\mathsf{r}^{n-1}$$

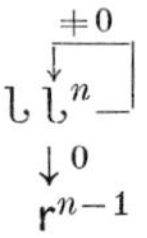

(2) The *equality relation* between numbers is Turing-decidable (the corresponding statement is naturally also valid for words over an arbitrary alphabet (cf. Exercise 2)). For this, we give an account of a machine which, if placed behind the last stroke of the tape expression $\cdots * z_1 * z_2 * \cdots$, stops operating after finitely many steps and does so over $*$ if $z_1 = z_2$, and over I if $z_1 \neq z_2$. The machine determined here removes (starting from the middle) alternatively a stroke of z_1 and z_2 and checks each time whether by the removal of the last stroke the argument in question is exhausted. The result $*$ appears at the end if and only if both arguments are erased at the same time (i.e. if and only if $z_1 = z_2$).

$$\mathsf{L}\, r * r \overset{1}{\to} \overset{\curvearrowleft}{\lambda} * \mathfrak{l} \overset{1}{\to} \rho$$
$$\overset{0}{\underset{\to \lambda\, \mathfrak{l}}{|}} \quad \overset{0}{\underset{\to \mathsf{I}}{|}}$$

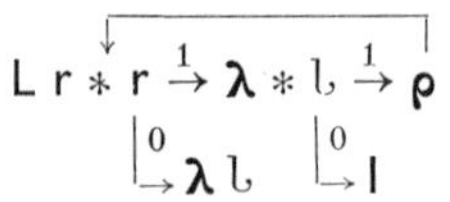

Exercise 1. Give a table for $\mathfrak{R}$ explicitly.

Exercise 2. Give an account of a machine which decides the equality relation between arbitrary words over an alphabet $\{a_1, \ldots, a_N\}$.

CHAPTER 3

μ-RECURSIVE FUNCTIONS

The concept of computable function was at first given intuitively (§ 2). We have, by virtue of an analysis of the behaviour of a calculator (§ 3), arrived at an exact definition of Turing-computability (§ 6). The direct connection with intuition, which is gained by this method, is without doubt a great advantage in realizing the meaning of the precise concepts obtained. On the other hand, the concept of Turing-computability, just as it stands, is not flexible enough for the work of the mathematician. When we want to consider the properties of computable functions we shall try to find, as all mathematicians would do, a new, to the original equivalent definition, which can more easily be handled mathematically. We know today several concepts which are equivalent to Turing-computability. Each one of these new concepts has an intuitive background as well. However, this background is on the whole not of the kind that we would be inclined to believe relatively quickly (as in the case of Turing-computability) that the precise replacement obtained on such a basis comprehends *all* possible computable functions. The fact that for every one of these concepts we can prove rigorously the equivalence to Turing-computability strengthens in any case the conviction that in all these investigations we are dealing with a quite fundamental concept.

A possible equivalent paraphrase of the concept of Turing-computability is the concept of *μ-recursiveness*. This chapter is devoted to a discussion of this concept.[1]

§ 10. Primitive Recursive Functions

As preparation and first step to the introduction of the concept of μ-recursive function we shall now discuss the concept of primitive recursive function. We arrive at this concept by an analysis of the methods used in mathematics to introduce fundamental functions familiar to every mathematician, like sum, product, etc. We have to note once and for all that we shall only consider functions whose domain of arguments is the domain of natural numbers (0, 1, 2, ...), and whose values are natural numbers as well. We require for a function of n argu-

[1] Cf. the historical remarks on p. 29. For other concepts equivalent to Turing-computability cf. §§ 19, 31, 33.

ments that it is defined for *every* n-tuple of natural numbers. Because of this we shall in some cases consider functions which are somewhat different from those which usually occur in mathematics. We shall for instance introduce instead of the difference $x - y$ a modified difference $x \dot{-} y$, which has the value 0 whenever $x - y$ is negative.

1. Definitions. The two processes which are mainly used in mathematical practice to define new functions are the processes of substitution and of inductive definition. Let us now consider these more closely.

Let $h_1, \ldots, h_r$ be functions of n arguments ($r \geqq 1$, $n \geqq 0$[1]) and g an r-ary function. Now, if we have that for arbitrary arguments[2] $\mathfrak{x}$

$$(*) \qquad f(\mathfrak{x}) = g(h_1(\mathfrak{x}), \ldots, h_r(\mathfrak{x})),$$

then we say that f is obtained from g by the *substitution* of $h_1, \ldots, h_r$. We call (*) the *substitution schema.* It is obvious that we can, if the functions $g, h_1, \ldots, h_r$ are given, conceive (*) as a possible *definition* of the function f.

Let g be a function of n arguments ($n \geqq 0$) and let h be a function of $n + 2$ arguments. Now, if we have that, for arbitrary arguments $\mathfrak{x}, y$,

$$(**) \qquad \begin{cases} f(\mathfrak{x}, 0) = g(\mathfrak{x}) \\ f(\mathfrak{x}, y') = h(\mathfrak{x}, y, f(\mathfrak{x}, y))^3, \end{cases}$$

then we say that f is *defined inductively* by the equations (**) by the help of g and h. We call (**) an *induction (recursion) schema.* Obviously, if the functions g and h are given, then the function f is unambiguously determined by the schema (**).

We may now be interested in functions which we could obtain, starting from certain functions which suggest themselves, by iterated applications of the substitution and induction processes. We take as *initial functions*

(a) the *successor function* S which, for an argument x, has the value $S(x) = x' = x + 1$,

(b) the *identity functions* U_n^i ($n \geqq 1$, $1 \leqq i \leqq n$) which are defined by $U_n^i(x_1, \ldots, x_n) = x_i$ for all $x_1, \ldots, x_n$,

(c) the 0-*ary constant* C_0^0.[4]

Definition. A function is called *primitive recursive* if it is one of the above mentioned initial functions, or if it is obtainable from one of these

---

[1] For the concept of 0-ary function cf. § 6.1.

[2] We write here and in the rest of the book $\mathfrak{x}$ for $x_1, \ldots, x_n$. Similarly we use $\mathfrak{y}$, $\mathfrak{z}$, etc.

[3] We write, as usual, y' for the successor $y + 1$.

[4] C_0^0 has the value 0. Cf. § 6.1.

initial functions by finitely many applications of the substitution and induction processes.

The initial functions are (intuitively speaking) computable. It is immediately clear that the processes of substitution and inductive definition lead from computable functions to computable functions again (i.e. f is computable in case of the substitution schema (*) if $g, h_1, \ldots, h_r$ are computable, and f is computable in case of the induction schema (**), if g and h are computable). *This means that all primitive recursive functions are computable*[1].

2. *More general substitution processes.* We required in (*) that all of the functions $h_1, \ldots, h_r$ be of the same number of variables and furthermore that they are functions of the same variables. Thus in the case of (*) we are dealing with a rather special substitution process. But we can easily see that even more general substitution processes lead from primitive recursive functions to primitive recursive functions again. For this we require the help of the functions U_n^i. We show the method by three examples.

(a) *Identification of two variables.* Let $1 \leq i \leq n$. Let, for all $x_1, \ldots, x_n$,

$$f(x_1, \ldots, x_n) = g(x_1, \ldots, x_k, x_i, x_{k+1}, \ldots, x_n).$$

Then, if g is primitive recursive, so is f.

In agreement with mathematical usage we can express this assertion in short by saying that if g is primitive recursive, then so is the function $g(x_1, \ldots, x_k, x_i, x_{k+1}, \ldots, x_n)$. We shall occasionally use a similar way of expression in other cases.[2]

The proof for the above assertion follows from the representation

$$f(x_1, \ldots, x_n) = g(U_n^1(x_1, \ldots, x_n), \ldots, U_n^k(x_1, \ldots, x_n), U_n^i(x_1, \ldots, x_n), U_n^{k+1}(x_1, \ldots, x_n), \ldots, U_n^n(x_1, \ldots, x_n)).$$

(b) *Permutation of variables.* Let π be a permutation of $1 \ldots n$. Let $f = g_\pi$, i.e., for all $x_1, \ldots, x_n$,

$$f(x_1, \ldots, x_n) = g(x_{\pi(1)}, \ldots, x_{\pi(n)}).$$

Then, if g is primitive recursive, so is f. We have, as a matter of fact, that

$$f(x_1, \ldots, x_n) = g(U_n^{\pi(1)}(x_1, \ldots, x_n), \ldots, U_n^{\pi(n)}(x_1, \ldots, x_n)).$$

---

[1] In these last remarks we have dealt with *intuitive* computability. The *Turing*-computability of the primitive recursive functions, and more generally of all μ-recursive functions, will be proved in § 16.

[2] It is true that this way of expression is usual in mathematics, however, it is not always recommendable. Cf. the remarks in § 31.1. There we shall give an account of a better notation $\lambda x_1 \ldots x_n g(x_1, \ldots, x_k, x_i, x_{k+1}, \ldots, x_n)$.

(c) *Substitution for a single variable.* Let, for all $x_1, \ldots, x_n$ and for all $y_1, \ldots, y_m$ (abbreviated by $\mathfrak{y}$),

$$f(x_1, \ldots, x_n, \mathfrak{y}) = g(x_1, \ldots, x_k, h(\mathfrak{y}), x_{k+1}, \ldots, x_n).$$

Then, if g and h are primitive recursive, so is f. To prove this we introduce first the primitive recursive function H by

$$H(x_1, \ldots, x_n, \mathfrak{y}) = h(U_{n+m}^{n+1}(x_1, \ldots, x_n, \mathfrak{y}), \ldots, U_{n+m}^{n+m}(x_1, \ldots, x_n, \mathfrak{y})).$$

Then, we have that

$$f(x_1, \ldots, x_n, \mathfrak{y}) = g(U_{n+m}^1(x_1, \ldots, x_n, \mathfrak{y}), \ldots, U_{n+m}^k(x_1, \ldots, x_n, \mathfrak{y}), \\ H(x_1, \ldots, x_n, \mathfrak{y}), U_{n+m}^{k+1}(x_1, \ldots, x_n, \mathfrak{y}), \ldots, U_{n+m}^n(x_1, \ldots, x_n, \mathfrak{y})).$$

3. Other induction procedures. The induction schema (**) is very special. It only allows inductions over the *last* variable. Furthermore, it is required that the function g is a function of n variables and that the function h is a function of $n+2$ variables, and finally that the value $f(\mathfrak{x}, y)$ is the last argument in h. Just as in the case of substitution, we can show here that other simple induction procedures lead from primitive recursive functions to primitive recursive functions again. We give two examples.

(a) Let f be introduced by the equations

$$f(\mathfrak{x}, 0) = g(\mathfrak{x})$$
$$f(\mathfrak{x}, y') = h(f(\mathfrak{x}, y)).$$

Then, if g and h are primitive recursive, so is f. To prove this we introduce the primitive recursive function $H(\mathfrak{x}, y, z)$ by the definition

$$H(\mathfrak{x}, y, z) = h(U_{n+2}^{n+2}(\mathfrak{x}, y, z)).$$

We can now replace the second line of the definition of f by

$$f(\mathfrak{x}, y') = H(\mathfrak{x}, y, f(\mathfrak{x}, y)).$$

(b) Let f be introduced by

$$f(0, \mathfrak{x}) = g(\mathfrak{x})$$
$$f(y', \mathfrak{x}) = h(y, f(y, \mathfrak{x}), \mathfrak{x}).$$

Then, if g and h are primitive recursive, so is f. For this, we first consider the functions F and H which are defined by

$$H(\mathfrak{x}, y, z) = h(y, z, \mathfrak{x})$$
$$F(\mathfrak{x}, 0) = g(\mathfrak{x})$$
$$F(\mathfrak{x}, y') = H(\mathfrak{x}, y, F(\mathfrak{x}, y)).$$

H is primitive recursive by Section 2 (b). Therefore F is also primitive recursive. We can easily show by induction on y that for all $\mathfrak{x}, y$

$$f(y, \mathfrak{x}) = F(\mathfrak{x}, y).$$

From this follows, again by Section 2 (b), that f is also primitive recursive.

Remark 1. Instead of the induction schema (**) we could have used also the schema discussed just now for the definition of primitive recursive functions. To show the equivalence of the two definitions we would have to show that the schema (**) does not lead outside the class of the functions defined by our new definition. This proof can be carried out completely analogously to the proof given last.

Remark 2. In § 12 we shall deal with more general induction (recursion) schemata which lead from primitive recursive functions to primitive recursive functions again. However, we must not believe that this is valid for *every* recursion schema. Cf. § 13.

4. Examples of primitive recursive functions. In order to establish the primitive recursiveness of functions we shall in this section frequently state definitions of these functions which are generalizations of the schemata (*) and (**) respectively. The primitive recursiveness of these functions follows either immediately from the results of the last two sections or from completely analogous considerations. — For the sake of the completeness of our enumeration we begin with the initial functions.

(1) $S(x)$ (Initial function)

(2) $U_n^i(x_1, \ldots, x_n)$ (Initial functions)

(3) C_0^0 (Initial function)

(4) *The functions* $C_n^k(x_1, \ldots, x_n)$ $(n \geqq 0)$ which have the constant value k.[1] These can be introduced for $n = 0, 1, 2, \ldots$ successively.

$n = 0$: C_0^0 is an initial function.

$$C_0^{k'} = S(C_0^k)$$

$n = 1$:

$$C_1^k(0) = C_0^k$$

$$C_1^k(y') = C_1^k(y)$$

$n = 2$:

$$C_2^k(x, 0) = C_1^k(x)$$

$$C_2^k(x, y') = C_2^k(x, y)$$

etc.

---

[1] Cf. § 7.4.

(5) *The sum* $x + y$ $$x + 0 = x$$ $$x + y' = (x + y)'.$$

(6) *The product* $x \cdot y$ $$x \cdot 0 = 0$$ $$x \cdot y' = xy + x.$$

(7) *The power* x^y $$x^0 = 1$$ $$x^{y'} = x^y \cdot x.$$ [1]

(8) *The factorial* $x!$ $$0! = 1$$ $$y'! = y! \cdot y'.$$

We also quote another few functions. For these we give the most obvious definition on the left and a hint on how the primitive recursiveness can be proved on the right. The proofs here and in similar considerations of the following paragraphs are in each case constructive. By that we mean here that we can read off from these proofs how the corresponding functions can be obtained from the initial functions by substitutions and inductive definitions (the two basic processes for obtaining primitive recursive functions).

(9) *The predecessor function*

$$V(x) = \begin{cases} 0 & \text{for } x = 0 \\ x - 1 & \text{for } x \neq 0 \end{cases} \qquad \begin{aligned} V(0) &= 0 \\ V(y') &= y. \end{aligned}$$

(10) *The modified difference*

$$x \dot{-} y = \begin{cases} 0 & \text{for } x < y \\ x - y & \text{for } y \leqq x \end{cases} \qquad \begin{aligned} x \dot{-} 0 &= x \\ x \dot{-} y' &= V(x \dot{-} y). \end{aligned}$$

(11) *The absolute difference*

$$|x - y| = \begin{cases} x - y \text{ for } y \leqq x \\ y - x \text{ for } x < y \end{cases} \qquad |x - y| = (x \dot{-} y) + (y \dot{-} x).$$

(12) *The signum function*

$$sg(x) = \begin{cases} 0 \text{ for } x = 0 \\ 1 \text{ for } x > 0 \end{cases} \qquad \begin{aligned} sg(0) &= 0 \\ sg(y') &= 1. \end{aligned}$$

(13) *The* $\overline{sg}$*-function*

$$\overline{sg}(x) = \begin{cases} 1 \text{ for } x = 0 \\ 0 \text{ for } x > 0 \end{cases} \qquad \begin{aligned} \overline{sg}(0) &= 1 \\ \overline{sg}(y') &= 0. \end{aligned}$$

---

[1] According to this definition of $0^0 = 1$, although usually the explanation of 0^0 is waived.

(14) *The ε-function*

$$\varepsilon(x, y) = \begin{cases} 0 \text{ for } x = y \\ 1 \text{ for } x \neq y \end{cases} \qquad \varepsilon(x, y) = sg(|x - y|).$$

(15) *The δ-function*

$$\delta(x, y) = \begin{cases} 1 \text{ for } x = y \\ 0 \text{ for } x \neq y \end{cases} \qquad \delta(x, y) = \overline{sg}(|x - y|).$$

5. *The processes Σ and Π.* We show the

Theorem. If the function f is primitive recursive, and if

$$g(\mathfrak{x}, z) = \sum_{y=0}^{z} f(\mathfrak{x}, y, z),$$

$$h(\mathfrak{x}, z) = \prod_{y=0}^{z} f(\mathfrak{x}, y, z),$$

then the functions g and h are also primitive recursive.

Proof. First we consider the primitive recursive functions g^* and h^* which are defined by

$$\begin{aligned} g^*(\mathfrak{x}, 0, z) &= f(\mathfrak{x}, 0, z) \\ g^*(\mathfrak{x}, w', z) &= g^*(\mathfrak{x}, w, z) + f(\mathfrak{x}, w', z) \\ h^*(\mathfrak{x}, 0, z) &= f(\mathfrak{x}, 0, z) \\ h^*(\mathfrak{x}, w', z) &= h^*(\mathfrak{x}, w, z) \cdot f(\mathfrak{x}, w', z). \end{aligned}$$

Obviously,

$$g^*(\mathfrak{x}, w, z) = \sum_{y=0}^{w} f(\mathfrak{x}, y, z), \qquad h^*(\mathfrak{x}, w, z) = \prod_{y=0}^{w} f(\mathfrak{x}, y, z).$$

Finally, we obtain from this that

$$\begin{aligned} g(\mathfrak{x}, z) &= g^*(\mathfrak{x}, z, z) \\ h(\mathfrak{x}, z) &= h^*(\mathfrak{x}, z, z). \end{aligned}$$

Exercise. Show explicitly by reduction to the initial functions and the schemata (*) and (**) that the sum and the product are primitive recursive functions.

References

GÖDEL, K.: Über formal unentscheidbare Sätze der Principia Mathematica und verwandter Systeme I. Mh. Math. Phys. **38**, 173–198 (1931). (It is here that the primitive recursive functions occur the first time under the name "recursive functions", whereas today we denote by this expression a more comprehensive class of functions; cf. § 19.)

GÖDEL, K.: On Undecidable Propositions of Formal Mathematical Systems. Mimeographed. Institute for Advanced Study, Princeton, N. J. 1934. 30 pp. (First complete account of the initial functions.)

HILBERT, D., and P. BERNAYS: Grundlagen der Mathematik I, Berlin: J. Springer 1934. (Introduction of the concept of *primitive recursion;* p. 326).

KLEENE, S. C.: General Recursive Functions of Natural Numbers. Math. Ann. **112**, 727–742 (1936). (Introduction of the expression "primitive recursive functions".)

PÉTER, R.: Rekursive Funktionen. Budapest: Verlag der ungarischen Akademie der Wissenschaften 1957. (Detailed treatment of the primitive recursive functions.)

§ 11. Primitive Recursive Predicates

In the previous paragraph we showed the primitive recursiveness of a few functions. In the examples given last we were able to do this by replacing the original definitions of the functions in question by equivalent ones so that the primitive recursiveness is clearly demonstrated in the new definitions. Such paraphrasing can become quite troublesome in more complicated cases. It would be nice if we had a method at our disposal by the help of which we could decide the primitive recursiveness of a function directly from its definition. It is for this reason that we introduce the concept of primitive recursive predicate.

1. Definitions. An n-ary predicate ($n \geqq 1$) is an n-ary relation between natural numbers which is valid for certain (ordered) n-tuples of numbers. — The prime number predicate for example is a singulary predicate which is valid for 2, 3, 5, ..., and not for 0, 1, 4, ... The "less than" relation is a binary predicate which is valid for the ordered pair (4, 8), but is not valid for the pair (6, 3) or the pair (4, 4). We can also for instance consider a ternary predicate of betweenness which is valid e.g. for the triple (3, 8, 9), since $3 < 8 < 9$.

$P\mathfrak{x}$ shall mean that the predicate P is valid for the n-tuple $\mathfrak{x}$.

Definition. An n-ary *predicate* P ($n \geqq 1$) is called *primitive recursive,* if there exists a primitive recursive n-ary function such that, for an arbitrary n-tuple $\mathfrak{x}$ of numbers,

$$P\mathfrak{x} \text{ if and only if } f(\mathfrak{x}) = 0 .$$

We can find out by calculating the value of the function f for the argument $\mathfrak{x}$ whether or not $P\mathfrak{x}$. This shows, since every primitive recursive function is computable, that every primitive recursive predicate is decidable.

In § 2.3 we introduced the concept of the characteristic function of a set. We can, more generally, speak of the characteristic function of a predicate.

Definition. The n-ary function f is called *characteristic function* of the n-ary predicate P if and only if for all $\mathfrak{x}$[1]

$$(P\mathfrak{x} \leftrightarrow f(\mathfrak{x}) = 0) \wedge (\neg P\mathfrak{x} \leftrightarrow f(\mathfrak{x}) = 1).$$

Every predicate has exactly one characteristic function. We have the

Theorem. A predicate P is primitive recursive if and only if the characteristic function of P is primitive recursive.

To prove this we need only show that the characteristic function of a primitive recursive predicate P is primitive recursive. There exists a primitive recursive function f (depending on P) such that

$$P\mathfrak{x} \leftrightarrow f(\mathfrak{x}) = 0 \qquad \text{for all } \mathfrak{x}.$$

Let $g(\mathfrak{x}) = sg(f(\mathfrak{x}))$. Then g is primitive recursive and it is the characteristic function of P.

2. *Processes for the generation of predicates.* In this section we shall derive a few operations in the domain of predicates. We shall show in Section 3 that several of these operations *(not all)* lead from primitive recursive predicates to primitive recursive predicates again.

(a) Let Q be an n-ary predicate. The predicate P is called *the negation* or *complement of* Q, if P and Q have the same number of arguments and if for all $\mathfrak{x}$

$$P\mathfrak{x} \leftrightarrow \neg Q\mathfrak{x}.$$

(b) Let Q be an n-ary and R be an m-ary predicate. The predicate P is called *the conjunction of* Q *and* R, if P is $(n+m)$-ary and if for all $\mathfrak{x}$ and $\mathfrak{y}$

$$P\mathfrak{x}\mathfrak{y} \leftrightarrow Q\mathfrak{x} \wedge R\mathfrak{y}.$$

(c) Let Q be an n-ary predicate. Let π be a permutation of $1, \ldots, n$. The predicate P is called the *π-permutation of* Q, if P is n-ary and if for all $x_1, \ldots, x_n$

$$Px_1 \ldots x_n \leftrightarrow Qx_{\pi(1)} \ldots x_{\pi(n)}.$$

---

[1] In this book we make use of the logical symbols $\neg$ (not), $\wedge$ (and), $\vee$ (or), $\rightarrow$ (if, then), $\leftrightarrow$ (if and only if), $\bigwedge_x$ (for all x), $\bigvee_x$ (there exists an x). If $\mathfrak{x}$ is an n-tuple, $\mathfrak{x} = (x_1, \ldots, x_n)$, then $\bigwedge_{\mathfrak{x}}$ shall be an abbreviation for $\bigwedge_{x_1} \ldots \bigwedge_{x_n}$, and $\bigvee_{\mathfrak{x}}$ shall be an abbreviation for $\bigvee_{x_1} \ldots \bigvee_{x_n}$. Finally we use the bounded *quantifiers* $\bigwedge_{x=0}^{n}$ (for all x between 0 and n including 0 and n), and $\bigvee_{x=0}^{n}$ (there exists an x between n and 0 including 0 and n). $\rightarrow$ and $\leftrightarrow$ will rank ahead of $\wedge$ and $\vee$, so that e.g. $p \wedge q \rightarrow r$ will stand for $(p \wedge q) \rightarrow r$ (omitting of brackets).

(d) Let Q be an n-ary predicate ($n \geqq 2$). Let $1 \leqq i < k \leqq n$. Then P is called the *(i, k)-identification of Q*, if P is $(n-1)$-ary and if for all $x_1, \ldots, x_{k-1}, x_{k+1}, \ldots, x_n$

$$P x_1 \ldots x_{k-1}\, x_{k+1} \ldots x_n \leftrightarrow Q x_1 \ldots x_{k-1}\, x_i\, x_{k+1} \ldots x_n .$$

(e) We shall call the predicates P which can be obtained from the predicates Q and R by conjunction and subsequent permutations and identifications *generalized conjunctions* of Q and R. P is for example a generalized conjunction of Q and R if for all x_1, x_2, x_3

$$P x_1 x_2 x_3 \leftrightarrow Q x_2 x_3 \wedge R x_1 x_2$$

or if for all $\mathfrak{x}$

$$P\mathfrak{x} \leftrightarrow Q\mathfrak{x} \wedge R\mathfrak{x} .$$

(f) Let Q be an n-ary predicate. Let $1 \leqq i \leqq n$. The $(n-1)$-ary predicate P is called *the ith generalization of Q*, if for all $x_1, \ldots, x_{i-1}, x_{i+1}, \ldots, x_n$

$$P x_1 \ldots x_{i-1}\, x_{i+1} \ldots x_n \leftrightarrow \bigwedge_{x_i} Q x_1 \ldots x_n .$$

(g) Let Q be an n-ary predicate. Let $1 \leqq i \leqq n$. Then the n-ary predicate P is called *the ith bounded generalization of Q*, if for all $x_1, \ldots, x_{i-1}, x_{i+1}, \ldots, x_{n+1}$

$$P x_1 \ldots x_{i-1}\, x_{i+1} \ldots x_n x_{n+1} \leftrightarrow \bigwedge_{x_i=0}^{x_{n+1}} Q x_1 \ldots x_n .$$

(h) There is a line of further processes which we only mention by way of suggestion because they can be traced back to the processes already known. We can introduce an *alternative* of the predicates Q and R by the definition $P\mathfrak{x}\mathfrak{y} \leftrightarrow Q\mathfrak{x} \vee R\mathfrak{y}$. The alternative can be traced back to negations and conjunctions since $Q\mathfrak{x} \vee R\mathfrak{y} \leftrightarrow \neg(\neg Q\mathfrak{x} \wedge \neg R\mathfrak{y})$. We can deal similarly with the *implication* of Q and R, because $(Q\mathfrak{x} \rightarrow R\mathfrak{y}) \leftrightarrow \neg Q\mathfrak{x} \vee R\mathfrak{y}$. Just as in the case of conjunction we can also speak of generalized alternatives and implications.

We call the predicate P the *i^{th} particularization of Q* if for all $x_1, \ldots, x_{i-1}, x_{i+1}, \ldots, x_n$

$$P x_1 \ldots x_{i-1}\, x_{i+1} \ldots x_n \leftrightarrow \bigvee_{x_i} Q x_1 \ldots x_n .$$

Since $\bigvee_{x_i} Q x_1 \ldots x_n \leftrightarrow \neg \bigwedge_{x_i} \neg Q x_1 \ldots x_n$ we have that the ith particularization can be traced back to the ith generalization. A similar result applies in the case of the ith *bounded particularization* (which is analogous to the ith bounded generalization), for

$$\bigvee_{x_i=0}^{x_{n+1}} Qx_1 \ldots x_n \leftrightarrow \neg \bigwedge_{x_i=0}^{x_{n+1}} \neg Qx_1 \ldots x_n .$$

(i) Let Q be an n-ary predicate and f be an m-ary function. Let $1 \leq i \leq n$. We say that we obtain the $(n-1+m)$-ary predicate P by *substitution of f in the ith place of P*, if for all $x_1, \ldots, x_{i-1}, x_{i+1}, \ldots, x_n, \mathfrak{y}$

$$Px_1 \ldots x_{i-1}\, x_{i+1} \ldots x_n \mathfrak{y} \leftrightarrow Qx_1 \ldots x_{i-1}\, f(\mathfrak{y})\, x_{i+1} \ldots x_n .$$

(j) Finally we introduce the n-ary *empty predicate* O_n, and the n-ary *universal predicate* A_n which we define, for all $x_1, \ldots, x_n$, by

$$O_n x_1 \ldots x_n \leftrightarrow x_1 \neq x_1 \wedge \ldots \wedge x_n \neq x_n, \quad A_n x_1 \ldots x_n \leftrightarrow x_1 = x_1 \wedge \ldots \wedge x_n = x_n .$$

We shall, following mathematical usage, frequently make use of abbreviated ways of expression and speak for instance of the *predicate*

$$\bigvee_{x_2} (f(x_1) = x_2 \wedge Qx_2 y) .$$

By this we understand the binary predicate P which is valid for x_1 and y (in this sequence!) if and only if the condition described is fulfilled. This abbreviated notation presupposes that the natural sequence of the variables occurring in it is given.[1]

3. Application of the generating processes to primitive recursive predicates.

Theorem 1. The operations of negation, of π-permutation, of generalized conjunctions and alternatives and of the bounded quantifications lead from primitive recursive predicates to primitive recursive predicates again.

Proof. Let Q and R be primitive recursive predicates. Thus, there exist primitive recursive functions g and h such that, for all $\mathfrak{x}$, $\mathfrak{y}$,

$$Q\mathfrak{x} \leftrightarrow g(\mathfrak{x}) = 0, \qquad R\mathfrak{y} \leftrightarrow h(\mathfrak{y}) = 0 .$$

(1) Let P be the negation of Q according to Section 2 (a). Then we have for each $\mathfrak{x}$

$$\begin{aligned} P\mathfrak{x} &\leftrightarrow \neg Q\mathfrak{x} \\ &\leftrightarrow g(\mathfrak{x}) \neq 0 \\ &\leftrightarrow \overline{sg}(g(\mathfrak{x})) = 0 . \end{aligned}$$

$\overline{sg}(g(\mathfrak{x}))$ is primitive recursive. Thus, so is P.

---

[1] A better, however longer notation for the predicate P is

$$\hat{x}_1 \hat{y} \left(\bigvee_{x_2} (f(x_1) = x_2 \wedge Qx_2 y)\right),$$

where it is expressed that the variables x_1 and y are really bounded and that they should be considered in the given sequence. Cf. the corresponding note in § 10.2.

(2) Let P be the conjunction of Q and R according to Section 2 (b). Then we have for all $\mathfrak{x}$, $\mathfrak{y}$

$$\begin{aligned} P\mathfrak{x}\mathfrak{y} &\leftrightarrow Q\mathfrak{x} \wedge R\mathfrak{y} \\ &\leftrightarrow g(\mathfrak{x}) = 0 \wedge h(\mathfrak{y}) = 0 \\ &\leftrightarrow g(\mathfrak{x}) + h(\mathfrak{y}) = 0, \end{aligned}$$

for a sum of natural numbers is zero if and only if all terms of the sum are equal to zero. Since $g(\mathfrak{x}) + h(\mathfrak{y})$ is primitive recursive, P is also primitive recursive.

(3) Let P be a π-permutation of Q according to Section 2 (c). Then we have for all $x_1, \ldots, x_n$

$$\begin{aligned} P x_1 \ldots x_n &\leftrightarrow Q x_{\pi(1)} \ldots x_{\pi(n)} \\ &\leftrightarrow g(x_{\pi(1)}, \ldots, x_{\pi(n)}) = 0 \\ &\leftrightarrow g_\pi(x_1, \ldots, x_n) = 0. \end{aligned}$$

g_π is primitive recursive (by § 10.2(b)). Thus, so is P.

(4) Let P be the (i, k)-identification of Q according to Section 2 (d). Then we have, for all $x_1, \ldots, x_{k-1}, x_{k+1}, \ldots, x_n$,

$$\begin{aligned} P x_1 \ldots x_{k-1}\, x_{k+1} \ldots x_n &\leftrightarrow Q x_1 \ldots x_{k-1}\, x_i\, x_{k+1} \ldots x_n \\ &\leftrightarrow g(x_1, \ldots, x_{k-1}, x_i, x_{k+1}, \ldots, x_n) = 0. \end{aligned}$$

The function $g(x_1, \ldots, x_{k-1}, x_i, x_{k+1}, \ldots, x_n)$ is primitive recursive (by § 10.2 (a)). Thus, so is P.

(5) It follows from (2), (3) and (4) that every generalized conjunction of Q and R is primitive recursive (by Section 2 (e)). This, together with (1), implies that every generalized alternative of Q and R is also primitive recursive.

(6) Let P be the ith bounded generalization of Q according to Section 2 (g). Then we have, for all $x_1, \ldots, x_{i-1}, x_{i+1}, \ldots, x_n, x_{n+1}$,

$$\begin{aligned} P x_1 \ldots x_{i-1}\, x_{i+1} \ldots x_n\, x_{n+1} &\leftrightarrow \bigwedge_{x_i=0}^{x_{n+1}} Q x_1 \ldots x_n \\ &\leftrightarrow \bigwedge_{x_i=0}^{x_{n+1}} g(x_1, \ldots, x_n) = 0 \\ &\leftrightarrow g(x_1, \ldots, 0, \ldots, x_n) = 0 \\ &\quad \wedge g(x_1, \ldots, 1, \ldots, x_n) = 0 \\ &\quad \wedge \ldots \\ &\quad \wedge g(x_1, \ldots, x_{n+1}, \ldots, x_n) = 0 \end{aligned}$$

$$\begin{aligned}\leftrightarrow\, & g(x_1, \ldots, 0, \ldots, x_n) \\ & + g(x_1, \ldots, 1, \ldots, x_n) \\ & + \cdots \\ & + g(x_1, \ldots, x_{n+1}, \ldots, x_n) = 0 \\ \leftrightarrow\, & \sum_{x_i=0}^{x_{n+1}} g(x_1, \ldots, x_n) = 0 .\end{aligned}$$

Now we see (cf. Theorem in § 10.5) that in the last line there is a primitive recursive function on the right. This shows that the ith bounded generalization of Q leads to a primitive recursive predicate P. The same applies in case of the ith bounded particularization, for this can be traced back (according to Section 2(h)) to the ith bounded generalization and the negation.

Remark. We see from (2) that the *conjunction* of predicates corresponds to the *addition* of the characteristic functions. In a similar way we obtain that

$$Q\mathfrak{x} \vee R\mathfrak{y} \leftrightarrow g(\mathfrak{x}) = 0 \vee h(\mathfrak{y}) = 0 \leftrightarrow g(\mathfrak{x}) \cdot h(\mathfrak{y}) = 0.$$

(For a product of natural numbers is equal to zero if and only if at least one factor is equal to zero.) Thus the *alternative* of predicates corresponds to the *product* of their characteristic functions.

Theorem 2. Let P be obtained from the primitive recursive predicate Q by substitution of the primitive recursive function f in the ith place. Then P is primitive recursive.

Proof. There exists a primitive recursive function g such that, for all $\mathfrak{x}$, $Q\mathfrak{x} \leftrightarrow g(\mathfrak{x}) = 0$. Then, we have (cf. Section 2 (i)) that

$$\begin{aligned}P x_1 \ldots x_{i-1}\, x_{i+1} \ldots x_n \mathfrak{y} & \leftrightarrow Q x_1 \ldots x_{i-1}\, f(\mathfrak{y})\, x_{i+1} \ldots x_n \\ & \leftrightarrow g(x_1, \ldots, x_{i-1}, f(\mathfrak{y}), x_{i+1}, \ldots, x_n) = 0 .\end{aligned}$$

In the last line we have a primitive recursive function on the right (by § 10.2(c)). This shows that P is primitive recursive.

4. *Remark on the unbounded quantifications.* We have just shown in Theorem 1 that the *bounded* quantifications lead from primitive recursive predicates to primitive recursive predicates again. On the other hand *the unbounded quantifications* $\bigwedge_x$ *and* $\bigvee_x$ *lead in general outside the range of the primitive recursive predicates.* We justify this statement as follows. Let us take for example the particularisator $\bigvee$. When we start with an arbitrary primitive recursive function $f(\mathfrak{x}, y)$ and consider the predicate P which is defined (by the help of the unbounded particularisator) by the stipulation that, for all $\mathfrak{x}$,

$$P\mathfrak{x} \leftrightarrow \bigvee_y f(\mathfrak{x}, y) = 0 .$$

Now, it is true that we can calculate for every $\mathfrak{x}$ and y the value $f(\mathfrak{x}, y)$ of the function. However, this is not sufficient to decide whether, for a given $\mathfrak{x}$, there exists an y for which $f(\mathfrak{x}, y) = 0$. Namely, if there exists no such y, then we are not able to determine this by working out for this $\mathfrak{x}$ and for all y the values of $f(\mathfrak{x}, y)$, because that is not a finite process. Thus we cannot just assume, without further investigation, that P is decidable or, what is more, that it is primitive recursive.[1]

5. *Further primitive recursive predicates.* We shall apply the Theorems of Section 3 to show the primitive recursiveness for a list of predicates.

Theorem 3. The predicates $=, \leqq, <, \geqq, >, /$ (divides), E (is even), D (is odd), Pr (is a prime number), O_n, A_n (cf. Section 2(j)) are primitive recursive.

Proof. We show this step by step by means of the following relations, which are valid for arbitrary x and y.

$$x = y \leftrightarrow \varepsilon(x, y) = 0$$

$$x \leqq y \leftrightarrow \bigvee_{z=0}^{y} (x + z = y)$$

$$x < y \leftrightarrow x \leqq y \wedge x \neq y$$

$$x \geqq y \leftrightarrow y \leqq x$$

$$x > y \leftrightarrow y < x$$

$$x/y \leftrightarrow \bigvee_{z=0}^{y} xz = y$$

$$Ex \leftrightarrow 2/x$$

$$Dx \leftrightarrow \neg Ex$$

$$Pr\ x \leftrightarrow x \neq 0 \wedge x \neq 1 \wedge \bigwedge_{z=0}^{x} (z/x \rightarrow z = 1 \vee z = x)$$

For O_n and A_n the assertion follows immediately from the definition of these predicates, because $=$ is primitive recursive.

In the cases of the relations given for $\leqq$ and $/$ we should keep in in mind that the usual definitions $x \leqq y \leftrightarrow \bigvee_z x + z = y$ and $x/y \leftrightarrow \bigvee_z xz = y$ respectively are not sufficient for our purpose. They do not show the primitive recursiveness, for *unbounded* particularisators appear in them. However, we see at once that we may confine ourselves to $z = 0, \ldots, y$. Similar assertions are valid in connection with the definition of Pr and with several definitions of the following paragraphs.

6. *Definition by cases.* Now we shall deal with a procedure, the so-called definition by cases, which is often used to define a function f

[1] For an exact proof cf. § 22.3.

by the help of the functions $g_1, \ldots, g_m$ and predicates $P_1, \ldots, P_m$ (all g_i's and P_i's already known). Such a definition looks like this:

$$f(\mathfrak{x}) = \begin{cases} g_1(\mathfrak{x}), & \text{if } P_1\mathfrak{x} \\ \ldots\ldots, & \ldots\ldots \\ \ldots\ldots, & \ldots\ldots \\ g_m(\mathfrak{x}), & \text{if } P_m\mathfrak{x}, \end{cases}$$

where it is presumed that for every $\mathfrak{x}$ *exactly one of the predicates* $P_1, \ldots, P_m$ *is valid.* We have here

Theorem 4. If $g_1, \ldots, g_m$ are primitive recursive functions and $P_1, \ldots, P_m$ are primitive recursive predicates, then f is also a primitive recursive function.

Proof. We have for all $\mathfrak{x}$ and every r $(r = 1, \ldots, m)$ that $P_r\mathfrak{x} \leftrightarrow h_r(\mathfrak{x}) = 0$, where $h_1, \ldots, h_m$ are primitive recursive functions. Then we have

$$f(\mathfrak{x}) = g_1(\mathfrak{x}) \cdot \overline{sg}(h_1(\mathfrak{x})) + \cdots + g_m(\mathfrak{x}) \cdot \overline{sg}(h_m(\mathfrak{x})).$$

This is so, since if P_r is the only predicate that is valid for $\mathfrak{x}$, then $h_r(\mathfrak{x}) = 0$ and the other $h_i(\mathfrak{x}) \neq 0$. Thus $\overline{sg}(h_r(\mathfrak{x})) = 1$ and all other $\overline{sg}(h_i(\mathfrak{x})) = 0$, so that the right hand side of the equation coincides with $g_r(\mathfrak{x})$. The given representation shows the primitive recursiveness of f.

Corollary 1. The schema

$$f(\mathfrak{x}) = \begin{cases} g_1(\mathfrak{x}), & \text{if } P_1\mathfrak{x} \\ \ldots\ldots, & \ldots\ldots \\ g_{m-1}(\mathfrak{x}), & \text{if } P_{m-1}\mathfrak{x} \\ g_m(\mathfrak{x}), & \text{otherwise} \end{cases}$$

defines a primitive recursive function f, provided $g_1, \ldots, g_m, P_1, \ldots, P_{m-1}$ are primitive recursive and $P_1, \ldots, P_{m-1}$ are mutually exclusive. This follows from the fact that the case "otherwise" occurs if and only if $\neg P_1\mathfrak{x} \wedge \ldots \wedge \neg P_{m-1}\mathfrak{x}$ and this defines according to Section 3 a primitive recursive predicate.

Corollary 2. A predicate P which is valid for only finitely many n-tuples of numbers is primitive recursive.

Proof. The theorem is true for an empty predicate P by Theorem 3 (Section 5). Let P be non-empty and let it be valid for the n-tuples $\mathfrak{x}_1, \ldots, \mathfrak{x}_s$. Then we can define the characteristic function f of P by definition by cases as follows.

$$f(\mathfrak{x}) = \begin{cases} 0 \text{ for } \mathfrak{x} = \mathfrak{x}_1 \vee \ldots \vee \mathfrak{x} = \mathfrak{x}_s \\ 1 \text{ otherwise.} \end{cases}$$

This representation shows that f and with it P are primitive recursive.

Theorem 5. The functions $\max(x_1, \ldots, x_n)$ and $\min(x_1, \ldots, x_n)$ are primitive recursive.

Proof. First, for $n = 2$, the primitive recursiveness of max follows from the representation

$$\max(x_1, x_2) = \begin{cases} x_1, \text{ if } x_1 \geqq x_2 \\ x_2, \text{ if } x_1 < x_2. \end{cases}$$

Starting from this, the assertion follows step by step for n by virtue of the representation

$$\max(x_1, \ldots, x_{n+1}) = \max(\max(x_1, \ldots, x_n), x_{n+1}).$$

For min we prove the theorem by similar considerations.

Theorem 6. If f is primitive recursive, then so is the function

$$g(\mathfrak{x}, z) = \max_{y=0}^{z} f(\mathfrak{x}, y).$$

Proof. We have

$$g(\mathfrak{x}, 0) = f(\mathfrak{x}, 0)$$
$$g(\mathfrak{x}, z') = \max(g(\mathfrak{x}, z), f(\mathfrak{x}, z')).$$

§ 12. The μ-Operator

μ-operators can be applied to predicates and turn them into functions. By the help of μ-operators we shall introduce (in § 14) the concept of μ-recursive function.

1. *The unbounded μ-operator.* Let P be an $(n+1)$-ary predicate for natural numbers $(n \geqq 0)$. If for $\mathfrak{x}$ there exists a y such that $P\mathfrak{x}y$, then for this $\mathfrak{x}$ there exists an unambiguously determined *smallest* y such that $P\mathfrak{x}y$. We shall denote this smallest y, which depends on $\mathfrak{x}$, by $\mu y P\mathfrak{x}y$ (μ originates from μικρός, which means *small*). If, on the other hand, there is no y for $\mathfrak{x}$ such that $P\mathfrak{x}y$, then we define $\mu y P\mathfrak{x}y = 0$. Thus, $\mu y P\mathfrak{x}y$ is unambiguously defined for every predicate P. μ is called *the unbounded μ-operator*. By the help of μ we can associate with each $(n+1)$-ary predicate P an n-ary function

$$f(\mathfrak{x}) = \mu y P\mathfrak{x}y.$$

We may want to know whether the computability of f follows from the decidability of P. If P is decidable, then we can certainly compute $f(\mathfrak{x})$, provided there exists a y for $\mathfrak{x}$ such that $P\mathfrak{x}y$. For, we can decide then one by one whether $P\mathfrak{x}0$, $P\mathfrak{x}1$, $P\mathfrak{x}2$, ... until we come across a y for the first time for which $P\mathfrak{x}y$. This y is equal, according to definition, to $f(\mathfrak{x})$.

If however there exists no y for $\mathfrak{x}$ such that $P\mathfrak{x}y$, then the procedure just described does not give a computation of $f(\mathfrak{x})$, for it does not terminate after finitely many steps. So it is plausible that there are decidable predicates P for which the corresponding functions f are not computable. As a matter of fact it is obvious that

$$\bigvee_y P\mathfrak{x}y \leftrightarrow f(\mathfrak{x}) \neq 0 \vee (f(\mathfrak{x}) = 0 \wedge P\mathfrak{x}0).$$

If f were computable, then the right hand side would be decidable, and so $\bigvee_y P\mathfrak{x}y$ also. However, we have already worked out in a similar case in § 11.4 that we cannot expect this if we do not know more about P than that it is decidable.

We shall say that *a predicate P is regular* if for *every* $\mathfrak{x}$ there exists a y for which $P\mathfrak{x}y$. In this case the procedure given above carries out satisfactorily the computation of $f(\mathfrak{x})$ for each $\mathfrak{x}$. *Thus, in the case of decidable predicates which are regular the application of the μ-operator leads to computable functions.*

Let us assume that for each $\mathfrak{x}$ there exists exactly one y such that $P\mathfrak{x}y$. We can in this case denote *the y for which $P\mathfrak{x}y$* by $\mu y P\mathfrak{x}y$.

2. *The bounded μ-operator.* We could on account of these facts believe that the application of the μ-operator to a regular primitive recursive predicate always leads to a primitive recursive function. However, this is not the case as we shall see in § 13. On the other hand, we obtain a similar result if we (analogously to the case of quantifiers) turn from the *unbounded* μ-operator μy considered until now to a *bounded* μ-operator $\overset{y}{\underset{z=0}{\mu}}$. First we give the

Definition of the bounded μ-operator.

$$\overset{y}{\underset{z=0}{\mu}} P\mathfrak{x}z = \begin{cases} \text{the smallest } z \text{ between } 0 \text{ and } y \text{ (including the limits } 0 \text{ and } y) \\ \quad \text{for which } P\mathfrak{x}z, \text{ if such } z \text{ exists at all,} \\ 0, \text{ if no such } z \text{ exists.} \end{cases}$$

(The first case is obviously more interesting; that we have defined the value of the function to be 0 in the second case proves to be convenient in many applications.) We should keep in mind that the application of the bounded μ-operator to an $(n+1)$-ary predicate P leads to an $(n+1)$-ary function, since the value of the function depends also on the upper limit y. We have here the

Theorem. Let P be a primitive recursive predicate. Let

$$f(\mathfrak{x}, y) = \overset{y}{\underset{z=0}{\mu}} P\mathfrak{x}z.$$

Then f is a primitive recursive function.

Proof. (Sketch.) First we verify the two equations

$$f(\mathfrak{x}, 0) = 0$$

$$f(\mathfrak{x}, y') = \begin{cases} f(\mathfrak{x}, y), \text{ if there exists a } z \text{ between } 0 \text{ and } y \text{ (including the limits) such that } P\mathfrak{x}z, \\ y', \text{ if the first case does not apply, but } P\mathfrak{x}y', \\ 0, \text{ otherwise.} \end{cases}$$

We introduce the function h by the definition

$$h(\mathfrak{x}, y, t) = \begin{cases} t, \text{ if there exists a } z \text{ between } 0 \text{ and } y \text{ (including the limits) such that } P\mathfrak{x}z, \\ y', \text{ if the first case does not apply, but } P\mathfrak{x}y', \\ 0, \text{ otherwise.} \end{cases}$$

We see at once, especially by the help of § 11.6, Corollary 1, that h is primitive recursive. Now the primitive recursiveness of f is evident, since the two equations stated above can be written in the form (defining f)

$$f(\mathfrak{x}, 0) = 0$$
$$f(\mathfrak{x}, y') = h(\mathfrak{x}, y, f(\mathfrak{x}, y)).$$

Remark. We shall often use expressions of the form

$$\overset{y}{\underset{z=0}{\mu}} P\mathfrak{x}yz.$$

We shall understand by such an expression that we first form $\overset{y}{\underset{z=0}{\mu}} P\mathfrak{x}uz$ and then identify u with y afterwards. The reader should convince himself that this leads to the same result as if we had defined $\overset{y}{\underset{z=0}{\mu}} P\mathfrak{x}yz$ as above from the outset.

3. *Further primitive recursive functions.* We shall show by the help of the bounded μ-operator that a few functions, which we shall use later on, are primitive recursive.

First we introduce the *quotient* $\frac{x}{y}$ by the definition

$$\frac{x}{y} = \overset{x}{\underset{z=0}{\mu}} yz' > x.$$

If $y \neq 0$, then $\frac{x}{y}$ is the largest number t such that $ty \leq x$. If $y \neq 0$ and y divides x, then we have the ordinary quotient. If $y = 0$, then $\frac{x}{y} = 0$.

Further we consider the *prime number function* $p(n)$ or in short p_n, which determines the nth prime number (thus $p(0) = 2$, $p(1) = 3$, $p(2) = 5, \ldots$). We have that

$$p(0) = 2$$

$$p(n') = \mu_{z=0}^{p(n)!+1} (Pr\, z \wedge p(n) < z).$$

In the determination of the upper limit of the μ-operator we have made use of the fact that there is always a prime number between p and $p! + 1$.[1]

Further we introduce the *exponent function* $\exp(n, x)$, which determines the highest possible exponent of the prime number $p(n)$ in the prime decomposition of the number x.[2] If we take into consideration that for $x \neq 0$ the number $\exp(n, x)$ is always smaller than x, then we can write

$$\exp(n, x) = \mu_{z=0}^{x} \neg\, p(n)^{z+1}/x.$$

Finally we shall call the largest n such that $p(n)$ divides x the *length* $l(x)$ of the number x. This can be applied for $x > 1$. We define further $l(0) = l(1) = 0$. For $x \neq 0$ we have $l(x) < x$. Then

$$l(x) = \mu_{z=0}^{x} \bigwedge_{w=0}^{x} (w > z \rightarrow \neg\, p(w)/x).$$

4. The σ-functions. We shall frequently have the task to characterize number pairs, number triples, etc. by numbers. This is a case of *Gödel numbering* (cf. § 1.3). We start with number pairs.

Every (natural) number $z \geqq 1$ can be represented in the form $z = 2^x(2y + 1)$. Here x and y are unambiguously determined. Because of this we know that every number $z \geqq 0$ can be represented in the form $z = 2^x(2y + 1) \dot{-} 1$, where x and y are unambiguously determined for each z. Now, if we associate with each number pair x, y a Gödel number by the function

$$\sigma_2(x, y) = 2^x(2y + 1) \dot{-} 1,$$

---

[1] We should make certain that this definition falls under the schema (**) of § 10.1. If we use abbreviations Qwz for $Pr\, z \wedge w < z$, $k(w, y)$ for $\mu_{z=0}^{y} Qwz$ and $h(n, x)$ for $k(U_2^2(n, x), U_2^2(n, x)! + 1)$, then we see that Q, k and h are primitive recursive and that $p(n') = h(n, p(n))$.

[2] We define $\exp(n, 0) = 0$.

then we have a one-one mapping of the number pairs onto the natural numbers. The inverse functions are given by

$$\sigma_{21}(z) = \exp(0, z+1)$$

$$\sigma_{22}(z) = \frac{\dfrac{z+1}{2^{\exp(0,z+1)}} \dot{-} 1}{2}.$$

σ_2, σ_{21}, σ_{22} are primitive recursive functions. $\sigma_{21}(z)$ and $\sigma_{22}(z)$ are the first and the second component respectively of the number pair whose Gödel number is z. Thus, we have that

$$\sigma_{21}(\sigma_2(x, y)) = x$$
$$\sigma_{22}(\sigma_2(x, y)) = y$$
$$\sigma_2(\sigma_{21}(z), \sigma_{22}(z)) = z.$$

By the help of σ_2, σ_{21}, σ_{22} we can obtain one by one the mappings σ_3, σ_4, ... of triples, quadruples, ... of natural numbers together with the corresponding inverse mappings. For this purpose we define σ_{n+1}, $\sigma_{n+1\,1}, \ldots, \sigma_{n+1\,n+1}$ by the help of the functions $\sigma_n, \sigma_{n1}, \ldots, \sigma_{nn}$, which we assume to be known, as follows

$$\sigma_{n+1}(x_1, \ldots, x_{n+1}) = \sigma_2(\sigma_n(x_1, \ldots, x_n), x_{n+1})$$
$$\sigma_{n+1\,j}(z) = \sigma_{nj}(\sigma_{21}(z)) \qquad (j = 1, \ldots, n)$$
$$\sigma_{n+1\,n+1}(z) = \sigma_{22}(z).$$

Thus, $\sigma_{n+1}(x_1, \ldots, x_{n+1})$ is obtained as the Gödel number of the pair $(\sigma_n(x_1, \ldots, x_n), x_{n+1})$. Especially, we have for $n = 3$ that

$$\sigma_3(x, y, z) = \sigma_2(\sigma_2(x, y), z)$$
$$\sigma_{31}(z) = \sigma_{21}(\sigma_{21}(z))$$
$$\sigma_{32}(z) = \sigma_{22}(\sigma_{21}(z))$$
$$\sigma_{33}(z) = \sigma_{22}(z).$$

All functions σ_n, σ_{nj} are primitive recursive.

5.[1] *An inductive definition, where we substitute in the parameter.* In the ordinary induction schema (§ 10.1 (**)) $f(\mathfrak{x}, y')$ is traced back to $f(\mathfrak{x}, y)$. Thus, the parameter $\mathfrak{x}$ occurs unaltered on the right hand side.

---

[1] The rest of this paragraph can be omitted at the first reading. We shall prove here a few theorems about primitive recursive predicates. We shall use these theorems in § 21. We do not aim at a systematical completeness in stating and proving these theorems. The methods used in the proofs are characteristic of the work with primitive recursive functions and predicates. More about this can be found in the book by R. PÉTER.

We find a different situation for instance in the case of the following definition schema (we confine ourselves in this schema to *one* parameter x)

$$\text{(i)}\begin{cases} f(x, 0) = g(x) \\ f(x, y') = h(x, y, f(x, y), f(\exp(1, x), y), \sum_{k=0}^{x} f(\exp(k, x), y) \cdot H(k, x, y)). \end{cases}$$

Here, for the computation of $f(x, y')$ not only the knowledge of $f(x, y)$ but also that of other values $f(i, y)$ is required. In the special case considered here such i is always less than or equal to x (because $i = \exp(k, x) \leqq x$). (An inductive definition schema in which we also allow $i > x$ will be considered in Section 7.) We assert the

Theorem. If g, h, H are primitive recursive and if F satisfies the conditions (i), then f is also primitive recursive.

Proof. We consider the function F which is defined by

$$F(x, y) = p_0^{f(0,0)} \dots p_x^{f(x,0)} p_{x+1}^{f(0,1)} \dots p_{2x+1}^{(x,1)} p_{2(x+1)}^{f(0,2)} \dots p_{y(x+1)}^{f(0,y)} \dots p_{y(x+1)+x}^{f(x,y)}.$$

For $i \leqq x$ we have

$$\text{(i')} \qquad f(i, y) = \exp(y(x+1) + i, F(x, y)).$$

Further, we introduce the function G by

$$\begin{cases} G(0) = p_0^{g(0)} & (= p_0^{f(0,0)}) \\ G(x') = G(x) p_{x'}^{g(x')} & (= G(x) p_{x'}^{f(x',0)}). \end{cases}$$

G is primitive recursive. We see at once that $G(x) = F(x, 0)$. We have that

$$F(x, y') = F(x, y) \cdot p_{y'(x+1)}^{f(0,y')} \dots p_{y'(x+1)+x}^{f(x,y')} = F(x, y) \prod_{l=0}^{x} p_{y'(x+1)+l}^{f(l,y')}.$$

Now, we can transform here $f(l, y')$ by virtue of the second equation of (i). If we introduce the function F again by (i'), then we obtain that

$$\text{(i'')} \qquad \begin{cases} F(x, 0) = G(x) \\ F(x, y') = F(x, y) \cdot \prod_{l=0}^{x} p_{y'(x+1)+l}^{h^*(l,x,y,F(x,y))} \end{cases}$$

where the primitive recursive function h^* is defined by

$$h^*(l, x, y, z) = h(l, y, \exp(y(x+1) + l, z), \exp(y(x+1) + \exp(1, l), z),$$
$$\sum_{k=0}^{l} \exp(y(x+1) + \exp(k, l), z) H(k, l, y)).$$

Now the relation (i″) shows that F is primitive recursive (for Σ and Π cf. § 10.5). From (i′) we obtain for $i = x$ the representation

$$f(x, y) = \exp(y(x+1) + x, F(x, y)),$$

by which the primitive recursiveness of f is proved as well.

6. A further inductive definition with substitution in the parameter.

Theorem. Let B and R be primitive recursive predicates, f a primitive recursive function and M a natural number. Let us assume that for all r-tuples $\mathfrak{x}, \mathfrak{y}, \mathfrak{z}$ and for all m

(1) $$B\mathfrak{x} \to \max \mathfrak{x} \leqq M,$$

(2) $$R\mathfrak{x}\mathfrak{y}\mathfrak{z}m \to \max \mathfrak{x} \leqq f(\mathfrak{y}, \mathfrak{z}, m).$$

Let us introduce the predicate A by induction on m by the relations

(3) $$A\mathfrak{x}0 \leftrightarrow B\mathfrak{x},$$

(4) $$A\mathfrak{x}m' \leftrightarrow A\mathfrak{x}m \vee \bigvee_{\mathfrak{y}} \bigvee_{\mathfrak{z}} (A\mathfrak{y}m \wedge A\mathfrak{z}m \wedge R\mathfrak{x}\mathfrak{y}\mathfrak{z}m).$$

Then A is also primitive recursive.

To *prove* this we first transform the unbounded particularisators in (4) into bounded particularisators. We introduce the primitive recursive function h by the inductive definition

$$\begin{cases} h(0) = M \\ \vdots \\ h(m') = \max\left(h(m), \max\limits_{y_1=0}^{h(m)} \dots \max\limits_{y_r=0}^{h(m)} \max\limits_{z_1=0}^{h(m)} \dots \max\limits_{z_r=0}^{h(m)} f(\mathfrak{y}, \mathfrak{z}, m)\right). \end{cases}$$

For this h we have the estimate

(5) $$A\mathfrak{x}m \to \max \mathfrak{x} \leqq h(m).$$

We show this by induction. If $A\mathfrak{x}0$, then $B\mathfrak{x}$, and so $\max \mathfrak{x} \leqq M$ ($M = h(0)$) according to (1). If $A\mathfrak{x}m'$, then we have two possibilities:

(α) $A\mathfrak{x}m$. Then $\max \mathfrak{x} \leqq h(m) \leqq h(m')$ by induction hypothesis.

(β) There exists $\mathfrak{y}$ and $\mathfrak{z}$ with $A\mathfrak{y}m$, $A\mathfrak{z}m$, $R\mathfrak{x}\mathfrak{y}\mathfrak{z}m$. Because of the induction hypothesis $\max \mathfrak{y} \leqq h(m)$, $\max \mathfrak{z} \leqq h(m)$. This, together with $R\mathfrak{x}\mathfrak{y}\mathfrak{z}m$ implies (by the help of (2)) that

$$\max \mathfrak{x} \leqq f(\mathfrak{y}, \mathfrak{z}, m) \leqq \max_{y_1=0}^{h(m)} \dots \max_{z_r=0}^{h(m)} f(\mathfrak{y}, \mathfrak{z}, m) \leqq h(m').$$

By the help of the estimate (5) we can tighten up (4) so that

(4′) $$A\mathfrak{x}m' \leftrightarrow A\mathfrak{x}m \vee \bigvee_{y_1=0}^{h(m)} \dots \bigvee_{z_r=0}^{h(m)} (A\mathfrak{y}m \wedge A\mathfrak{z}m \wedge R\mathfrak{x}\mathfrak{y}\mathfrak{z}m).$$

Let a, b, r be the characteristic functions of the predicates A, B, R. b and r are primitive recursive by hypothesis. We have to show that this is true for a as well. This follows from the following representation for a, which is obtained directly from (3) and (4').

$$\begin{cases} a(\mathfrak{x}, 0) = b(\mathfrak{x}) \\ a(\mathfrak{x}, m') = a(\mathfrak{x}, m) \prod_{y_1=0}^{h(m)} \cdots \prod_{z_r=0}^{h(m)} sg(a(\mathfrak{y}, m) + a(\mathfrak{z}, m) + r(\mathfrak{x}, \mathfrak{y}, \mathfrak{z}, m)). \end{cases}$$

7. *Course of values recursion.* In the induction schema (**) of § 10.1 $f(\mathfrak{x}, y')$ is traced back to $f(\mathfrak{x}, y)$. More generally we could trace back $f(\mathfrak{x}, y')$ to the previous values $f(\mathfrak{x}, 0), \ldots, f(\mathfrak{x}, y)$, i.e. to the whole previous "course of values" of the function f. We speak in this case of a *course of values recursion.* As a typical example we shall deal with the case of the definition schema

(6) $$f(\mathfrak{x}, 0) = g(\mathfrak{x}),$$

(7) $$f(\mathfrak{x}, y') = h(\mathfrak{x}, y, \prod_{i=0}^{y} p_i^{f(\mathfrak{x}, G(\mathfrak{x}, y, i)) H(\mathfrak{x}, y, i)}),$$

where we *assume that* $G(\mathfrak{x}, y, i) \leqq y$ for $i \leqq y$. *We shall show that if* g, h, G, H *are primitive recursive, then so is* f. For this purpose we introduce a function F by

(8) $$F(\mathfrak{x}, y) = p_0^{f(\mathfrak{x}, 0)} p_1^{f(\mathfrak{x}, 1)} \cdots p_y^{f(\mathfrak{x}, y)}.$$

It is obvious that

(9) $$f(\mathfrak{x}, i) = \exp(i, F(\mathfrak{x}, y)) \quad \text{for} \quad i \leqq y$$

and thus, especially, that

(10) $$f(\mathfrak{x}, y) = \exp(y, F(\mathfrak{x}, y)).$$

After this it is sufficient to show that F is primitive recursive. This follows at once from the two equations

(11) $$F(\mathfrak{x}, 0) = p_0^{f(\mathfrak{x}, 0)} = p_0^{g(\mathfrak{x})},$$

$$F(\mathfrak{x}, y') = F(\mathfrak{x}, y)\, p_{y'}^{f(\mathfrak{x}, y')}$$

$$= F(\mathfrak{x}, y)\, p_{y'}^{h(\mathfrak{x}, y, \prod_{i=0}^{y} p_i^{f(\mathfrak{x}, G(\mathfrak{x}, y, i)) H(\mathfrak{x}, y, i)})}$$

(12) $$= F(\mathfrak{x}, y)\, p_{y'}^{h(\mathfrak{x}, y, \prod_{i=0}^{y} p_i^{\exp(G(\mathfrak{x}, y, i), F(\mathfrak{x}, y)) H(\mathfrak{x}, y, i)})}.$$

Finally, we shall trace back to the schema (6), (7) a further definition schema, which we shall need later on (in § 21.3). This is given in an unessentially modified form

$$f(\mathfrak{x}, 0) = 0, \tag{13}$$

$$f(\mathfrak{x}, y) = K(\mathfrak{x}, y), \quad \textit{if } y \textit{ is odd}, \tag{14}$$

$$f(\mathfrak{x}, p_0^{\nu_0} \cdots p_r^{\nu_r}) = p_0^{\nu_0} p_1^{f(\mathfrak{x}, \nu_1)} \cdots p_r^{f(\mathfrak{x}, \nu_r)} \textit{ for } \nu_0 > 0. \tag{15}$$

Obviously (13), (14), (15) define a function f unambiguously. We shall show that if K is primitive recursive, then so is f. To do this we start with an even number y'. We can write

$$y' = p_0^{\nu_0} p_1^{\nu_1} \cdots p_r^{\nu_r}, \text{ and even } y' = p_0^{\nu_0} p_1^{\nu_1} \cdots p_y^{\nu_y}.$$

The right hand side of (15) becomes

$$p_0^{\exp(0, y')} p_1^{f(\mathfrak{x}, \exp(1, y'))} \cdots p_y^{f(\mathfrak{x}, \exp(y, y'))}.$$

If we define $G(\mathfrak{x}, y, i) = \exp(i, y')$, then obviously $G(\mathfrak{x}, y, i) \leqq y$ for $i \leqq y$. If further we define $H(\mathfrak{x}, y, i) = sg(i)$, then we can write the last product in the form

$$p_0^{\exp(0, y')} \prod_{i=0}^{y} p_i^{f(\mathfrak{x}, G(\mathfrak{x}, y, i)) H(\mathfrak{x}, y, i)}.$$

Let us now combine the equations (14), (15) into the equation

$$f(\mathfrak{x}, y') = \overline{sg}(\exp(0, y')) \cdot K(\mathfrak{x}, y') +$$
$$+ sg(\exp(0, y')) \cdot p_0^{\exp(0, y')} \prod_{i=0}^{y} p_i^{f(\mathfrak{x}, G(\mathfrak{x}, y, i)) H(\mathfrak{x}, y, i)}.$$

This is indeed a special case of the formula (7), and thus the proof of the primitive recursiveness of f is complete.

§ 13. Example of a Computable Function which is not Primitive Recursive

In § 10 we introduced the concept of primitive recursive function. As the examples of the last paragraph show, many functions of the mathematical praxis (whose arguments and values are natural numbers) are primitive recursive. From this we could surmise that every computable function is primitive recursive. This problem was formulated by Hilbert in 1926. Ackermann showed in 1928 by a counter example that this conjecture is not valid.

1. The essence of Ackermann's proof of the existence of a computable function which is not primitive recursive consists in defining a computable function which increases in a certain sense faster than any primitive recursive function. It is known that the sequence "sum, product, power" leads to faster and faster increasing functions. Since the power is obtained from the product similarly to how the product is obtained from the sum, we can continue this process further. We obtain in this way a hyperpower, etc. Let us pursue this more rigorously. For $n = 1, 2, 3$ let $f_n(x, y)$ be the sum, the product and the power respectively. We add to this sequence the function $f_0(x, y) = S(x)$. Then, we have that

$$\begin{cases} f_1(x, 0) = x \\ f_1(x, y') = f_0(f_1(x, y), x) \end{cases}$$

$$\begin{cases} f_2(x, 0) = 0 \\ f_2(x, y') = f_1(f_2(x, y), x) \end{cases}$$

$$\begin{cases} f_3(x, 0) = 1 \\ f_3(x, y') = f_2(f_3(x, y), x). \end{cases}$$

We see that these definitions (except for the initial function $f_0(x, y)$) fall under the schema

$$\begin{cases} f_{n'}(x, 0) = g_{n'}(x) \\ f_{n'}(x, y') = f_n(f_{n'}(x, y), x). \end{cases}$$

If we take for $g_{n'}(x)$ suitable primitive recursive functions, then we arrive at a sequence $f_n(x, y)$ of functions $(n = 0, 1, 2, \ldots)$ which has the sequence "successor function, sum, product, power" as initial subsequence. Every such function $f_n(x, y)$ is primitive recursive.

The decisive step consists now in replacing the *infinite sequence* $f_n(x, y)$ of functions of two arguments by *one* function $f(n, x, y)$ of three arguments. In other words, the n which was an *index* until now should be used as an *argument*. Thus, we define

$$f(n, x, y) = f_n(x, y).$$

$f(n, x, y)$ is obviously computable. $f(n, x, y)$ satisfies the functional equation

$$(0) \qquad f(n', x, y') = f(n, f(n', x, y), x),$$

which is still to be supplemented for the cases when the first or the third argument is zero. This functional equation is a kind of inductive definition. It is, however, of a more general type than the inductive definition which we encountered in the definition of primitive recursive functions. At the end of § 12 we reduced more general schemata to the induction schema occurring in the definition of primitive recursive functions. However, this is not possible for (0); we can, as a matter of fact, show that $f(n, x, y)$ is not primitive recursive.

2. Definition of Ackermann's function. We shall carry out the proof for the existence of a computable but not primitive recursive function not with the aid of the function $f(n, x, y)$ discussed just now, but on the basis of a simpler function. In the equation (0) the variable x, which appears throughout as parameter, obviously plays a less essential role than the variables n and y, the successors of which (n' and y') also occur. We shall, therefore, discard the variable x of (0) altogether. Then, instead of n we shall again use the letter x. Doing this we obtain the third equation of the following schema. Since this equation is obtained from equation (0) we shall carry on referring to the function discussed as Ackermann's function.[1] The other two equations are chosen in a simple way to make the following considerations possible. Now we give the

Definition:

(1) $$f(0, y) = y',$$

(2) $$f(x', 0) = f(x, 1),$$

(3) $$f(x', y') = f(x, f(x', y)).$$

We see at once by induction on x that $f(x, y)$ is unambiguously determined, for every x, y, by these equations and that it is computable. Thus, there exists one and only one function which satisfies these equations, and this function is computable.

3. The outline of the proof. We want to show that Ackermann's function is not primitive recursive. For this purpose we shall use the *Lemma* which will be proved in Section 5:

Lemma. For every primitive recursive function $g(x_1, \ldots, x_n)$ there exists a number c such that, for all $x_1, \ldots, x_n$,

(*) $$g(x_1, \ldots, x_n) < f(c, x_1 + \ldots + x_n).$$

If g is a function of *zero* variables, then (*) is to mean that

$$g < f(c, 0).$$

Now, if Ackermann's function $f(x, y)$ is primitive recursive, then so is the function $g(x) = f(x, x)$. Then there exists, according to the Lemma, a constant c such that, *for all* x,

$$g(x) < f(c, x).$$

[1] Thus, the function that is examined here is not the function originally suggested by ACKERMANN. For the definition of the original function and for the proof that it is not primitive recursive see the cited paper of ACKERMANN.

This is true especially for $x = c$. And so we obtain the contradiction

$$g(c) < f(c, c) = g(c).$$[1]

4. Estimates for f. As a preparation for the proof of the Lemma we shall derive a few estimates.

$$(4) \qquad y < f(x, y).$$

We show by induction on x that $y < f(x, y)$ for every y. For $x = 0$, $y < y' = f(0, y)$. Let us assume now that the estimate is already proved for one x and every y (induction hypothesis). Now, we have to prove the estimate for x' and all y by induction on y. For x' and $y = 0$, $1 < f(x, 1)$ by the induction hypothesis, and so $0 < 1 < f(x, 1) = f(x', 0)$. As second induction hypothesis we have the validity of (4) for x' and a certain y. We have to prove the estimate for x' and y'. First we have by the second induction hypothesis that

$$y < f(x', y),$$

and then, by the first induction hypothesis $(y < f(x, y))$, if we take there $f(x', y)$ for y, that $f(x', y) < f(x, f(x', y))$. Thus, by (3)

$$f(x', y) < f(x', y').$$

From the last two inequalities we obtain $y' < f(x', y')$; q.e.d.

(5) $f(x, y) < f(x, y')$. *Monotony in the second argument.*

We show this by induction on x. We have that $f(0, y) = y' < y'' = f(0, y')$. Finally, by (4), $f(x', y) < f(x, f(x', y)) = f(x', y')$; q.e.d.

$$(6) \qquad f(x, y') \leqq f(x', y).$$

Proof by induction on y. By (2), we have that $f(x, 1) = f(x', 0)$. Further, $y' < f(x, y')$ by (4), and so $y'' \leqq f(x, y') \leqq f(x', y)$ by induction hypothesis. From this follows, by the help of (5), that

$$f(x, y'') \leqq f(x, f(x', y)) = f(x', y'); \text{ q.e.d.}$$

(7) $f(x, y) < f(x', y)$. *Monotony in the first argument.*

$f(x, y) < f(x, y')$ by (5) and $f(x, y') \leqq f(x', y)$ by (6); q.e.d.

We shall now represent $f(1, y)$ and $f(2, y)$ by elementary functions.

$$(8) \qquad f(1, y) = y + 2.$$

Proof by induction on y: $f(1, 0) = f(0, 1) = 2$.

$$f(1, y') = f(0, f(1, y)) = f(0, y + 2) = y + 3 = y' + 2; \text{ q.e.d.}$$

---

[1] We speak in this example of a "diagonal procedure", for we use the values of the function $f(x, y)$ obtained on the diagonal $x = y = c$.

(9) $$f(2, y) = 2y + 3.$$

Proof by induction on y: $f(2, 0) = f(1, 1) = 3$ by (8).

$$f(2, y') = f(1, f(2, y)) = f(1, 2y + 3) = 2y + 5 = 2y' + 3; \text{ q.e.d.}$$

Finally, we need the estimate:

(10) For arbitrary $c_1, \ldots, c_r$ there exists a c, such that, for all x,

$$\sum_{j=1}^{r} f(c_j, x) \leqq f(c, x).$$

It is obviously sufficient to prove this assertion for $r = 2$. Let $d = \max(c_1, c_2)$ and $c = d + 4$. Then, we have that

$$\begin{aligned} f(c_1, x) + f(c_2, x) &\leqq f(d, x) + f(d, x) && \text{by (7)} \\ &< 2f(d, x) + 3 \\ &= f(2, f(d, x)) && \text{by (9)} \\ &< f(d + 2, f(d + 3, x)) && \text{by (5), (7)} \\ &= f(d + 3, x') \\ &\leqq f(d + 4, x) && \text{by (6)} \\ &= f(c, x). \end{aligned}$$

5. *Proof of the Lemma* (*) *of Section 3.* We prove (*) first for the initial functions and after that, we show that the estimate of the Lemma is preserved for functions which are obtained by the substitution and induction processes.

(11) $$S(x) < f(1, x),$$

because $S(x) = f(0, x) < f(1, x)$ by (7).

(12) $$U_n^i(x_1, \ldots, x_n) < f(0, x_1 + \cdots + x_n),$$

for $U_n^i(x_1, \ldots, x_n) = x_i < (x_1 + \cdots + x_n)' = f(0, x_1 + \cdots + x_n)$.

(13) $$C_0^0 < f(0, 0),$$

for $f(0, 0) = 1$.

(14) The substitution process. Given the functions $g, g_1, \ldots, g_n$ let the numbers $c, c_1, \ldots, c_n$ be such that

$$g(x_1, \ldots, x_n) < f(c, x_1 + \cdots + x_n)$$
$$g_j(y_1, \ldots, y_r) < f(c_j, y_1 + \cdots + y_r) \quad (j = 1, \ldots, n).$$

Let

$$h(y_1, \ldots, y_r) = g(g_1(y_1, \ldots, y_r), \ldots, g_n(y_1, \ldots, y_r)).$$

Then there exists a d such that, for all $y_1, \ldots, y_r$,

$$h(y_1, \ldots, y_r) < f(d, y_1 + \cdots + y_r).$$

Proof.

$$\begin{aligned} h(y_1, \ldots, y_r) &= g(g_1(y_1, \ldots, y_r), \ldots, g_n(y_1, \ldots, y_r)) \\ &< f(c, g_1(y_1, \ldots, y_r) + \cdots + g_n(y_1, \ldots, y_r)) \\ &< f(c, f(c_1, y_1 + \cdots + y_r) + \cdots + f(c_n, y_1 + \cdots + y_r)) \quad \text{by (5)} \\ &\leqq f(c, f(c^*, y_1 + \cdots + y_r)) \text{ with suitable } c^* \text{ by (10)} \\ &\leqq f(c + c^*, f(c + c^* + 1, y_1 + \cdots + y_r)) \quad \text{by (5), (7)} \\ &= f(c + c^* + 1, y_1 + \cdots + y_r + 1) \\ &\leqq f(c + c^* + 2, y_1 + \cdots + y_r) \quad \text{by (6).} \end{aligned}$$

(15) The induction process. Given the functions g_1 and g_2 let the numbers c_1 and c_2 be such that

$$g_1(x_1, \ldots, x_n) < f(c_1, x_1 + \cdots + x_n) \quad \text{for all } x_1, \ldots, x_n$$

$$g_2(x_1, \ldots, x_n, y, z) < f(c_2, x_1 + \cdots + x_n + y + z) \quad \text{for all } x_1, \ldots, x_n, y, z.$$

Let the function h be given inductively by

$$h(x_1, \ldots, x_n, 0) = g_1(x_1, \ldots, x_n)$$

$$h(x_1, \ldots, x_n, y') = g_2(x_1, \ldots, x_n, y, h(x_1, \ldots, x_n, y)).$$

Then, there exists a constant c such that for all $x_1, \ldots, x_n, y$

$$h(x_1, \ldots, x_n, y) < f(c, x_1 + \cdots + x_n + y).$$

Instead of this assertion we prove the stronger statement: There exists a c such that, for arbitrary $x_1, \ldots, x_n, y$,

$$(**) \qquad h(x_1, \ldots, x_n, y) + x_1 + \cdots + x_n + y < f(c, x_1 + \cdots + x_n + y).$$

For this purpose we show first that there exists a c_1^* such that

$$g_1(x_1, \ldots, x_n) + x_1 + \cdots + x_n < f(c_1^*, x_1 + \cdots + x_n) \text{ for all } x_1, \ldots, x_n.$$

This follows in view of (12) and (10) from the following.

$$\begin{aligned} &g_1(x_1, \ldots, x_n) + x_1 + \cdots + x_n \\ &\quad = g_1(x_1, \ldots, x_n) + U_n^1(x_1, \ldots, x_n) + \cdots + U_n^n(x_1, \ldots, x_n) \\ &\quad < f(c_1, x_1 + \cdots + x_n) + f(0, x_1 + \cdots + x_n) + \cdots + f(0, x_1 + \cdots + x_n) \\ &\quad < f(c_1^*, x_1 + \cdots + x_n) \quad \text{with suitable } c_1^*. \end{aligned}$$

Similarly, we can show that there exists a constant c_2^* such that, for all $x_1, \ldots, x_n, y, z$,

$$g_2(x_1, \ldots, x_n, y, z) + x_1 + \cdots + x_n + y + z < f(c_2^*, x_1 + \cdots + x_n + y + z).$$

Now we shall, by induction on y, prove (**) with

$$c = \max(c_1^*, c_2^*) + 1.$$

We easily obtain (**) for $y = 0$ by the help of (7). Further we have that

$$\begin{aligned}
& h(x_1, \ldots, x_n, y') + x_1 + \cdots + x_n + y' \\
&\quad = g_2(x_1, \ldots, x_n, y, h(x_1, \ldots, x_n, y)) + x_1 + \cdots + x_n + y' \\
&\quad < f(c_2^*, x_1 + \cdots + x_n + y + h(x_1, \ldots, x_n, y)) + 1 \\
&\quad < f(c_2^*, f(c, x_1 + \cdots + x_n + y)) + 1 \\
&\qquad \text{by the induction hypothesis and together with (7)} \\
&\quad \leqq f(c-1, f(c, x_1 + \cdots + x_n + y)) + 1 \\
&\quad = f(c, x_1 + \cdots + x_n + y') + 1.
\end{aligned}$$

If we note that in the estimate the "less than" sign appears twice, then we see that

$$h(x_1, \ldots, x_n, y') + x_1 + \cdots + x_n + y' < f(c, x_1 + \cdots + x_n + y'),$$

q.e.d.

Exercise 1. Strengthen the Lemma by showing that for every primitive recursive function $g(x_1, \ldots, x_n)$ there exists a number c such that, for all $x_1, \ldots, x_n$,

$$g(x_1, \ldots, x_n) \leqq f(c, \max(x_1, \ldots, x_n)).$$

Hint. Form the function $G(x) = \max\limits_{\substack{x_1 \leqq x \\ \cdots\cdots \\ x_n \leqq x}} g(x_1, \ldots, x_n)$.

Exercise 2. Show that not all (in the intuitive sense) computable functions are primitive recursive by giving an effective enumeration of all primitive recursive functions and then applying a diagonal procedure.

Reference

Ackermann, W.: Zum Hilbertschen Aufbau der rellen Zahlen. Math. Ann. **99**, 118–133 (1928).

§ 14. μ-Recursive Functions and Predicates

The example of Ackermann's function, which we dealt with in the last paragraph, shows that the operations of substitution and inductive definition are not sufficient to obtain all computable functions, if we start with the initial functions S, U_n^i and C_0^0 (cf. § 10.1). The addition of the μ-operator to these operators is an obvious extension. As we have already seen in § 12.1 $\mu y P\mathfrak{x}y$ *(the smallest y such that $P\mathfrak{x}y$)* is a computable function of $\mathfrak{x}$, provided that P is decidable and that for every $\mathfrak{x}$ there exists at least one y such that $P\mathfrak{x}y$ (regular predicate).

We shall say that $f(\mathfrak{x})$ *is obtained from the regular function* $g(\mathfrak{x}, y)$ *by application of the μ-operator,* if for every $\mathfrak{x}$ there exists a y such that $g(\mathfrak{x}, y) = 0$ and if $f(\mathfrak{x})$ is the smallest such y.

If, in this case, g is computable, then $g(\mathfrak{x}, y) = 0$ is decidable and by the previous considerations f is computable. Thus, the application of the μ-operator to a regular function does not lead outside the domain of computable functions.

1. μ-recursive functions.

Definition. A function is called *μ-recursive*[1], if it is generable, starting from the initial functions S, U_n^i and C_0^0, by means of the following operations:

(1) Substitution (cf. § 10.1),

(2) Inductive definition (cf. § 10.1),

(3) Application of the μ-operator to regular functions.

It is clear that every μ-recursive function is computable, and that every primitive recursive function is μ-recursive.

2. μ-recursive predicates. We can introduce the concept of μ-recursive predicate analogously to the concept of primitive recursive predicate (cf. § 11.1) by the

Definition. An n-ary *predicate* P $(n \geqq 1)$ is called *μ-recursive* if there exists a μ-recursive n-ary function f such that, for arbitrary n-tuples of numbers $\mathfrak{x}$,

$$P\mathfrak{x} \textit{ if and only if } f(\mathfrak{x}) = 0.$$

Every μ-recursive predicate is decidable.

The theorems which we derived in § 11, Sections 3 and 5, for *primitive recursive* predicates and functions are valid mutatis mutandis for the μ-recursive predicates and functions, with the same proofs. We confine ourselves to making a note of the result.

Theorem. The operations of negation, of π-permutation, of generalized conjunctions and alternatives, and also of the bounded quantifications lead from μ-recursive predicates to μ-recursive predicates again. The same applies for the substitution of a μ-recursive function into a μ-recursive predicate. A function defined by cases by the help of μ-recursive functions and μ-recursive predicates is μ-recursive.

3. μ-recursiveness of Ackermann's function. That the domain of μ-recursive functions is larger than the domain of primitive recursive functions is shown by the

---

[1] This concept originates from KLEENE.

Theorem. Ackermann's function $f(x, y)$ is μ-recursive.

For *proof* we follow up the computation of $f(x, y)$ by means of the defining equations (1), (2), (3) of the last paragraph. In the schedule given below the calculation of $f(2, 1)$ is represented. The calculation is carried out in steps (column 1). It is completed after 14 steps and provides the value 5. We see (column 2) that we achieve our aim if we paraphrase each time the expression $f(n, m)$ which is enclosed between the innermost brackets. For this, exactly one of the equations (1), (2), (3) comes into consideration. In this way we obtain an unambiguous calculating procedure. We see that the writing down of the symbols "f" and of the parentheses is superfluous in the calculation, for all the right parentheses are at the right end of the expression. It is entirely sufficient to write down the arguments only (column 3).

The calculation itself is completed after the 14th step. However, we have continued the third column further down by repeated reproductions of the value of the function, 5.

The calculation of $f(2, 1)$ for Ackermann's function.

step	calculation	abbreviated notation	Gödel numbers		
0	$f(2, 1)$	2, 1	$2^3\,3^2$	=	72
1	$f(1, f(2, 0))$	1, 2, 0	$2^2\,3^3\,5^1$	=	540
2	$f(1, f(1, 1))$	1, 1, 1	$2^2\,3^2\,5^2$	=	900
3	$f(1, f(0, f(1, 0)))$	1, 0, 1, 0	$2^2\,3^1\,5^2\,7^1$	=	2100
4	$f(1, f(0, f(0, 1)))$	1, 0, 0, 1	$2^2\,3^1\,5^1\,7^2$	=	2940
5	$f(1, f(0, 2))$	1, 0, 2	$2^2\,3^1\,5^3$	=	1500
6	$f(1, 3)$	1, 3	$2^2\,3^4$	=	324
7	$f(0, f(1, 2))$	0, 1, 2	$2^1\,3^2\,5^3$	=	2250
8	$f(0, f(0, f(1, 1)))$	0, 0, 1, 1	$2^1\,3^1\,5^2\,7^2$	=	7350
9	$f(0, f(0, f(0, f(1, 0))))$	0, 0, 0, 1, 0	$2^1\,3^1\,5^1\,7^2\,11^1$	=	16170
10	$f(0, f(0, f(0, f(0, 1))))$	0, 0, 0, 0, 1	$2^1\,3^1\,5^1\,7^1\,11^2$	=	25410
11	$f(0, f(0, f(0, 2)))$	0, 0, 0, 2	$2^1\,3^1\,5^1\,7^3$	=	10290
12	$f(0, f(0, 3))$	0, 0, 3	$2^1\,3^1\,5^4$	=	3750
13	$f(0, 4)$	0, 4	$2^1\,3^5$	=	486
14	5	5	2^6	=	64
15		5	2^6	=	64
16		5	2^6	=	64
17		5	2^6	=	64
.		.	.		.
.		.	.		.
.		.	.		.

In the fourth column we represent each sequence of natural numbers occuring in the third column by a single number, i.e. we carry out a

Gödel numbering. In this Gödel numbering we associate with the (finite) sequence

$$r_0, r_1, \ldots, r_k$$

of natural numbers ($k \geqq 0$, $r_j \geqq 0$) the number

$$n = p(0)^{r_0+1}\, p(1)^{r_1+1} \ldots p(k)^{r_k+1}$$

as the Gödel number of the sequence, where $p(k)$ runs through the sequence of the prime numbers (cf. § 12.3). In this representation it is essential that one can reconstruct the original sequence from the Gödel number by factorizing n into prime numbers and considering the sequence of exponents each diminished by one.[1]

If the Gödel number n is given, then the length k of the sequence $r_0, \ldots, r_k$ represented by n is equal to $l(n)$ (cf. § 12.3), and we have, for $0 \leqq j \leqq l$,

$$(*) \qquad r_j = V(\exp(j, n))$$

(where V is the predecessor function introduced in § 10.4 (9)).

If we now consider the dependence of the Gödel numbers (occurring in the fourth column) of the sequences (occurring in the third column) on the step number z (first column), then we obtain a function $g(z)$. More explicitly, we shall write in our case $g(2, 1, z)$, for we start from the argument pair 2, 1 for which we want to find the value of f. If we start from an arbitrary argument pair x, y, then we obtain a function $g(x, y, z)$ of three variables. As long as (with fixed x and y) the calculation of $f(x, y)$ is not yet finished, $g(x, y, z)$ assumes, for increasing z, different values only, for otherwise the calculation procedure of f would be circular, and would not come to an end. As soon as the calculation of $f(x, y)$ is finished, $g(x, y, z)$ remains constant. For the step number u which denotes the end of the calculation of $f(x, y)$, we obviously have (cf. the example above)

$$u = \mu z(g(x, y, z) = g(x, y, z')).$$

We associate with u the Gödel number $g(x, y, u)$. This represents a sequence with one term only. The term of this sequence is equal to the value $f(x, y)$ which we are trying to find. By this we have, because of (*), that $f(x, y) = V(\exp(0, g(x, y, u)))$. If we finally write $\varepsilon(g(x, y, z), g(x, y, z')) = 0$ for $g(x, y, z) = g(x, y, z')$, then we have the representation

$$(**) \qquad f(x, y) = V(\exp(0, g(x, y, \mu z(\varepsilon(g(x, y, z), g(x, y, z')) = 0)))).$$

[1] If for the definition of the Gödel number of the sequence $r_0, r_1, \ldots, r_k$ we took the exponent of $p(j)$ to be equal to r_j (and not $r_j + 1$), then the sequences 4, 3, 6 and 4, 3, 6, 0 for instance, would have the same number. We should keep in mind that in the Gödel numbering given in the text not every number n is the Gödel number of a sequence.

In the next section we shall show that g is primitive recursive. Since the calculation, for arbitrary x, y, comes to an end after finitely many steps, there certainly exists a z such that $g(x, y, z) = g(x, y, z')$, and so $\varepsilon(g(x, y, z), g(x, y, z')) = 0$. That means that the μ-operator is applied to a regular function. *Thus, the representation* (**) *shows that* f *is a* μ*-recursive function.*

It should be noted that the μ-operator occurs *only once*. That this is not a special case will follow from § 18 where we shall show that in fact *every* μ-recursive function has a representation which has not more than one μ-operator in it.[1]

4. *Proof of the primitive recursiveness of* g. If we observe how $g(x, y, z)$ is obtained in each step, then it seems obvious to try to define g by induction on z. The initial step is easy, for $g(x, y, 0)$ is the Gödel number of the argument pair x, y.

(i) $$g(x, y, 0) = 2^{x+1}\, 3^{y+1}.$$

The induction step proceeds in four different ways, according to whether $g(x, y, z)$ characterizes a sequence, for the further paraphrasing of which we have to refer to the 1st, 2nd or 3rd defining equation of Ackermann's function, or a sequence of one term, whose term is, therefore, already equal to the Gödel number of the value $f(x, y)$ of the function. In order to carry out this more precisely we consider first four singulary predicates P_j $(j = 1, 2, 3, 4)$ by the help of which we can distinguish between the cases mentioned above, and four singulary functions h_j $(j = 1, 2, 3, 4)$ which bring about the passage from $g(x, y, z)$ to $g(x, y, z')$.

We define

$$P_1 n \leftrightarrow l(n) > 0 \wedge \exp(l(n) \dot{-} 1, n) = 1$$
$$P_2 n \leftrightarrow l(n) > 0 \wedge \exp(l(n) \dot{-} 1, n) > 1 \wedge \exp(l(n), n) = 1$$
$$P_3 n \leftrightarrow l(n) > 0 \wedge \exp(l(n) \dot{-} 1, n) > 1 \wedge \exp(l(n), n) > 1$$
$$P_4 n \leftrightarrow \neg P_1 n \wedge \neg P_2 n \wedge \neg P_3 n.$$

We easily see that $P_1, \ldots, P_4$ are primitive recursive predicates and that for each n one and only one of these predicates is valid. If n is the Gödel number of a sequence which can be altered by means of the defining equation (j) $(j = 1, 2, 3)$ of Ackermann's function, then $P_j n$. If n is the Gödel number of a one-term sequence, then $P_4 n$.

Now we must define functions h_j such that $h_j(n)$ is the Gödel number of a new sequence which is obtained from the sequence which has the

[1] The proof, given in this paragraph, of the μ-recursiveness of Ackermann's function contains already the essence of the considerations by the help of which we shall show in § 18 that every Turing-computable function is μ-recursive.

Gödel number n, in case $P_j\, n$ $(j = 1, \ldots, 4)$. The following functions are sufficient.

$$h_1(n) = \frac{n\,(p\,(l\,(n) \dot{-} 1))^{\exp(l(n),n)+1}}{(p\,(l\,(n)))^{\exp(l(n),n)}}$$

$$h_2(n) = \frac{n\,p\,(l\,(n))}{p\,(l\,(n) \dot{-} 1)}$$

$$h_3(n) = \frac{n\,(p\,(l\,(n) + 1))^{\exp(l(n),n) \dot{-} 1} \cdot (p(l(n)))^{\exp(l(n) \dot{-} 1,n)}}{p\,(l\,(n) \dot{-} 1) \cdot (p(l(n)))^{\exp(l(n),n)}}$$

$$h_4(n) = n.$$

We confine ourselves to an explanation for $h_2(n)$. In this case $P_2(n)$ is valid and the sequence is of the form $\ldots, t', 0$. The new resulting sequence is then of the form $\ldots, t, 1$. The Gödel number of this sequence is obtained from the Gödel number n of the initial sequence by multiplying n by $p\,(l\,(n))$ and dividing by $p\,(l\,(n) \dot{-} 1)$.

Finally we put

$$h(n) = \begin{cases} h_1(n), & \text{if } P_1 n \\ h_2(n), & \text{if } P_2 n \\ h_3(n), & \text{if } P_3 n \\ h_4(n), & \text{if } P_4 n, \end{cases}$$

so $h(n)$ is, in every case, the Gödel number of the new sequence obtained from the sequence with Gödel number n and thus

(ii) $$g(x, y, z') = h(g(x, y, z)).$$

(i) and (ii) give an inductive definition of the function g. This definition shows that g is primitive recursive.

Reference

Kleene, S. C.: General Recursive Functions of Natural Numbers. Math. Ann. **112**, 727–742 (1936).

CHAPTER 4

THE EQUIVALENCE OF TURING-COMPUTABILITY AND μ-RECURSIVENESS

We have already emphasized in the preface that the equivalence of the suggested precise replacements of the intuitive concept of computable function can be shown by purely mathematical considerations. We shall do this in this chapter for the concept of Turing-computable function and the concept of μ-recursive function. (Cf. also Chapter 5 and § 30.)

An equivalence proof of this kind generally leads to a standard representation of computable functions. Thus, we shall obtain (in § 18) Kleene's normal form theorem.

§ 15. Survey. Standard Turing-Computability

In this paragraph we shall make some preliminary remarks on the theorems of the following two paragraphs. These theorems show that the class of μ-recursive functions coincides with that of Turing-computable functions. In showing this we shall always assume, in this and in the following two paragraphs, that the *functions mentioned are defined* for *all n-tuples of natural numbers and that the values are natural numbers again.* Thus, we are dealing with the proofs of the following theorems.

Theorem A. Every μ-recursive function is Turing-computable.

Theorem B. Every Turing-computable function is μ-recursive.

Theorem A is *intuitively* clear. Every μ-recursive function is computable (§ 14.1), and every computable function is Turing-computable (§ 6, Introduction). We shall, however, give a rigorous proof for Theorem A which is independent of the intuitive concept of computability. Theorem B is not immediately obvious intuitively.

We can deduce from Theorems A and B the following

Theorem A'. Every μ-recursive predicate is Turing-decidable.

Theorem B'. Every Turing-decidable predicate is μ-recursive.

Proof of Theorem A'. Let P be a μ-recursive predicate. According to § 14.2 there exists a μ-recursive function f such that, for all $\mathfrak{x}$, $f(\mathfrak{x}) = 0$ if and only if $P\mathfrak{x}$. f is Turing-computable by Theorem A. Let M be a Turing machine which computes f in the sense of § 6.1. Then the machine $\mathsf{M}\mathfrak{l}^2$, placed behind $\mathfrak{x}$, stops operating after finitely many steps and, indeed, over $*$ if $P\mathfrak{x}$ holds and over I if $P\mathfrak{x}$ does not hold (the natural numbers are represented by sequences of strokes as we have described in § 1.4). This shows that P is Turing-decidable. (Cf. § 6.3.)

Proof of Theorem B'. Let P be Turing-decidable. Let M be a Turing machine which decides P in the sense that if M is placed behind $\mathfrak{x}$ it stops operating over $*$ if $P\mathfrak{x}$ holds and over I otherwise (cf. also Exercise 2 in § 7). Then the Turing machine

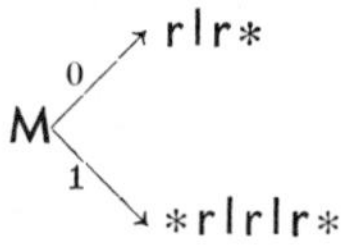

obviously computes a function f which is such that $f(\mathfrak{x}) = 0$ if and only if $P\mathfrak{x}$ holds. Thus, f is Turing-computable. According to Theorem B

f is μ-recursive. Hence it follows from the Definition of § 14.2 that P is μ-recursive.

1. Standard Turing-computability. It is easier to prove Theorem A first in a modified form. For this purpose we introduce the concept of "standard" Turing-computability. This concept differs from the concept of Turing-computability introduced in § 6.1. First of all it deals with functions whose values and arguments are natural numbers (i.e. sequences of strokes), while for Turing-computability we have allowed arbitrary non-empty words to be the values and arguments. For the Turing machines discussed in this section the symbol | must in any case belong to their alphabet $\mathfrak{A}$. Moreover, we shall even put $\mathfrak{A} = \{|\}$. Besides, we have two differences worthy of remark. (1) The original scanned square is defined, which is a simplification. (2) A list of aggravating conditions is given. We shall deal with them immediately. Such aggravating conditions have the advantage, as we shall see later on, that we can easily construct machines which compute more complicated functions from machines which satisfy these conditions. That we shall not assume any longer that the computing tape is empty at the beginning of the computation except for the given argument is an important limitation. Furthermore we shall allow that *on the left* of the given arguments (with a certain gap in between) arbitrary inscriptions may be printed on the tape, while we shall however assume (as before) that *on the right* of the arguments the tape is empty. For a more convenient formulation of the definition it is recommendable to introduce two concepts.

(a) A *half tape* is determined by a square. It consists of this square (as first square) and all squares *on the right* of it.

(b) By an *argument strip* of an n-ary function we understand a finite part of the tape on which $* W_1 * W_2 \ldots * W_n$ is printed. Thus, the first square of the argument strip is empty, the last square has the last symbol of W_n printed in it. This applies if $n \geqq 1$. An argument strip for a 0-ary function will contain no squares at all.

We shall assume that at the beginning of a computation of a function the given arguments are on an argument strip, and that the half tape beginning immediately on the right of this argument strip is empty. On the other hand the tape can be marked in an arbitrary manner on the left of this argument strip. Naturally we shall have to see to it that the inscription on the left of the argument strip does not disturb the computation of the function. We shall therefore assume that during the computation all the squares which become scanned are situated in the argument strip or in the half tape mentioned above. More precisely we have the

Definition. An n-ary function ($n \geqq 1$) is called *standard Turing-computable* if there exists a Turing machine M over an alphabet $\{|\}$ of one element which has the following property. If we print the n-tuple $\mathfrak{x}$ of arguments onto the computing tape in the usual manner (cf. § 6.1), and if the half tape H whose first square is the square immediately on the right of the argument strip is empty, then the machine placed over the first square of H stops operating after finitely many steps, and at the end of the computation we have that

(0) the arguments given are at the same place as at the beginning,

(1) the value of the function starts at the second square of H, thus there is a one-gap space between the value of the function and the arguments,

(2) M is over the square immediately behind the last stroke of the value of the function,

(3) H is empty except for the value of the function.

Furthermore we have that

(4) during the computation only the squares of the argument strip, determined by the arguments, and the squares of H are scanned.

In the case of a 0-*ary function* we shall modify the definition of standard computability in the following way.

Definition. A 0-ary function is called *standard Turing-computable* if there exists a Turing machine M over the alphabet $\{|\}$ with the following property. If we place the machine over an arbitrary square A of the computing tape, and if the half tape H, whose first square is A, is empty, then M stops operating after finitely many steps and at the end of the computation we have that

(1′) the value of the function begins at the second square of H,

(2′) M is over the square immediately behind the last stroke of the value of the function,

(3′) H is empty except for the value of the function.

Furthermore we have that

(4′) during the computation only the squares of H are scanned.

2. *Turing-computability and standard Turing-computability.* A standard Turing-computable function is obviously also Turing-computable. This follows directly from § 9.1, Theorem 1. But the converse assertion is also valid. For in the next paragraph we shall show

Theorem A_0. Every μ-recursive function is standard Turing-computable.

Theorem A_0 obviously implies Theorem A.

Then, together with Theorem B, this gives the

Corollary. Every Turing-computable function is standard Turing-computable.

Thus, to compute computable functions we can make do with a machine which (apart from the empty symbol) uses only *one* symbol (cf § 1.3). Besides, the Corollary shows that to compute a computable function we can make do with a computing tape which reaches out to infinity on one side only (cf. the second note of § 3.3).

3. Normal form of μ-recursive functions. We shall prove Theorem B also in a modified and, indeed, strengthened form (Theorem B_0). We shall apply a method which is in principle the same as the one which we made use of in § 14.3 to show the μ-recursiveness of Ackermann's function $f(x, y)$. There we have obtained the representation

$$f(x, y) = V(\exp(0, g(x, y, \mu z(\varepsilon(g(x, y, z), g(x, y, z')) = 0)))),$$

where $g(x, y, z)$ was a primitive recursive function. It is worthy of remark that in this representation the μ-operator occurs only once. We shall show that one can give an account of a representation with only one μ-operator for *every* Turing-computable function. That in addition to this an extensive standardisation can be carried out is shown in

Theorem B_0. There exists a singulary primitive recursive function U and, for each n, an $(n+2)$-ary primitive recursive predicate T_n, with the property that for every n-ary Turing-computable function f there exists a number k such that

(i) for every $\mathfrak{x}$, there exists a y with $T_n k \mathfrak{x} y$,

(ii) $f(\mathfrak{x}) = U(\mu y T_n k \mathfrak{x} y)$ for every $\mathfrak{x}$.

(ii) shows the μ-recursiveness of f, for the μ-operator is applied to a regular predicate (because of (i)). Thus, we can obtain, just by varying k, *all* n-ary Turing-computable functions in the described form.[1] — For the proof of Theorem B_0 cf. § 18.

By the help of Theorems A and B_0 we obtain as *Corollary*

Kleene's normal form theorem. There exists a primitive recursive function U and, for every n ($n \geqq 0$), an $(n+2)$-ary primitive recursive predicate T_n with the following property. For every n-ary μ-recursive function f there exists a number k such that[1]

(i) For every $\mathfrak{x}$ there exists a y with $T_n k \mathfrak{x} y$.

(ii) $f(\mathfrak{x}) = U(\mu y T_n k \mathfrak{x} y)$ for every $\mathfrak{x}$.

Reference

KLEENE, S. C.: Introduction to Metamathematics. Amsterdam: North-Holland Publishing Company 1952, 2nd reprint 1959.

[1] We should keep in mind that it is not required that (i) is valid for *every* k.

§ 16. The Turing-Computability of μ-Recursive Functions

We shall prove in this paragraph Theorem A_0, which we formulated in the last paragraph and from which follows, as we have seen, the Turing-computability of μ-recursive functions. We carry out the proof by induction, in which we show the assertion for the initial functions (in Section 1), and that the property of standard Turing-computability is invariant under substitution, under inductive definition and under application of the μ-operator to regular functions (in Sections 2, 3, 4 respectively).

1. The standard Turing-computability of the initial functions. The *successor function* is standard Turing-computed by $\mathsf{K|r}$. To show that the condition § 15.1 (4) for standard Turing-computability is satisfied we should bear in mind that K does not move any further to the left than the square which is in front of the first stroke to be copied. (Cf. for this what is said at the end of the remark of the Summary in § 8. The same applies for the machines which will be discussed later on in this paragraph, but we shall not draw attention to it any more.)

The function U_n^i ($n \geqq 1$, $1 \leqq i \leqq n$) can be standard Turing-computed by means of K_{n+1-i}.

The function C_0^0 is standard Turing-computed by $\mathsf{r|r}$.

2. The standard Turing-computability is invariant under substitution. Let g be a function of r variables ($r \geqq 1$) and $h_1, \ldots, h_r$ functions of n variables ($n \geqq 0$). Let the function f of n variables be defined by

$$f(x_1, \ldots, x_n) = g(h_1(x_1, \ldots, x_n), \ldots, h_r(x_1, \ldots, x_n)).$$

Let the functions $g, h_1, \ldots, h_r$ be standard Turing-computable by the help of the machines $\mathsf{M}, \mathsf{M}_1, \ldots, \mathsf{M}_r$. Then f is standard Turing-computable. The computation is performed by

$$\mathsf{r|r\,K}_{n+1}^{n}\,\mathsf{L}^n\mathsf{\mathfrak{l}}{*}\mathfrak{R}\,\mathsf{M}_1\,\mathsf{K}_{n+1}^{n}\,\mathsf{M}_2 \ldots \mathsf{K}_{n+1}^{n}\,\mathsf{M}_r\,\mathsf{K}_{r+(r-1)n}\,\mathsf{K}_{r+(r-2)n} \ldots \mathsf{K}_r\mathsf{MC}.$$

To show this we follow up the computing process in detail. We assume that $n \geqq 1$. (We can easily convince ourselves that the assertion is also valid in the case $n = 0$.) We denote the argument strip (cf. § 15) together with the arguments printed on it by $\mathfrak{x}$. ($\mathfrak{x}$ also denotes, as always, the n-tuple $x_1, \ldots, x_n$).)

Thus, at the beginning we have on the computing tape

$$*\mathfrak{x}\underline{*}\cdots.$$

The computation proceeds so that at first the values $h_j(\mathfrak{x})$ are computed by the help of M_j, and then the value $f(\mathfrak{x})$ is computed from these $h_j(\mathfrak{x})$ by the help of M. Finally the whole computational procedure with the exception of the arguments and the value has to be erased. This is done by the cleaning up machine C. In order to be able to apply C it is necessary to produce first a gap of three squares behind the arguments. However, the arguments $\mathfrak{x}$ must be at our disposal (on the right of this gap) for further computations. Hence, we shall first build, by the help of $\mathsf{r|r}$, a "bridge"

$$*\mathfrak{x}*|\underline{*}\ldots,$$

across which we shall then transport the arguments by means of K_{n+1}^{n}:

$$*\mathfrak{x}*|*\mathfrak{x}\underline{*}\cdots.$$

After the completion of this operation we shall pull down the bridge by means of $\mathsf{L}^{n}\mathfrak{L}*$

$$*\mathfrak{x}*\underline{*}*\mathfrak{x}*\cdots$$

and we shall go (by $\mathfrak{R}$) behind the last stroke of the copied argument:

$$*\mathfrak{x}***\mathfrak{x}\underline{*}\cdots.$$

Now we can use the machine M_1 which standard computes $h_1(\mathfrak{x})$ (we should keep in mind that, because of the conditions for *standard* computation, the machine M_1 does not go any further to the left than the square on the left of the copied $\mathfrak{x}$, so that the original argument $\mathfrak{x}$ remains unaltered and the computation of $h_1(\mathfrak{x})$ is not disturbed). We obtain

$$*\mathfrak{x}***\mathfrak{x}*h_1\underline{*}\cdots,$$

(where h_1 is an abbreviation of $h_1(\mathfrak{x})$). In order to compute $h_2(\mathfrak{x})$ the argument $\mathfrak{x}$ must first be carried over (by the help of K_{n+1}^{n}) to the right hand end:

$$*\mathfrak{x}***\mathfrak{x}*h_1*\mathfrak{x}\underline{*}\cdots.$$

Now M_2 computes without difficulty the value $h_2 = h_2(\mathfrak{x})$:

$$*\mathfrak{x}***\mathfrak{x}*h_1*\mathfrak{x}*h_2\underline{*}\cdots.$$

In the same way we carry over the arguments once more, compute h_3, etc. After the computation of h_r by means of M_r we have on the tape

$$*\mathfrak{x}***\mathfrak{x}*h_1*\mathfrak{x}*h_2\cdots*\mathfrak{x}*h_r\underline{*}\cdots.$$

Now, in order to compute $g(h_1, \ldots, h_r)$ we must first have the values $h_1, \ldots, h_r$ at our disposal at the right hand end. h_1 is carried over by $\mathsf{K}_{r+(r-1)n}$, then h_2 by $\mathsf{K}_{r+(r-2)n}$, $\ldots$, and finally h_r by $\mathsf{K}_{r+(r-r)n}$. Then we have

$$*\mathfrak{x}***\mathfrak{x}*h_1*\mathfrak{x}*h_2\cdots*\mathfrak{x}*h_r*h_1*h_2\cdots*h_r\underline{*}\cdots.$$

Now, $f = g(h_1, \ldots, h_r) = g(h_1(\mathfrak{x}), \ldots, h_r(\mathfrak{x}))$ can be computed straightaway by the help of M:

$$*\mathfrak{x}***\mathfrak{x}*h_1*\mathfrak{x}*h_2 \cdots *\mathfrak{x}*h_r*h_1*h_2 \cdots *h_r*f\underline{*} \cdots.$$

Now the cleaning up machine C solves the task of erasing the intermediate computations, and brings back the value f of the function to the original arguments:

$$*\mathfrak{x}*f\underline{*} \cdots.$$

This step concludes the standard Turing-computation of f.

3. *The standard Turing-computability is invariant under inductive definition.* Let the $(n+1)$-ary function f $(n \geqq 0)$ be defined inductively by the help of the two equations

$$f(\mathfrak{x}, 0) = g_1(\mathfrak{x})$$
$$f(\mathfrak{x}, y') = g_2(\mathfrak{x}, y, f(\mathfrak{x}, y)),$$

where g_1 and g_2 are standard Turing-computable by means of M_1 and M_2 respectively. Then f is also standard Turing-computable, and it is standard Turing-computed by

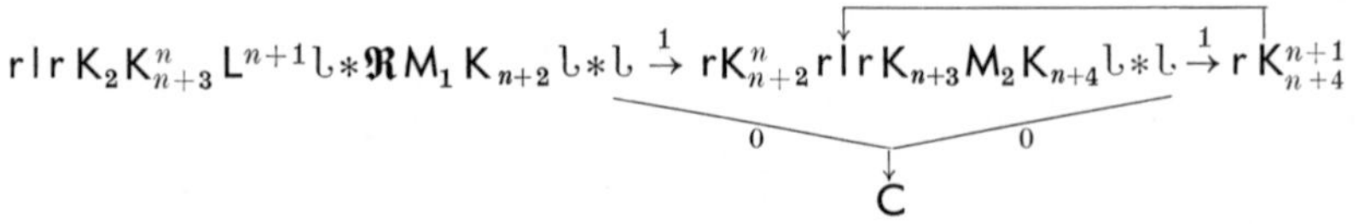

Let the arguments $\mathfrak{x}$, y be given on the computing tape at the beginning. We give, in the following list, the essential stages of the computation and explain them afterwards. In the list we use the abbreviation f_y for $f(\mathfrak{x}, y)$. (We again assume that $n > 0$, the machine works however for $n = 0$ as well, as can be verified easily.)

(a) $*\mathfrak{x}*y\underline{*} \cdots$

(b) $*\mathfrak{x}*y***y*\mathfrak{x}\underline{*} \cdots$

(c) $*\mathfrak{x}*y***y*\mathfrak{x}*f_0\underline{*} \cdots$

(d) $*\mathfrak{x}*y***y*\mathfrak{x}*f_0*y-1\underline{*} \cdots$

(e) $*\mathfrak{x}*y***y*\mathfrak{x}*f_0*y-1*\mathfrak{x}*0*f_0*f_1\underline{*} \cdots$

(f) $*\mathfrak{x}*y***y*\mathfrak{x}*f_0*y-1*\mathfrak{x}*0*f_0*f_1*y-2*\mathfrak{x}*1*f_1*f_2\underline{*} \cdots$

(g) $*\mathfrak{x}*y***$ as above until finally $*y-y*\mathfrak{x}*y-1*f_{y-1}*f_y\underline{*} \cdots$

(h) $*\mathfrak{x}*y*f_y\underline{*} \cdots.$

(a) denotes the initial situation. By the help of the bridge building procedure carried out in Section 2 we come to (b) by means of

$$\mathsf{r\,l\,r\,K_2\,K^n_{n+3}\,L^{n+1}}\,\mathfrak{l}*\mathfrak{R}$$

and from there to (c), i.e. to the value f_0 of the function, by

$$\mathsf{M}_1 .$$

Now, we copy y, erase the last stroke of the copy, and go back a square by means of

$$\mathsf{K}_{n+2}\mathfrak{l}*\mathfrak{l}.$$

Now, if the machine is over an empty square, then $y = 0$, and the computation is essentially at the end after the production of f_0; the cleaning up will be performed by C.

However, if there is a stroke in the square scanned, then the computation is not yet finished; and we go by r one square to the right, and stay in (d) behind $y - 1$.

Now we want to compute $f_1, f_2, \ldots, f_y$ successively by the help of the induction process. For this purpose we first copy $\mathfrak{x}$, print a 0 (which is represented by a stroke) behind it and finally copy f_0. Then we are staying on the right of $\mathfrak{x}*0*f_0$, i.e. behind the arguments from which f_1 can be computed directly. All this is done by

$$\mathsf{K}^n_{n+2}\,\mathsf{r}\,|\,\mathsf{r}\,\mathsf{K}_{n+3}\mathsf{M}_2 ,$$

after which we reach the stage (e). Now we copy $y - 1$, erase the last stroke, and go back one square by the help of

$$\mathsf{K}_{n+4}\mathfrak{l}*\mathfrak{l}.$$

Now, if the machine is over an empty square, then $y - 1 = 0$, $y = 1$ and so we have already computed f_y except for the cleaning up which is done by C. However, if the machine is over a stroke, then $y - 1 \neq 0$ and we must continue with the process. We obtain the arguments $\mathfrak{x}$, 0 by

$$\mathsf{r}\,\mathsf{K}^{n+1}_{n+4} .$$

The argument 0 is increased now by one by $|\,\mathsf{r}$ and f_1 is copied by means of K_{n+3}, after which the process considered last is repeated. Obviously, we can couple back at $|\,\mathsf{r}$ already. By this we arrive at (f) and finally at (g) and (h).

4. The standard Turing-computability is invariant under the application of the μ-operator to regular functions. Let g be a function of $n + 1$ variables ($n \geqq 0$). Assume that

$$\bigwedge_{\mathfrak{x}} \bigvee_{y} g(\mathfrak{x}, y) = 0.$$

Let the function f be defined by

$$f(\mathfrak{x}) = \mu y\, g(\mathfrak{x}, y) = 0.$$

If g is standard Turing-computable by means of M, then f can be standard Turing-computed by

$$\overset{\overset{0}{\downarrow}}{\mathsf{r}}\,|\,\mathsf{r}\mathsf{M}\mathsf{l}*\mathsf{l} \xrightarrow{1} \underset{\underset{1}{\curvearrowleft}}{*}\,\mathsf{l}.$$

The computation proceeds in the following way (where we assume that the machine is placed behind the last argument $\mathfrak{x}$ at the beginning and that the adjoining right half tape is empty (cf. § 15.1)).

(1) The machine moves one square to the right and proceeds further according to (2).

(2) The machine prints a stroke and moves a square further to the right. Now we have the arguments $\mathfrak{x}, y$ for the computation of g (initially $y = 0$). The machine computes $g(\mathfrak{x}, y)$ by means of M. It erases the last stroke of this function value by means of $\mathsf{l}*$ and goes (by the help of l) a square further to the left. According to whether the machine is now over a marked or an empty square, it proceeds further according to (3) or (4) respectively.

(3) In this case $g(\mathfrak{x}, y) \neq 0$, and so we must try out the next number y. By the help of $\overset{\overset{1}{\curvearrowleft}}{*\,\mathsf{l}}$ the machine erases the rest of $g(\mathfrak{x}, y)$. Then it is coupled back to (2) and so the procedure is continued with $y + 1$ instead of y.

(4) In this case $g(\mathfrak{x}, y) = 0$, so that the smallest y for which $g(\mathfrak{x}, y) = 0$ is immediately in front of the square scanned just now. Since we also know that the computing tape is completely empty on the right, the computation is complete.

Since we assumed that we apply the μ-operator to *regular functions* only, there must be a y such that $g(\mathfrak{x}, y) = 0$. This shows that the computation must finally terminate with the procedure part (4). We note for an application later on (in § 18.6) that the machine described here never stops operating in the case when *there exists no* y *such that* $g(\mathfrak{x}, y) = 0$. It will keep writing down the new arguments y and computing the value $g(\mathfrak{x}, y)$.

5. *Final remark.* The proof that every μ-recursive function f can be standard Turing-computed is *constructive* in the sense that for every μ-recursive function for which we have a chain of substitutions, inductive definitions, and μ-operations we can give *effectively* (on the basis of the above mentioned proof) an account of a Turing machine which standard computes f.

§ 17. Gödel Numbering of Turing Machines

As preparation for the considerations in the next paragraph we shall carry out several Gödel numberings. We shall namely (1) denote the squares of the computing tape by natural numbers and characterize on this basis the configurations of a Turing machine by numbers, and (2) form a one-one mapping between arbitrary Turing machines and some numbers. On this basis it will become possible to introduce functions by the help of which we can determine the numbers of the consecutive configuration from the number of a given configuration.

1. Numbering of the squares of a computing tape. We shall select an arbitrary square of the computing tape of a Turing machine and give it the number 0. We shall number the other squares according to the following schema.[1]

...	9	7	5	3	1	0	2	4	6	8	10	...

We shall speak, in the sense of this numbering, in short of the "square x". On the right of the square x is the square $R(x)$, on the left the square $L(x)$. We have[2]

$$R(x) = \begin{cases} x + 2, & \text{if } Ex \\ 0, & \text{if } x = 1 \\ x \dot{-} 2, & \text{if } Dx \wedge x \neq 1, \end{cases}$$

$$L(x) = \begin{cases} x + 2, & \text{if } Dx \\ 1, & \text{if } x = 0 \\ x \dot{-} 2, & \text{if } Ex \wedge x \neq 0. \end{cases}$$

$R(x)$ and $L(x)$ are primitive recursive functions.

Rxy will mean that the square x is on the right of the square y. We have

$$Rxy \leftrightarrow (Ex \wedge Ey \wedge x > y) \vee (Ex \wedge Dy) \vee (Dx \wedge Dy \wedge x < y).$$

This definition shows that R *is a primitive recursive predicate.*

Finally, we need one more function $Z(x, y)$ which is the number of the squares lying between x and y (including the square x but not the

[1] We have already developed a method (in § 5.2) for denoting the squares of a computing tape by numbers. However, we cannot use the method applied there for we want to use natural numbers only.

[2] Ex (resp. Dx) means that x is even (resp. odd). Cf. § 11.5.

square y), provided that y is on the left of x. $Z(x, y)$ *is a primitive recursive function* because

$$Z(x, y) = \begin{cases} \frac{x \dot{-} y}{2}, & \text{if } Ex \wedge Ey \\ \frac{y \dot{-} x}{2}, & \text{if } Dx \wedge Dy \\ \frac{(x+y)+1}{2}, & \text{otherwise.} \end{cases}$$

2. *Characterization of tape expressions by numbers.* The numbering of the squares discussed just now makes it possible to characterize a tape expression by a number b in a simple way. We speak, in this sense, in short of the "tape expression b". Let the square j contain the letter $a_{\beta(j)}$. Then we define

$$b = \prod_{j=0}^{\infty} p_j^{\beta(j)}.$$

We should keep in mind that an empty square has the letter a_0 printed in it, hence empty squares provide the factor 1 for the product, which is therefore infinite in a formal sense only. $b = 1$ denotes an empty tape. If the tape has the expression b, then the square j has the letter $a_{\exp(j,b)}$ printed in it.

Now we assume that a computation of a function is completed. Then we find on the tape the expression b. Let the square a be the last (empty) scanned square. According to § 6.1 the word which represents the value of the function ends immediately in front of the square a. The value w of the function is equal to the number of strokes which form this word, diminished by one.[1] w is determined by a and b. We want to describe a function $W_0(a, b)$ for which $w = W_0(a, b)$. For this purpose we first describe the left end square $E_l(a, b)$ and the right end square $E_r(a, b)$ of the word in question. The right end square is obviously

$$E_r(a, b) = L(a).$$

The left end square is unambiguously characterized by two conditions. (1) The square lying on its left is empty, and (2) every square which lies between the square lying on its left and the square a carries the symbol a_1. These conditions give (cf. § 12.1)

$$E_l(a, b) = \mu x [\exp(L(x), b) = 0 \wedge \bigwedge_y (Ry L(x) \wedge Ray \rightarrow \exp(y, b) = 1)].$$

[1] We can, in view of the application later on, confine ourselves to the case when the value of the function is given by sequences of strokes. If, more generally, the value of the function may be an arbitrary word, then the following formulae have to be slightly modified.

E_r and E_l are primitive recursive. This needs only to be shown for E_l. For this we must give bounds for the μ-operator and the generalizator occurring in the definition. To find x we need only consider such numbers for which the square x is marked. Thus, p_x must divide b. This makes certain that $x \leqq b$. — For y we can choose $\max(L(x), a)$ to be the bound. We need as a matter of fact consider only those y's for which $RyL(x)$ and Ray. If $RyL(x)$, then y is even or $y < L(x)$. If $y < L(x)$, then $y \leqq \max(L(x), a)$. If, however, y is even, then it follows from Ray that $y < a$. So, in every case, $y \leqq \max(L(x), a)$. This finally shows that[1]

$$E_l(a, b) = \mu_{x=0}^{b}\left[\exp(L(x), b) = 0 \wedge \bigwedge_{y=0}^{\max(L(x), a)} (RyL(x) \wedge Ray \rightarrow \exp(y, b) = 1)\right].$$

Now we obviously have that

$$(*) \qquad W_0(a, b) = Z(E_r(a, b), E_l(a, b)).$$

This representation shows that W_0 is *primitive recursive.*

3. *The Gödel number t of a Turing machine* **M**. **M** is given by a table. **M** has $m + 1$ states $0, \ldots, m$ $(m \geqq 0)$ and works with the symbols $a_0, \ldots, a_N$. In the third column are the instructions, in the fourth column the new states.

It is clear that we need only know the number N and the last two columns, for we can put in the first two columns without difficulty. We shall replace the symbols of the third column, which contains the instruction by numbers in the following way:

$$l, r, h, a_0, \ldots, a_N$$

is replaced by $\quad 1, 2, 3, 4, \ldots, N + 4$, respectively.

After this alteration we have the number matrix A_{ij} with $(N+1)(m+1)$ lines $(i = 1, \ldots, (N + 1)(m + 1))$ and two columns $(j = 3, 4)$. This we characterize by the number[2]

$$t = p_0^N p_1^m \prod_{i=1}^{(N+1)(m+1)} \prod_{j=3}^{4} p_{\sigma_2(i,j)}^{A_{ij}}.$$

t is called *the Gödel number* of **M**.

From t we can again obtain the table for **M**. First of all we have that $\sigma_2(i, j) > 1$ for $j = 3$ or $j = 4$. This gives, obviously, that

$$N = \exp(0, t)$$
$$m = \exp(1, t).$$

[1] We should keep in mind that $E_l(a, b)$ does not *always* have the above meaning. In spite of this the function is defined for *all a* and *b*.

[2] For σ_2 cf. § 12.4.

Further we have that, for $i = 1, \ldots, (N+1)(m+1)$ and $j = 3, 4$,

$$A_{ij} = \exp(\sigma_2(i, j), t).$$

Thus, the $N+1$ lines of the table of **M** which belong to the state c ($c = 0, \ldots, m$) look like this (if we represent the third column by the numbers characterizing it):

$$\begin{array}{llll} c & a_0 & \exp(\sigma_2((N+1)\,c+1, 3), t) & \exp(\sigma_2((N+1)\,c+1, 4), t) \\ \ldots\ldots & & \ldots\ldots & \ldots\ldots \\ c & a_N & \exp(\sigma_2((N+1)\,c+N+1, 3), t) & \exp(\sigma_2((N+1)\,c+N+1, 4), t). \end{array}$$

If we define the abbreviation

$$h(p, q, c, t) = \exp(\sigma_2((N+1)\,c+q+1, p), t),$$

then the line of the table of **M** which begins with $c\ a_q$ will be

$$(**) \qquad c \quad a_q \quad h(3, q, c, t) \quad h(4, q, c, t).$$

It should be noticed that the function h is primitive recursive.

The procedure given just now provides at the same time a decision procedure to determine whether or not an arbitrarily given number t is the Gödel number of a Turing machine. For this we compute first, as above, the numbers N and m from t. Then we produce, according to (**), a matrix of $(m+1)(N+1)$ lines and four columns. Now we must check whether the following conditions are satisfied:

$$(1) \quad N \geqq 1,$$
$$(2) \quad 1 \leqq h(3, q, c, t) \leqq N+4,$$
$$(3) \quad h(4, q, c, t) \leqq m.$$

If these conditions are not satisfied, then t is certainly not the Gödel number of a Turing machine. On the other hand, if the conditions are satisfied, then the matrix produced just now gives a Turing machine. For this Turing machine we must now calculate the Gödel number t_0 belonging to it according to the instruction at the beginning of this section. Now it is obvious that t is a Gödel number of a Turing machine if and only if $t = t_0$.[1]

4. The functions A, B, C. Let t be the Gödel number of a Turing machine. Let us consider the configuration (cf. § 5.3) denoted by the

[1] That it is necessary to check whether $t = t_0$ can be shown as follows. If we multiply the Gödel number t_0 of a Turing machine M_0 by a prime number which is not a factor of t_0, then we obtain a number t which is certainly not the Gödel number of a Turing machine. On the other hand, the procedure described above, if applied to t, will obviously produce the table of M_0 again.

scanned square a, the tape expression b, and the state c. If the machine does not stop operating at this configuration we obtain a consecutive configuration denoted by a new scanned square, which is determined unambiguously by t, a, b, c and so can be written in the form $A(t, a, b, c)$, a new tape expression $B(t, a, b, c)$, and a new state $C(t, a, b, c)$. We shall give an account of the functions A, B, C explicitly.

Originally the scanned square a contains the symbol $a_{\exp(a,b)}$. Thus, the line which is decisive for the next step of the machine is the one beginning with $c\, a_{\exp(a,b)}$, which according to Section 3 (**) is

$$(***) \qquad c \quad a_{\exp(a,b)} \quad h(3, \exp(a, b), c, t) \qquad h(4, \exp(a, b), c, t).$$

This provides at once the new state

$$C(t, a, b, c) = h(4, \exp(a, b), c, t).$$

The new scanned square is on the left or on the right of the old scanned square, if $h(3, \exp(a, b), c, t) = 1$ or 2 respectively. Otherwise the scanned square remains where it was. So we have that

$$A(t, a, b, c) = \begin{cases} L(a), & \text{if } h(3, \exp(a, b), c, t) = 1 \\ R(a), & \text{if } h(3, \exp(a, b), c, t) = 2 \\ a, & \text{otherwise.} \end{cases}$$

The tape expression is altered only if another symbol is printed in the scanned square. The alteration will be described by multiplication or division by a suitable power of p_a. We can write

$$B(t, a, b, c) = \begin{cases} \dfrac{b \cdot p_a^{h(3, \exp(a, b), c, t)}}{p_a^{\exp(a, b)}}, & \text{if } 4 \leq h(3, \exp(a, b), c, t) \\ b, & \text{if } 4 > h(3, \exp(a, b), c, t). \end{cases}$$

The given definitions show that A, B, C are *primitive recursive functions*.

Naturally, the values $A(t, a, b, c)$, $B(t, a, b, c)$, $C(t, a, b, c)$ of the functions have the given meaning only if (a, b, c) is not a terminal configuration.

Finally, we shall consider one more *predicate* $E_0 t a b c$ which, under the assumption that t is the Gödel number of a Turing machine **M**, says that the configuration (a, b, c) is a *terminal configuration of* **M**. This is the case if and only if the line of the table which begins with $c\, a_{\exp(a,b)}$ has the symbol h in the third column. We have represented this symbol by the number 3. This gives, according to (***), that

$$E_0 t a b c \leftrightarrow h(3, \exp(a, b), c, t) = 3.$$

This representation shows that E_0 *is primitive recursive.*

§ 18. The μ-Recursiveness of Turing-Computable Functions. Kleene's Normal Form

After the preparations of the last paragraph we shall now prove Theorem B_0 of § 15.3. From this theorem follows, as we have seen there, the μ-recursiveness of Turing-computable functions and also Kleene's normal form theorem for μ-recursive functions.

1. Gödel numbers of configurations and of functions and predicates connected with them. Let M be an arbitrary Turing machine. A configuration of M is given by a number triple (a, b, c). We can characterize this configuration unambiguously by the number $\sigma_3(a, b, c)$ which we shall call the *Gödel number of the configuration*. We speak in short of the *configuration* k. Naturally,

$$a = \sigma_{31}(k)$$
$$b = \sigma_{32}(k)$$
$$c = \sigma_{33}(k).$$

Provided t is the Gödel number of a Turing machine, let Etk mean that k is the Gödel number of a *terminal* configuration of the Turing machine with the number t. According to § 17.4 we have that

$$Etk \leftrightarrow E_0 t \sigma_{31}(k)\, \sigma_{32}(k)\, \sigma_{33}(k).$$

Now we assume that k has a consecutive configuration. The Gödel number of this configuration is given by a function $F(t, k)$ for which we have, according to § 17.4, that

$$\begin{aligned} F(t, k) = \sigma_3\big(&A(t, \sigma_{31}(k), \sigma_{32}(k), \sigma_{33}(k)),\\ &B(t, \sigma_{31}(k), \sigma_{32}(k), \sigma_{33}(k)),\\ &C(t, \sigma_{31}(k), \sigma_{32}(k), \sigma_{33}(k))\big). \end{aligned}$$

Finally let us consider a configuration k in which the scanned square is empty and is immediately behind a sequence of strokes, in front of which we have an empty square. This sequence of strokes represents a natural number (e.g. a value of a function). This number is given by (according to § 17.2 (*))

$$W(k) = W_0(\sigma_{31}(k), \sigma_{32}(k)).$$

It is immediately clear that E, F and W are primitive recursive.

2. The function $K(t, \mathfrak{x}, z)$. We start from an *arbitrary* Turing machine M with the Gödel number t. Further let $\mathfrak{x}$ be an arbitrary n-tuple of arguments. We print $\mathfrak{x}$ onto the otherwise empty tape. We choose as the square with the number 0 the square which is immediately in front of the arguments (thus, the first square of the argument strip) (an arbitrary square, if $n = 0$). Consequently we obtain a sequence of configurations which perhaps terminates at its last term. The numbers of these configurations form a sequence $K(t, \mathfrak{x}, z)$, $z = 0, 1, 2, \ldots$. If the sequence of configurations terminates after the step z_0, then $K(t, \mathfrak{x}, z)$ is determined for $z \leqq z_0$ only. We shall however define in this case $K(t, \mathfrak{x}, z)$ also for $z > z_0$ by the stipulation $K(t, \mathfrak{x}, z) = K(t, \mathfrak{x}, z_0)$.

We note two properties of K.

(1) *If* $K(t, \mathfrak{x}, z) = K(t, \mathfrak{x}, z')$, *then* $K(t, \mathfrak{x}, z) = K(t, \mathfrak{x}, z') = K(t, \mathfrak{x}, z'') = \ldots$. To prove this we distinguish between two cases. (a) $K(t, \mathfrak{x}, z)$ is not a terminal configuration. Then, since $K(t, \mathfrak{x}, z) = K(t, \mathfrak{x}, z')$, the configuration $K(t, \mathfrak{x}, z)$ is its own consecutive configuration. Because of the uniqueness of the consecutive configuration all the configurations following must coincide with $K(t, \mathfrak{x}, z)$. (b) $K(t, \mathfrak{x}, z)$ is a terminal configuration. This case can only occur if M stops operating after finitely many, say z_0, steps. Then no configuration $K(t, \mathfrak{x}, u)$ with $u < z_0$ is a terminal configuration, and so $z_0 \leqq z$. But, according to the definition of the function K we have for *every* $z \geqq z_0$ that $K(t, \mathfrak{x}, z) = K(t, \mathfrak{x}, z_0)$, from which the assertion follows.

(2) *If* M *stops operating after* z_0 *steps, then*

$$z_0 = \mu z K(t, \mathfrak{x}, z) = K(t, \mathfrak{x}, z').$$

We put $z_1 = \mu z\, K(t, \mathfrak{x}, z) = K(t, \mathfrak{x}, z')$. From the definition of K immediately follows that $z_1 \leqq z_0$. If $z_1 < z_0$, then, according to (1), $K(t, \mathfrak{x}, z_1) = K(t, \mathfrak{x}, z_1') = \cdots = K(t, \mathfrak{x}, z_0)$ is a terminal configuration, in contradiction with the fact that $K(t, \mathfrak{x}, z_0)$ is the first terminal configuration of the sequence of configurations $K(t, \mathfrak{x}, z)$. So, $z_1 = z_0$.

Now, we set ourselves the task to give an inductive definition of $K(t, \mathfrak{x}, z)$.

First of all $K(t, \mathfrak{x}, 0)$ is the number of the initial configuration. Here $c = 0$. The initial tape expression consists of the argument $\mathfrak{x}$. If we notice that all squares on the right of 0 have even numbers associated with them, then we easily see that the initial tape expression is given by

$$b_0(\mathfrak{x}) = \frac{\prod\limits_{j=0}^{x_1 + \cdots + x_n + 2n} p(2j)}{p(2(x_1 + \cdots + x_n + 2n))\, p(2(x_1 + \cdots + x_{n-1} + 2(n-1))) \cdots p(2(x_1 + 2))\, p(0)}.$$

We have chosen the scanned square for the initial configuration at the beginning of this section. This square has the number

$$a_0(\mathfrak{x}) = 2(x_1 + \cdots + x_n + 2n).$$

If $n = 0$, then we put $a_0(\mathfrak{x}) = 0$ and $b_0(\mathfrak{x}) = 1$.

Now we have

$$(*) \qquad K(t, \mathfrak{x}, 0) = \sigma_3(a_0(\mathfrak{x}), b_0(\mathfrak{x}), 0).$$

Furthermore, as we have seen earlier on, $K(t, \mathfrak{x}, z') = K(t, \mathfrak{x}, z)$ if $EtK(t, \mathfrak{x}, z)$, and $K(t, \mathfrak{x}, z') = F(t, K(t, \mathfrak{x}, z))$, otherwise. If we temporarily introduce, for abbreviation, a primitive recursive function φ by

$$\varphi(t, y) = \begin{cases} y, & \text{if } Ety \\ F(t, y), & \text{otherwise,} \end{cases}$$

then we have, obviously, that

$$(**) \qquad K(t, \mathfrak{x}, z') = \varphi(t, K(t, \mathfrak{x}, z)).$$

The equations (*), (**) show that the function K *is primitive recursive.*[1]

3. *The predicate* T_n, which is $(n+2)$-ary $(n \geqq 0)$, is defined by

$$(***) \quad T_n t\mathfrak{x}y \leftrightarrow K(t, \mathfrak{x}, \sigma_{21}(y)) = K(t, \mathfrak{x}, (\sigma_{21}(y))') \wedge \sigma_{22}(y) = K(t, \mathfrak{x}, \sigma_{21}(y)).$$

This definition shows immediately that T_n *is primitive recursive.* We show that

(3) *If there exists a z such that* $K(t, \mathfrak{x}, z) = K(t, \mathfrak{x}, z')$, *then there exists a y such that* $T_n t\mathfrak{x}y$. Let $K(t, \mathfrak{x}, z) = K(t, \mathfrak{x}, z')$. We put $y = \sigma_2(z, K(t, \mathfrak{x}, z))$. Then $z = \sigma_{21}(y)$, and so $K(t, \mathfrak{x}, \sigma_{21}(y)) = K(t, \mathfrak{x}, (\sigma_{21}(y))')$. Further $K(t, \mathfrak{x}, \sigma_{21}(y)) = K(t, \mathfrak{x}, z) = \sigma_{22}(y)$.

(4) *If there exists a y such that $T_n t\mathfrak{x}y$, then there exists also a z such that* $K(t, \mathfrak{x}, z) = K(t, \mathfrak{x}, z')$, namely $z = \sigma_{21}(y)$.

(5) (*Here we assume that t is the Gödel number of a Turing machine* **M**.) *If* **M** *stops operating after z_0 steps and if $T_n t\mathfrak{x}y$, then* $\sigma_{22}(y) = K(t, \mathfrak{x}, z_0)$. Because $T_n t\mathfrak{x}y$ we have $K(t, \mathfrak{x}, \sigma_{21}(y)) = K(t, \mathfrak{x}, (\sigma_{21}(y))')$; on the other hand, according to (2), z_0 is the smallest number z for which $K(t, \mathfrak{x}, z) = K(t, \mathfrak{x}, z')$. This gives $z_0 \leqq \sigma_{21}(y)$. Together with (1) this implies that $K(t, \mathfrak{x}, z_0) = K(t, \mathfrak{x}, z_0) = \cdots = K(t, \mathfrak{x}, \sigma_{21}(y))$, and $K(t, \mathfrak{x}, \sigma_{21}(y)) = \sigma_{22}(y)$ because $T_n t\mathfrak{x}y$.

4. *Kleene's normal form.* Now we assume that the n-ary function f is computed by the machine **M** with the Gödel number t. We shall represent f by the help of T_n. Let z_0 be the number of steps after which **M**, placed behind $\mathfrak{x}$, stops operating. According to (3) there exists a y such that $T_n t\mathfrak{x}y$. Therefore, $\mu y T_n t\mathfrak{x}y$ is an application of the μ-operator to a *regular* function. Naturally, $T_n t\mathfrak{x}(\mu y T_n t\mathfrak{x}y)$. According to (5), from this follows that $\sigma_{22}(\mu y T_n t\mathfrak{x}y) = K(t, \mathfrak{x}, z_0) =$ Gödel number of the terminal

[1] It should be noticed that the function K is defined by (*), (**) *for all t* (not only for those t's which are Gödel numbers of Turing machines).

configuration of M. From this we obtain, according to Section 1, the value of the function $f(\mathfrak{x}) = W(K(t, \mathfrak{x}, z_0)) = W(\sigma_{22}(\mu y T_n t \mathfrak{x} y))$. Finally, we introduce a *primitive recursive* function U by the definition

$$(\substack{**\\**}) \qquad U(u) = W(\sigma_{22}(u)),$$

and so we obtain the final representation

$$(\substack{***\\**}) \qquad f(\mathfrak{x}) = U(\mu y T_n t \mathfrak{x} y).$$

Thus we have shown that every Turing-computable function is μ-recursive. In the representation $(\substack{***\\**})$ the μ-operator is applied only once. *Thus, Theorem B_0 is proved.*

Since we already know that every μ-recursive function is Turing-computable we obtain the

Theorem of Kleene's normal form for μ-recursive functions. For the singulary primitive recursive function defined in $(\substack{**\\**})$ and for the $(n+2)$-ary primitive recursive predicate T_n introduced in (***) the following statement is valid. For every μ-recursive n-ary function there exists at least one number t such that, for every n-tuple of arguments $\mathfrak{x}$,

(a) *there exists a y such that $T_n t \mathfrak{x} y$,*

(b) $f(\mathfrak{x}) = U(\mu y T_n t \mathfrak{x} y)$.

5. *Two remarks.*

(1) In the proof given above we have proved in fact more than the existence of a number t with the given properties. As a matter of fact if we are given any μ-recursive function f explicitly, so that it is shown how f is obtained from the initial functions by the help of substitutions, inductive definitions and μ-operations (which are applied to regular functions), then we can give an account, explicitly as done in § 16, of a Turing machine M which computes f. Now we can put t equal to the Gödel number of M, i.e. we can give t effectively.

(2) The reader should keep in mind that the statement (a) is not asserted for arbitrary t. It would be false to do so. Let us consider for instance the Turing machine $\mathsf{M} = \mathsf{r}\overleftarrow{\mathsf{l}}$. If we place M behind the argument 0, i.e. behind a stroke, then no two configurations following one another are the same. Therefore, according to (4), we cannot have a y such that $T_n t \mathfrak{x} y$. — Another proof to show that (a) is not valid for every t can be given as follows. If there exists a y, for every t and $\mathfrak{x}$, such that $T_n t \mathfrak{x} y$, then the function $\mu y T_n t \mathfrak{x} y$ is μ-recursive. However, we have the

Theorem. The function $\mu y T_n t \mathfrak{x} y$ is not μ-recursive.

Proof is by reductio ad absurdum. If $\mu y T_n t \mathfrak{x} y$ is μ-recursive, then so is the n-ary function

$$f(\mathfrak{x}) = U(\mu y T_n x_1 \mathfrak{x} y) + 1.$$

According to Kleene's normal form theorem there exists a t such that for all $\mathfrak{x}$

$$f(\mathfrak{x}) = U(\mu y T_n t \mathfrak{x} y).$$

If we choose an argument $\mathfrak{x}$ such that $x_1 = t$, then we obtain a contradiction (diagonal procedure!).

6. Kleene's enumeration theorem.[1] *For every* $(n+1)$*-ary* μ*-recursive predicate* $R\mathfrak{x}y$ *there exists a number* t *such that for all* $\mathfrak{x}$

$$\bigvee_y R\mathfrak{x}y \leftrightarrow \bigvee_y T_n t \mathfrak{x} y.$$

For *proof* we start with the *characteristic function*[2] $g(\mathfrak{x}, y)$ of the predicate $R\mathfrak{x}y$. g is μ-recursive and so, according to § 16, computable by a Turing machine **M**. Now, we consider the machine, constructed by the help of **M**, which is described in § 16.4. We call this machine M_0 and use it for the computation of $\mu y g(\mathfrak{x}, y) = 0$. Let t be the Gödel number of M_0. We assert that, for this t, the relation given in the Theorem is valid.

(a) Suppose $\bigvee_y R\mathfrak{x}y$. Then there exists a y such that $g(\mathfrak{x}, y) = 0$. The machine M_0, when used to compute $\mu y g(\mathfrak{x}, y) = 0$, stops operating after finitely many steps. Then according to Section 2 (2) and Section 3 (3) (for M_0 in place of **M**) there exists a y such that $T_n t \mathfrak{x} y$.

(b) Suppose there exists a y such that $T_n t \mathfrak{x} y$. According to Section 3 (4) there exists a z such that $K(t, \mathfrak{x}, z) = K(t, \mathfrak{x}, z')$. According to Section 2 (1) $K(t, \mathfrak{x}, z)$, considered as a function of z, is finally constant. If the assertion $\bigvee_y R\mathfrak{x}y$ is false, then there exists no y such that $g(\mathfrak{x}, y) = 0$. Then the machine M_0 prints, as it has been shown at the end of § 16.4, always new arguments y and computes the values $g(\mathfrak{x}, y)$. From this immediately follows that the configurations $K(t, \mathfrak{x}, z)$ cannot be finally constant as z increases. By reductio ad absurdum the assertion is proved.

Reference

KLEENE, S. C.: General Recursive Functions of Natural Numbers. Math. Ann. **112**, 727–742 (1936). (Normal form theorem.)

[1] The choice of this phrase will be justified by the considerations of § 28.
[2] Cf. § 11.1.

CHAPTER 5

RECURSIVE FUNCTIONS

In the last two chapters we considered the properties of μ-recursive functions. It was shown that the class of μ-recursive functions is the same as the class of Turing-computable functions and so the same as the class of the functions which are computable in the intuitive sense. Thus, we can say that the concept of μ-recursive function, just like that of Turing-computable function, is a precise replacement of the concept of computable function. Another concept which can be considered to be a precise replacement of the concept of computable function (and which historically precedes the concept of μ-recursive function) is the concept of recursive function (HERBRAND, GÖDEL, KLEENE). After the definition of recursiveness (in § 19) we shall show in the two following paragraphs that the class of μ-recursive functions coincides with the class of recursive functions.

Today we often denote by the name *recursive function* any function which is computable in the sense of a precise definition which is equivalent to the definition of recursive function in the strict sense. We frequently speak of *recursive predicates, recursive enumerability, recursive decidability,* etc. in the same sense.

§ 19. Definition of Recursive Functions

Before giving the exact definition (Section 4) we shall carry out some considerations to show that recursive functions form a class which has something in common with the class of computable functions. We shall see that this concept of recursive function is based upon a quite general idea. However, we shall have to admit that this generality is not sufficient to show *directly* (as in the case of Turing-computability) that the class of recursive functions contains *all* computable functions.

1. Heuristic considerations. We start out from the equations which are usually given to define the product $P(x, y)$ inductively (cf. § 10.4). In one of these equations we have the sum $S_u(x, y)$ as a "helpfunction". If we want to give a complete definition of the product, then we must also attach the equations which define $S_u(x, y)$. Then we obtain the following equation system for the definition of the product

$$(*)\qquad \begin{cases} S_u(x, 0) & = x \\ S_u(x, S(y)) & = S(S_u(x, y)) \\ P(x, 0) & = 0 \\ P(x, S(y)) & = S_u(P(x, y), x), \end{cases}$$

where 0 stands for zero and S denotes the successor function, so that the "numerals" $0, S(0), S(S(0)), S(S(S(0))), \ldots$ can be comprehended as symbols for the natural numbers. We shall not consider S to be a helpfunction.

Obviously, there exists exactly one pair (S_u, P) of functions which satisfies this definition, namely the pair (sum, product). Thus, the product is defined by (*).

In the considerations which led in § 10 to the concept of primitive recursive function it was important that (*) is a pair of ordinary inductive definitions. However, this is no longer of importance, since we have shown that the class of primitive recursive functions does not contain all computable functions (§ 13). In fact, the equation system by which we defined Ackermann's function $f(x, y)$ leads outside the scope of the ordinary inductive definition:

$$(**)\qquad \begin{cases} F(0, y) & = S(y) \\ F(S(x), 0) & = F(x, S(0)) \\ F(S(x), S(y)) & = F(x, F(S(x), y)). \end{cases}$$

We have seen that there exists one and only one function F which satisfies the equations (**). (Contrary to (*) no helpfunction is used here).

Now, it suggests itself to consider functions which (like in the examples above the product function and Ackermann's function) are unambiguously defined by a finite system of equations.

We may ask the question whether such functions are always computable in the intuitive sense. The answer is yes for functions which are defined by systems which we considered in the definition of primitive recursive functions, since the systems appearing there represent substitutions or inductive definitions. However, we can give examples of systems of equations (§ 21.7) which unambiguously define non-computable functions.

To exclude such cases from the start we shall say *how* the values of the function must be calculated from the system of equations. We shall give for this purpose certain obvious rules and demand that the value of the function can be obtained from the given equation *by the application of these rules only*. The reader ought to keep in mind that in the rest of this chapter we shall speak of derivations only in this restricted sense. Such rules are the following two, which we shall formulate precisely in Section 3.

(SR) *Substitution* of numerals for variables.

(RR) *Replacement* of an expression $h(\nu_1, \ldots, \nu_n)$ by a numeral ν, provided that the "numeral equation" $h(\nu_1, \ldots, \nu_n) = \nu$ has already been derived.

These rules are plausible, more precisely "correct" (cf. § 20.2) and they are, as experience shows, sufficient for the calculation of values of functions for many systems of equations which we would think of. This applies for instance in the cases of the above introduced systems (*) and (**). Let us calculate for instance the value $P(1, 1)$ from (*). In this calculation we shall number the equations which occur and give a hint on the rule applied on the right. (The reader should convince himself that the sequence of the lines in the proof can be altered to a certain extent.)

(1)	$S_u(x, 0) = x$	initial equation	
(2)	$S_u(x, S(y)) = S(S_u(x, y))$	initial equation	
(3)	$P(x, 0) = 0$	initial equation	
(4)	$P(x, S(y)) = S_u(P(x, y), x)$	initial equation	
(5)	$S_u(0, 0) = 0$	(SR)	(1)
(6)	$S_u(0, S(0)) = S(S_u(0, 0))$.	(SR)	(2)
(7)	$S_u(0, S(0)) = S(0)$	(RR)	(5), (6)
(8)	$P(S(0), 0) = 0$	(SR)	(3)
(9)	$P(S(0), S(0)) = S_u(P(S(0), 0), S(0))$	(SR)	(4)
(10)	$P(S(0), S(0)) = S_u(0, S(0))$	(RR)	(8), (9)
(11)	$P(S(0), S(0)) = S(0)$.	(RR)	(7), (10)

We shall in future consider a system $\mathfrak{S}$ of equations sufficient for the definition of a function φ if and only if (if we denote φ by the symbol F) an equation $F(\nu_1, \ldots, \nu_n) = \nu$ is derivable from $\mathfrak{S}$ if and only if the corresponding relation between the arguments and the values of the function holds.

It would be possible to place similar demands also on the symbols which, like the symbols S_u in (*), occur as "helpsymbols". However, this is not usual and we shall therefore forgo to do so. We shall not even demand that there exist functions for which all the equations of $\mathfrak{S}$ are valid (in the sense of § 20.1).

We could conjecture that the additional requirements on the systems of equations mentioned just now would reduce the class of functions definable by such a system. However, this is not the case, as we shall see from the theorems of § 20.3 and § 21.

2. *Terms, numerals, equations.* We start from the symbol 0, the number variables $x_0, x_1, x_2, \ldots,$ the function constant S and the function variables $F_0^0, F_0^1, F_0^2, \ldots, F_1^0, F_1^1, F_1^2, \ldots, F_2^0, F_2^1, F_2^2, \ldots, \ldots$. In the function variable F_j^n we call n the *place index* and j the *difference index*. We shall use 0 as a symbol for the number zero, and S as a symbol for the successor function. Further, we use parentheses (,) and the equality symbol $=$.

The *terms* are defined inductively by the following stipulations.

(a) 0 is a term.

(b) Every variable x_i is a term.

(c) Every 0-ary function variable F_i^0 is a term.

(d) If τ is a term, then so is $S(\tau)$.

(e) If F_j^n is an n-ary function variable $(n \geqq 1)$ and $\tau_1, \ldots, \tau_n$ are terms, then $F_j^n(\tau_1, \ldots, \tau_n)$ is also a term.

We call the terms formed by (a), (b), (c) *simple terms* and the terms formed by (d), (e) *compound terms.*

The *numerals* are special terms which are defined by the following.

(a') 0 is a numeral.

(b') If ν is a numeral, then so is $S(\nu)$.

The numerals are names for the natural numbers.

By an *F-term* we understand a term of the form $F_j^n(\nu_1, \ldots, \nu_n)$, where $\nu_1, \ldots, \nu_n$ are numerals. Especially, every term F_j^0 is an F-term.

Finally, an *equation* is a line of symbols of the form $\tau_1 = \tau_2$, where τ_1 and τ_2 are terms.

In order to describe the statement that the terms τ_1 and τ_2, considered as lines of symbols, are identical, we use the expression $\tau_1 \equiv \tau_2$.

3. The rules (SR) *and* (RR). Let x_i be a variable and τ_0 an arbitrary term. We associate with every term τ a term $\tau' = \tau^{x_i}/\tau_0$. We shall say that τ' is obtained from τ by the *substitution* of τ_0 for x_i. The substitution operation is defined inductively as follows.

(a) $0^{x_i}/\tau_0 \equiv 0$;

(b) $x_i{}^{x_i}/\tau_0 \equiv \tau_0$; $x_k{}^{x_i}/\tau_0 \equiv x_k$, if $k \neq i$;

(c) $F_j^0{}^{x_i}/\tau_0 \equiv F_j^0$

(d) $[S(\tau)]^{x_i}/\tau_0 \equiv S(\tau^{x_i}/\tau_0)$;

(e) $[F_j^n(\tau_1, \ldots, \tau_n)]^{x_i}/\tau_0 \equiv F_j^n(\tau_1{}^{x_i}/\tau_0, \ldots, \tau_n{}^{x_i}/\tau_0)$ for $n \geqq 1$.

By the help of this operation we formulate the two rules (SR) and (RR) which we have already mentioned in Section 1.

(SR) *Substitution rule.* If x_i is a variable and ν is a numeral, then we can proceed from the equation

$$\tau = \tilde{\tau}$$

by *substitution of ν for x_i* to the equation

$$\tau^{x_i}/\nu = \tilde{\tau}^{x_i}/\nu .$$

(RR) *Replacement rule.* Let τ' be an F-term, $\tilde{\tau}'$ be a numeral and x_i be a variable. Further let τ'', $\tilde{\tau}''$, τ_0, $\tilde{\tau}_0$, τ, $\tilde{\tau}$ be terms for which we have

$$\tau'' \equiv \tau_0^{x_i}/\tau', \qquad \tilde{\tau}'' \equiv \tilde{\tau}_0^{x_i}/\tau'$$
$$\tau \equiv \tau_0^{x_i}/\tilde{\tau}', \qquad \tilde{\tau} \equiv \tilde{\tau}_0^{x_i}/\tilde{\tau}'.$$

Then we can proceed from the equations

$$\tau' = \tilde{\tau}'$$
$$\tau'' = \tilde{\tau}''$$

to the equation

$$\tau = \tilde{\tau}.$$

We can say that $\tau = \tilde{\tau}$ is obtained from $\tau'' = \tilde{\tau}''$ by replacement of τ' by $\tilde{\tau}'$ at arbitrarily many places.[1]

The rules (SR) and (RR) define a calculus, which we shall call *equation calculus*.

4. *Recursive functions.* An n-ary function f ($n \geqq 0$) is called (*general*) *recursive*, if there exists a finite system $\mathfrak{S}$ of equations and a function variable F_j^n such that *for all* numerals $\nu_1, \ldots, \nu_n, \nu$ we have that if $\nu_1, \ldots, \nu_n, \nu$ denote the numbers $k_1, \ldots, k_n, k$, then

$$f(k_1, \ldots, k_n) = k \quad \text{if and only if}$$

$F_j^n(\nu_1, \ldots, \nu_n) = \nu$ is derivable from $\mathfrak{S}$ by means of (SR) and (RR).

We shall say in this case that $\mathfrak{S}$ defines the function f with regard to F_j^n.

Instead of recursive functions we often speak of *general recursive functions* in order to emphasize the difference from primitive recursive functions.

A function f which is given by a system $\mathfrak{S}$ of equations with regard to F_j^n is computable in the intuitive sense. We show this as follows. The set of valid[2] equations $F_j^n(\nu_1, \ldots, \nu_n) = \nu$ is enumerable. In order to find the value of the function for given arguments $\nu_1, \ldots, \nu_n$ we only need to continue systematically the enumeration procedure given by (SR) and (RR) for a sufficiently long time; we shall finally come across an equation of the form $F_j^n(\nu_1, \ldots, \nu_n) = \nu$ which gives the value of the function.

5. *Renaming of function variables.* Naturally the choice of the function variables which we use is not essential. In a system $\mathfrak{S}$ of equations,

---

[1] In this fashion we obtain from the equation $F_1^2(F_1^2(x_1, x_2), F_1^2(x_1, x_2)) = x_1$ the equation $F_1^2(F_1^2(x_1, x_2), x_2) = x_1$ by the replacement of $F_1^2(x_1, x_2)$ by x_2 at arbitrarily many places. In order to show this we have to put $\tau_0 \equiv F_1^2(F_1^2(x_1, x_2), x_3)$, $\tilde{\tau}_0 \equiv x_1$ and $x_i \equiv x_3$.

[2] An exact definition of the validity of an equation will be given in § 20.1.

which defines f with regard to F_j^n, may occur for instance (apart from S and F_j^n) the function variables $F_{j_1}^{n_1}, \ldots, F_{j_q}^{n_q}$. Now, if we change successively F_j^n into F_k^n, $F_{j_1}^{n_1}$ into $F_{k_1}^{n_1}, \ldots, F_{j_q}^{n_q}$ into $F_{k_q}^{n_q}$ (where different symbols are changed into different symbols again), then each term τ will be changed in a trivial way into a corresponding term τ' and finally $\mathfrak{S}$ into $\mathfrak{S}'$. It is clear that $\mathfrak{S}'$ defines the function f with regard to F_k^n. — We shall make use of this possibility of changing the notation repeatedly in the next paragraph.

6. Partially recursive functions. It is possible that, for a finite system $\mathfrak{S}$ of equations and for a function variable F_j^n, we have that if the equations of the form

$$F_j^n(\nu_1, \ldots, \nu_n) = \nu \qquad \text{and} \qquad F_j^n(\nu_1, \ldots, \nu_n) = \nu^*$$

are derivable from $\mathfrak{S}$ by means of (SR) and (RR), then in all cases $\nu \equiv \nu^*$. As against the situation which we considered in Section 4 in defining recursive functions it is not demanded here that for *every* n-tuple $\nu_1, \ldots, \nu_n$ of numerals at least one equation of the form $F_j^n(\nu_1, \ldots, \nu_n) = \nu$ is derivable.

A system of equations of the above kind calls for a definition of an n-ary function f which is defined for those n-tuples of arguments for whose numeral representation $\nu_1, \ldots, \nu_n$ an equation $F_j^n(\nu_1, \ldots, \nu_n) = \nu$ is derivable and whose value is represented by ν. Such functions are called *partially recursive functions*. For the properties of these functions consult the books by KLEENE and DAVIS mentioned at the end of the preface.

Exercise. In § 1.6 we introduced the concept of superimposed rule system. Describe the equation calculus in the form of a superimposed rule system. (The solution of this exercise is very difficult. Readers who have no practice in calculi ought not to attempt it. For hint cf. § 24.1.)

References

HERBRAND, J.: Sur la non-contradiction de l'Arithmétique. J. reine angew. Math. **166**, 1–8 (1931). (Idea of recursive function.)

GÖDEL, K.: On Undecidable Propositions of Formal Mathematical Systems. Mimeographed. Institute for Advanced Study, Princeton, N. J. 1934. 30 pp. (First account of rules of inference.)

KLEENE, S. C.: General Recursive Functions of Natural Numbers. Math. Ann. **112**, 727–742 (1936). (Introduction of the expression "recursive functions" for these functions, and their precise definition.)

§ 20. The Recursiveness of μ-Recursive Functions

In this paragraph we shall prove the theorem that every μ-recursive function is definable by a system of equations. We shall show this theorem in a strengthened form. First we begin with a few preliminary remarks intended to explain what this strengthening consists of.

That a function f is defined by a finite system $\mathfrak{S}$ with regard to the function variable F_j^n means according to the definition given in the last paragraph that an equation of the form

$$F_j^n(\nu_1, \ldots, \nu_n) = \nu$$

is derivable from $\mathfrak{S}$ if and only if the numbers $k_1, \ldots, k_n, k$, which correspond to the numerals $\nu_1, \ldots, \nu_n, \nu$, are in the relation $f(k_1, \ldots, k_n) = k$. The last part can also be expressed in short by saying that *the equation in question is valid for* f. In general, further function variables will also occur in $\mathfrak{S}$, e.g. F_i^r. However, it is not required in the definition of recursiveness that for F_i^r also there exists a function such that this function is defined by $\mathfrak{S}$ with regard to F_i^r. It is possible that we are not able to derive an equation of the form $F_i^r(\nu_1, \ldots, \nu_r) = \nu^*$ from $\mathfrak{S}$ for every $\nu_1, \ldots, \nu_r$. It is also possible that for some $\nu_1, \ldots, \nu_r$ there exist numerals $\nu^* \neq \nu^{**}$ for which the equations $F_i^r(\nu_1, \ldots, \nu_r) = \nu^*$ and $F_i^r(\nu_1, \ldots, \nu_r) = \nu^{**}$ are both derivable from $\mathfrak{S}$. Finally, it is possible that both these cases occur at the same time.[1]

We shall call a function f *standard definable* by a finite system $\mathfrak{S}$ of equations with regard to the function variable F_j^n, if for *all* function variables $F_{j_1}^{n_1}, \ldots, F_{j_q}^{n_q}$ occurring in $\mathfrak{S}$ there exist functions $f_{j_1}^{n_1}, \ldots, f_{j_q}^{n_q}$ of corresponding number of places (where, especially, the function f corresponds to the symbol F_j^n) of the kind that for every such $F_{j_i}^{n_i}$ an equation of the form $F_{j_i}^{n_i}(\nu_1, \ldots, \nu_{n_i}) = \nu$ is derivable from $\mathfrak{S}$ if and only if it is valid for the function $f_{j_i}^{n_i}$. The strengthening will consist in proving the standard definability of μ-recursive functions. In this case we have to determine suitable functions $f_{j_i}^{n_i}$ (we shall call the system of the function $f_{j_i}^{n_i}$ an "interpretation" of the system of the function variables $F_{j_i}^{n_i}$) and to show for every $F_{j_i}^{n_i}$ that

(a) If an equation $F_{j_i}^{n_i}(\nu_1, \ldots, \nu_{n_i}) = \nu$ is derivable from $\mathfrak{S}$, then it is valid for $f_{j_i}^{n_i}$. This we shall show by proving that

(a_1) every equation of $\mathfrak{S}$ is valid if the function variables $F_{j_i}^{n_i}$ are given the interpretation $f_{j_i}^{n_i}$,

(a_2) the rules (SR) and (RR) lead from equations which are valid under this interpretation to such equations again.

(b) If an equation $F_{j_i}^{n_i}(\nu_1, \ldots, \nu_{n_i}) = \nu$ is valid for the function $f_{j_i}^{n_i}$, then we can derive this equation from $\mathfrak{S}$. We shall prove this by giving an actual derivation.

[1] For an example cf. the last note of this paragraph, and also Exercise.

Until now we have defined what it means to be valid under an interpretation only for equations of the form $F_{j_1}^{n_1}(\nu_1, \ldots, \nu_{n_1}) = \nu$. In order to make sense of (b_1) we must extend this concept to arbitrary equations. This will be our next task.

1. Interpretations and validity. By an *interpretation* $\mathfrak{J}$ we shall understand a mapping which is defined for certain function variables and which associates with these functions of the same number of arguments over the domain of natural numbers.

We consider an interpretation $\mathfrak{J}$ and a mapping φ which is defined for certain variables and maps these onto natural numbers. Then we can associate with every term τ (for the function variables of which $\mathfrak{J}$ and for the variables of which φ is defined) a number $\tau^{\mathfrak{J}\varphi}$ according to the following instruction (which is to be applied only to those F_j^n and x_i for which $\mathfrak{J}$ and φ resp. are defined):

(a) $0^{\mathfrak{J}\varphi}$ = the number zero;

(b) $x_i^{\mathfrak{J}\varphi} = \varphi(x_i)$;

(c) $F_j^{0\mathfrak{J}\varphi}$ = the value of $\mathfrak{J}(F_j^0)$;

(d) $S(\tau)^{\mathfrak{J}\varphi}$ = the successor of $\tau^{\mathfrak{J}\varphi}$;

(e) $F_j^n(\tau_1, \ldots, \tau_n)^{\mathfrak{J}\varphi} = \mathfrak{J}(F_j^n)(\tau_1^{\mathfrak{J}\varphi}, \ldots, \tau_n^{\mathfrak{J}\varphi})$.

(a) says that 0 is comprehended as the name of the number zero. (d) says that S is used as the notation for the successor function. This means that the image $\nu^{\mathfrak{J}\varphi}$ is defined for every numeral ν in such a way that $\nu^{\mathfrak{J}\varphi}$ is the natural number corresponding to ν, as we can easily show by induction.

We shall say that *an equation $\tau_1 = \tau_2$ is valid under an interpretation $\mathfrak{J}$*, if $\mathfrak{J}$ is defined at least for all function variables which occur in τ_1 or τ_2, and if, for every mapping φ which is defined for all variables occurring in τ_1 or τ_2, the number $\tau_1^{\mathfrak{J}\varphi}$ coincides with $\tau_2^{\mathfrak{J}\varphi}$.[1] A (finite or infinite) set $\mathfrak{S}$ of equations is called valid under $\mathfrak{J}$ if every equation of $\mathfrak{S}$ is valid under $\mathfrak{J}$.

2. The correctness of the rules (SR) *and* (RR). The *correctness* of these rules is expressed by the following

Theorem. If an equation $\tau_1 = \tau_2$ is derivable from a system $\mathfrak{S}$ of equations by means of (SR) *and* (RR), *then if $\mathfrak{S}$ is valid under a given interpretation $\mathfrak{J}$, then the equation $\tau_1 = \tau_2$ is also valid under $\mathfrak{J}$.*

We should notice here that an equation $\tau_1 = \tau_2$ which is derivable from $\mathfrak{S}$ can only contain function variables which occur already in $\mathfrak{S}$. First we prove the

[1] The reader should convince himself that in the case of an equation of the form $F_{j_i}^{n_i}(\nu_1, \ldots, \nu_{n_i}) = \nu$ this definition of validity is in accordance with the way we have treated validity in the introduction to this paragraph.

Lemma. Let τ, τ_0 be terms and x_i a variable. Let $\mathfrak{J}$ be defined for all function variables occurring in τ or τ_0. Let φ be a mapping of the variables occurring in τ or τ_0 onto natural numbers. Let us define a mapping ψ for these variables by the stipulation that $\psi(x_k) = \varphi(x_k)$, if $k \neq i$, and $\psi(x_i) = \tau_0^{\mathfrak{J}\varphi}$. Then we have that

$$\tau^{\mathfrak{J}\psi} = [\tau^{x_i}/\tau_0]^{\mathfrak{J}\varphi}.$$

The proof is easily carried out by induction on the construction of τ:

(a) $[0^{x_i}/\tau_0]^{\mathfrak{J}\varphi} = 0^{\mathfrak{J}\varphi} = \text{zero} = 0^{\mathfrak{J}\psi}$;

(b) $[x_k^{x_i}/\tau_0]^{\mathfrak{J}\varphi} = x_k^{\mathfrak{J}\varphi} = \varphi(x_k) = \psi(x_k) = x_k^{\mathfrak{J}\psi}$, if $k \neq i$;
$[x_i^{x_i}/\tau_0]^{\mathfrak{J}\varphi} = \tau_0^{\mathfrak{J}\varphi} = \psi(x_i) = x_i^{\mathfrak{J}\psi}$;

(c) $[F_j^{0\,x_i}/\tau_0]^{\mathfrak{J}\varphi} = F_j^0{}^{\mathfrak{J}\varphi} = \text{value of } \mathfrak{J}(F_j^0) = F_i^0{}^{\mathfrak{J}\psi}$

(d) $[[S(\tau)]^{x_i}/\tau_0]^{\mathfrak{J}\varphi} = [S(\tau^{x_i}/\tau_0)]^{\mathfrak{J}\varphi}$
$\qquad\qquad\qquad = \text{successor of } [\tau^{x_i}/\tau_0]^{\mathfrak{J}\varphi}$
$\qquad\qquad\qquad = \text{successor of } \tau^{\mathfrak{J}\psi}$
$\qquad\qquad\qquad = [S(\tau)]^{\mathfrak{J}\psi}$;

(e) $[[F_j^n(\tau_1, \ldots, \tau_n)]^{x_i}/\tau_0]^{\mathfrak{J}\varphi} = [F_j^n(\tau_1^{x_i}/\tau_0, \ldots, \tau_n^{x_i}/\tau_0)]^{\mathfrak{J}\varphi}$
$\qquad\qquad\qquad = \mathfrak{J}(F_j^n)([\tau_1^{x_i}/\tau_0]^{\mathfrak{J}\varphi}, \ldots, [\tau_n^{x_i}/\tau_0]^{\mathfrak{J}\varphi})$
$\qquad\qquad\qquad = \mathfrak{J}(F_j^n)(\tau_1^{\mathfrak{J}\psi}, \ldots, \tau_n^{\mathfrak{J}\psi})$
$\qquad\qquad\qquad = [F_j^n(\tau_1, \ldots, \tau_n)]^{\mathfrak{J}\psi}$.

To prove the Theorem above it is sufficient to show that (SR) and (RR) lead from equations valid under $\mathfrak{J}$ to equations valid under $\mathfrak{J}$ again.

About (SR). If $\tau = \tilde{\tau}$ is valid under an interpretation $\mathfrak{J}$, then $\tau^{x_i}/v = \tilde{\tau}^{x_i}/v$ is also valid under $\mathfrak{J}$. Let φ be given. If we define ψ by $\psi(x_k) = \varphi(x_k)$ for $k \neq i$ and $\psi(x_i) = v^{\mathfrak{J}\varphi}$, we have by the Lemma above: $[\tau^{x_i}/v]^{\mathfrak{J}\varphi} = \tau^{\mathfrak{J}\psi}$ and $[\tilde{\tau}^{x_i}/v]^{\mathfrak{J}\varphi} = \tilde{\tau}^{\mathfrak{J}\psi}$. Thus $[\tau^{x_i}/v]^{\mathfrak{J}\varphi} = [\tilde{\tau}^{x_i}/v]^{\mathfrak{J}\varphi}$, because $\tau = \tilde{\tau}$ is valid under $\mathfrak{J}$.

About (RR). Let $\tau'' \equiv \tau_0^{x_i}/\tau'$, $\tilde{\tau}'' \equiv \tilde{\tau}_0^{x_i}/\tau'$, $\tau \equiv \tau_0^{x_i}/\tilde{\tau}'$, $\tilde{\tau} \equiv \tilde{\tau}_0^{x_i}/\tilde{\tau}'$. Now, if $\tau' = \tilde{\tau}'$ and $\tau'' = \tilde{\tau}''$ are valid under an interpretation $\mathfrak{J}$, then $\tau = \tilde{\tau}$ is also valid under $\mathfrak{J}$. Let φ be given. We introduce ψ by $\psi(x_k) = \varphi(x_k)$ for $k \neq i$ and $\psi(x_i) = \tau'^{\mathfrak{J}\varphi}$. Then $\psi(x_i) = \tilde{\tau}'^{\mathfrak{J}\varphi}$ since $\tau' = \tilde{\tau}'$ is valid under $\mathfrak{J}$. Then according to the Lemma we have that $\tau_0^{\mathfrak{J}\psi} = \tau''^{\mathfrak{J}\varphi}$, $\tilde{\tau}_0^{\mathfrak{J}\psi} = \tilde{\tau}''^{\mathfrak{J}\varphi}$, $\tau_0^{\mathfrak{J}\psi} = \tau^{\mathfrak{J}\varphi}$, $\tilde{\tau}_0^{\mathfrak{J}\psi} = \tilde{\tau}^{\mathfrak{J}\varphi}$. Now we get $\tau^{\mathfrak{J}\varphi} = \tau_0^{\mathfrak{J}\psi} = \tau''^{\mathfrak{J}\varphi} = \tilde{\tau}''^{\mathfrak{J}\varphi}$ (because $\tau'' = \tilde{\tau}''$ is valid under $\mathfrak{J}$) $= \tilde{\tau}_0^{\mathfrak{J}\psi} = \tilde{\tau}^{\mathfrak{J}\varphi}$.

3. Now we show as the *main result* of this paragraph the

Theorem. For every n-ary μ-recursive function f there exists a finite system $\mathfrak{S}$ of equations and a function variable F_j^n occurring in $\mathfrak{S}$ with the property that if $F_{j_1}^{n_1}, \ldots, F_{j_q}^{n_q}$ are the function variables occurring in $\mathfrak{S}$, *then there exist functions $f_{j_1}^{n_1}, \ldots, f_{j_q}^{n_q}$ with corresponding number of places (where especially $f = f_j^n$) such that the following statements are valid:*

(a) *$\mathfrak{S}$ is valid under every interpretation $\mathfrak{J}$ for which $\mathfrak{J}(F_{j_i}^{n_i}) = f_{j_i}^{n_i}$ $(i = 1, \ldots, q)$.*

(b) *For every $F_{j_i}^{n_i}$ all equations of the form $F_{j_i}^{n_i}(v_1, \ldots, v_{n_i}) = v$ which are valid for such an $\mathfrak{J}$ are derivable from $\mathfrak{S}$ using the rules* (SR) *and* (RR) *only $(i = 1, \ldots, q)$.*

We shall call a system of equations with this property a *regular system of equations associated with f* with regard to F_j^n.

From (a) follows, in view of the correctness of the *rules* (SR) and (RR), that every equation of the form $F_{j_i}^{n_i}(v_1, \ldots, v_{n_i}) = v$ which is derivable from $\mathfrak{S}$ is valid for $f_{j_i}^{n_i}$. This gives, in view of (b), the

Corollary 1. Every μ-recursive function f is standard definable by a system of equations.

This Corollary provides us directly with the

Corollary 2. Every μ-recursive function is recursive.

We carry out the proof of the Theorem stated above inductively. First we consider the initial functions (in Section 4). Then, in the following sections, we consider the processes which lead from these initial functions to μ-recursive functions. We shall make use in the proof (cf. § 19.6) of the following almost obvious remark. Let $\mathfrak{S}$ be a regular system of equations associated with f with regard to F_j^n. Let $\mathfrak{S}^*$ be obtained from $\mathfrak{S}$ by renaming the function variables occurring in $\mathfrak{S}$ so that the number of argument places is preserved and different function variables will again be different after the renaming. Let F_j^n be renamed as F_{j*}^n, say. Then $\mathfrak{S}^*$ is a regular system of equations associated with f with regard to F_{j*}^n.

4. We give an account of regular systems of equations which are associated with the *initial functions.* Each one of these systems consists of one equation only. These equations are for the *successor function S*

(*) $$F_0^1(x_0) = S(x_0),$$

for the *identity functions U_n^i* $(n = 1, 2, 3, \ldots; 1 \leqq i \leqq n)$

(**) $$F_1^n(x_1, \ldots, x_n) = x_i,$$

and for the *constant C_0^0*

(***) $$F_0^0 = 0.$$

The proof is carried out in all cases according to the same schema. We shall only give it for the successor function.

About (a). If we define $\mathfrak{F}(F_0^1)$ to be equal to the successor function, then for arbitrary φ

$$F_0^1(x_0)^{\mathfrak{J}\varphi} = \mathfrak{F}(F_0^1)\,(x_0^{\mathfrak{J}\varphi}) = (x_0^{\mathfrak{J}\varphi})' = S(x_0)^{\mathfrak{J}\varphi}.$$

(We represent here and in future the successor of a natural number by the symbol $'$.)

About (b). Let ν_1 represent the number k_1 and ν the number k. Further, we assume that $F_0^1(\nu_1) = \nu$ is valid if $\mathfrak{F}(F_0^1) = S$. From this follows that $k_1' = k$ and so that $\nu \equiv S(\nu_1)$. Thus we must prove that $F_0^1(\nu_1) = S(\nu_1)$ is derivable from (∗). This follows straightaway from (SR) by the substitution of ν_1 for x_0 in (∗).

5. Now we shall discuss the *substitution process*. We presume that the functions $g_1, \ldots, g_r, g_{r+1}$ are defined by the regular systems $\mathfrak{S}_1, \ldots, \mathfrak{S}_r, \mathfrak{S}_{r+1}$ of equations with regard to the function variables $F_1^n, \ldots, F_r^n, F_{r+1}^r$ respectively. We can choose these function variables without loss of generality since the renaming of function variables makes no essential difference (cf. Section 3). We can also assume that no function variable occurs in more than one $\mathfrak{S}_i$. Finally, we stipulate that F_0^n does not occur in any of these systems. Now let the function f be defined by

$$(\dagger) \qquad f(k_1, \ldots, k_n) = g_{r+1}(g_1(k_1, \ldots, k_n), \ldots, g_r(k_1, \ldots, k_n))$$

for arbitrary numbers $k_1, \ldots, k_n$. Then the following system of equations is a regular system associated with f with regard to F_0^n.

$$\left(\begin{smallmatrix}**\\**\end{smallmatrix}\right) \begin{cases} \mathfrak{S}_1 \\ \vdots \\ \mathfrak{S}_r \\ \mathfrak{S}_{r+1} \\ F_0^n(x_1, \ldots, x_n) = F_{r+1}^r(F_1^n(x_1, \ldots, x_n), \ldots, F_r^n(x_1, \ldots, x_n)). \end{cases}$$

We give interpretations to the function variables occurring in $\mathfrak{S}_1, \ldots, \mathfrak{S}_{r+1}$ in the way defined by the induction hypothesis. Thus especially, the variables $F_1^n, \ldots, F_r^n, F_{r+1}^r$ are given the interpretations $g_1, \ldots, g_r, g_{r+1}$. F_0^n is given the interpretation f. Now we see the following.

About (a). The sets $\mathfrak{S}_1, \ldots, \mathfrak{S}_{r+1}$ of equations are valid according to the induction hypothesis. The last equation of $\left(\begin{smallmatrix}**\\**\end{smallmatrix}\right)$ is valid because of (†).

About (b). Because of the induction hypothesis the assertion need only be shown for F_0^n. Thus, we need to show that every valid equation of the form $F_0^n(\nu_1, \ldots, \nu_n) = \nu$ is derivable from $\left(\begin{smallmatrix}**\\**\end{smallmatrix}\right)$. If $\nu_1, \ldots, \nu_n, \nu$

denote the numbers $k_1, \ldots, k_n, k$ respectively, then $f(k_1, \ldots, k_n) = k$. Thus there exist numbers $k_1^*, \ldots, k_r^*$ such that

$$g_1(k_1, \ldots, k_n) = k_1^*, \ldots, g_r(k_1, \ldots, k_n) = k_r^*, g_{r+1}(k_1^*, \ldots, k_r^*) = k.$$

Let the numbers $k_1^*, \ldots, k_r^*$ be represented by the numerals $v_1^*, \ldots, v_r^*$. Because of the presumed properties of $\mathfrak{S}_1$ the equation $F_1^n(v_1, \ldots, v_n) = v_1^*$ is derivable from $\mathfrak{S}_1$ and also from $(^{**}_{**})$. Similarly the equations $F_2^n(v_1, \ldots, v_n) = v_2^*, \ldots, F_r^n(v_1, \ldots, v_n) = v_r^*$ and $F_{r+1}^r(v_1^*, \ldots, v_r^*) = v$ are also derivable from $(^{**}_{**})$. From the last equation of $(^{**}_{**})$ we obtain by n times repeated application of the rule (SR) the equation

$$F_0^n(v_1, \ldots, v_n) = F_{r+1}^r(F_1^n(v_1, \ldots, v_n), \ldots, F_r^n(v_1, \ldots, v_n)).$$

From this we obtain (by the help of the equations proved to be derivable just now) by r times repeated application of the rule (RR)

$$F_0^n(v_1, \ldots, v_n) = F_{r+1}^r(v_1^*, \ldots, v_r^*).$$

Finally, using (RR) again, we get

$$F_0^n(v_1, \ldots, v_n) = v.$$

6. Now we turn our attention to the *induction*. Let f be defined by the stipulation that for all $k_1, \ldots, k_{n+1}$

$$\begin{cases} f(k_1, \ldots, k_n, 0) = g_1(k_1, \ldots, k_n) \\ f(k_1, \ldots, k_n, k'_{n+1}) = g_2(k_1, \ldots, k_n, k_{n+1}, f(k_1, \ldots, k_n, k_{n+1})), \end{cases}$$

where we assume that g_1 and g_2 are standard defined by the regular systems $\mathfrak{S}_1$ and $\mathfrak{S}_2$ of equations with regard to F_1^n and F_2^{n+2} respectively. Further, we shall assume that $\mathfrak{S}_1$ and $\mathfrak{S}_2$ have no function variables in common and that F_0^{n+1} does not occur in either of them.

Then f is defined by the regular system

$$(^{***}_{**}) \quad \begin{cases} \mathfrak{S}_1 \\ \mathfrak{S}_2 \\ F_0^{n+1}(x_1, \ldots, x_n, 0) = F_1^n(x_1, \ldots, x_n) \\ F_0^{n+1}(x_1, \ldots, x_n, S(x_{n+1})) \\ \qquad = F_2^{n+2}(x_1, \ldots, x_n, x_{n+1}, F_0^{n+1}(x_1, \ldots, x_n, x_{n+1})) \end{cases}$$

of equations with regard to F_0^{n+1}.

Here we give interpretations to the function variables occurring in $\mathfrak{S}_1$ and $\mathfrak{S}_2$ in the same way as they are given according to the induction hypothesis. Thus especially, F_1^n is given the interpretation g_1 and F_2^{n+2}

is given the interpretation g_2. Further, we give the interpretation f to F_0^{n+1}. Then the equations $(^{***}_{**})$ are obviously valid. The proof of the assertion that for all $k_1, \ldots, k_{n+1}, k$ with $f(k_1, \ldots, k_{n+1}) = k$ the corresponding equation for F_0^{n+1} is derivable from $(^{***}_{**})$ can easily be carried out by induction over k_{n+1}.

7. It follows from Sections 4, 5 and 6 that every *primitive recursive function* is standard definable by a system of equations. This justifies our making use of the functions sg, $\dot{-}$, $+$, $\cdot$ defined in § 10.4.

8. Application of the μ-operator to regular functions. We presume that for every n-tuple $k_1, \ldots, k_n$ there exists at least one number k_{n+1} for which $g(k_1, \ldots, k_n, k_{n+1}) = 0$. Let $f(k_1, \ldots, k_n)$ be the smallest k_{n+1} with this property. Let g be defined by an associated system of equations. We shall show that this is also possible for f.

First we introduce a function h^* by induction.

$$\begin{cases} h^*(k_1, \ldots, k_n, 0) = sg(g(k_1, \ldots, k_n, 0)) \\ h^*(k_1, \ldots, k_n, k'_{n+1}) = sg(g(k_1, \ldots, k_n, k'_{n+1}) \cdot h^*(k_1, \ldots, k_{n+1})). \end{cases}$$

The reader can easily convince himself that if $k_1, \ldots, k_n$ are fixed, the function h^* has the value 0 for all k_{n+1} which are greater than or equal to the smallest k_{n+1} for which $g(k_1, \ldots, k_n, k_{n+1}) = 0$, and the value 1 for all k_{n+1} which are smaller than this k_{n+1}. We have especially that

$$f(k_1, \ldots, k_n) = \mu k_{n+1}\, h^*(k_1, \ldots, k_{n+1}) = 0.$$

Further, we define a function h inductively by

$$\begin{cases} h(k_1, \ldots, k_n, 0) = h^*(k_1, \ldots, k_n, 0) \\ h(k_1, \ldots, k_n, k'_{n+1}) = (1 \dot{-} h^*(k_1, \ldots, k_n, k_{n+1})) + h^*(k_1, \ldots, k_n, k'_{n+1}). \end{cases}$$

If $h^*(k_1, \ldots, k'_{n+1}) = 1$, then also $h^*(k_1, \ldots, k_{n+1}) = 1$, and so $h(k_1, \ldots, k'_{n+1}) = 1$. If $h^*(k_1, \ldots, k'_{n+1}) = 0$ and $h^*(k_1, \ldots, k_{n+1}) = 1$, then $h(k_1, \ldots, k'_{n+1}) = 0$. Finally, if $h^*(k_1, \ldots, k'_{n+1}) = h^*(k_1, \ldots, k_{n+1}) = 0$, then $h(k_1, \ldots, k'_{n+1}) = 1$. From this follows that if $k_1, \ldots, k_n$ are fixed and k_{n+1} is increasing, then the function h has the value 0 once and only once and, indeed, it has this value at the first place where h^* has the value 0. Therefore we have that

(††) $\quad h(k_1, \ldots, k_{n+1}) = 0$ if and only if $f(k_1, \ldots, k_n) = k_{n+1}$.

If h has a value different from 0, then this value is 1.

h has been traced back by inductions to g by the help of primitive recursive functions. Therefore h is definable by a regular system of equations. Let $\mathfrak{S}$ be a regular system associated with h with regard to

F_0^{n+1}. Now we consider the function variables F_0^n and F_0^{n+2} (about which we may assume that they do not occur in $\mathfrak{S}$) and the system of equations

$$(\substack{***\\***})\quad \begin{cases} \mathfrak{S} \\ F_0^{n+2}(x_1,\ldots,x_n,x_{n+1},0) = x_{n+1} \\ F_0^{n+2}(x_1,\ldots,x_n,x_{n+1},S(x_{n+2})) = F_0^n(x_1,\ldots,x_n) \\ F_0^{n+2}(x_1,\ldots,x_n,x_{n+1},S(x_{n+2})) = F_0^{n+2}(x_1,\ldots,x_n,S(x_{n+1}), \\ \qquad F_0^{n+1}(x_1,\ldots,x_n,S(x_{n+1}))) \\ F_0^n(x_1,\ldots,x_n) = F_0^{n+2}(x_1,\ldots,x_n,0, \\ \qquad F_0^{n+1}(x_1,\ldots,x_n,0)). \end{cases}$$

We shall show that $(\substack{***\\***})$ is a regular system associated with f with regard to F_0^n. We give interpretations to the function variables occurring in $\mathfrak{S}$ in the way these are given according to the induction hypothesis. Thus F_0^{n+1} has the interpretation h. Further we give the interpretation f to F_0^n and finally the interpretation Φ to F_0^{n+2}, where the function Φ is given by

$$(\dagger\dagger\dagger)\qquad \Phi(k_1,\ldots,k_{n+2}) = \begin{cases} k_{n+1}, \text{ if } k_{n+2} = 0 \\ f(k_1,\ldots,k_n), \text{ if } k_{n+2} \neq 0. \end{cases}$$

About (a). Under this interpretation the equations of $\mathfrak{S}$ are obviously valid, and so are the two following equations.

In order to show that the penultimate equation is valid under $\mathfrak{J}$ we must show that for all $k_1,\ldots,k_n,k_{n+1},k_{n+2}$

$$\Phi(k_1,\ldots,k_n,k_{n+1},k'_{n+2}) = \Phi(k_1,\ldots,k_n,k'_{n+1},h(k_1,\ldots,k_n,k'_{n+1})).$$

The definition $(\dagger\dagger\dagger)$ of Φ shows that the left hand side of this equation has the value $f(k_1,\ldots,k_n)$. The same is true for the right hand side, provided

$$h(k_1,\ldots,k_n,k'_{n+1}) \neq 0.$$

However, if $h(k_1,\ldots,k_n,k'_{n+1}) = 0$, then according to $(\dagger\dagger\dagger)$ the right hand side is equal to k'_{n+1}. But then we have according to $(\dagger\dagger)$ that

$$k'_{n+1} = f(k_1,\ldots,k_n).$$

In order to see that the last equation is valid under $\mathfrak{J}$ we must show that, for every $k_1,\ldots,k_n$, $f(k_1,\ldots,k_n) = \Phi(k_1,\ldots,k_n,0,h(k_1,\ldots,k_n,0))$. This follows at once from $(\dagger\dagger\dagger)$, provided $h(k_1,\ldots,k_n,0) \neq 0$. On the other hand, if $h(k_1,\ldots,k_n,0) = 0$, then the right hand side has the value 0 because of $(\dagger\dagger\dagger)$. But then $f(k_1,\ldots,k_n) = 0$ because of $(\dagger\dagger)$.

About (b). Now we still have to show that all equations of the forms

$$F_0^n(\nu_1,\ldots,\nu_n) = \nu^* \quad\text{and}\quad F_0^{n+2}(\nu_1,\ldots,\nu_{n+2}) = \nu^{**}$$

which are valid under $\mathfrak{J}$ are derivable from $\binom{***}{***}$. For this we only have to show that for every $\nu_1, \ldots, \nu_{n+2}$ there exists at least one ν^* and ν^{**} such that the equations above are derivable.

That this simple assertion is sufficient follows from the following considerations. We have just seen that all equations of $\binom{***}{***}$ are valid under $\mathfrak{J}$. Then, according to Section 2, all equations which are derivable from $\binom{***}{***}$ are also valid under $\mathfrak{J}$. However, if $\nu_1, \ldots, \nu_{n+2}$ are given, then there exists only one equation of each of the above given forms which is valid under $\mathfrak{J}$. Thus, if we succeed in showing that for every $\nu_1, \ldots, \nu_{n+2}$ there exists at least one ν^* and ν^{**} such that the equations above are derivable, then we shall have shown that *all* equations of these forms which are valid under $\mathfrak{J}$ are derivable.

We need only show our reduced assertion for F_0^n, since the assertion follows for F_0^{n+2} from the third and fourth equations from the bottom of $\binom{***}{***}$ by suitable substitutions.

Let $\nu_1, \ldots, \nu_n$ stand for $k_1, \ldots, k_n$. For $k_1, \ldots, k_n$ there exists one and only one k for which $h(k_1, \ldots, k_n, k) = 0$. Let this k be represented by the numeral ν. Then, according to the hypothesis about $\mathfrak{S}$ the equation

(1) $$F_0^{n+1}(\nu_1, \ldots, \nu_n, \nu) = 0$$

is derivable from $\mathfrak{S}$ and thus also from $\binom{***}{***}$. From the last equation of $\binom{***}{***}$ we obtain by substitution that

(2) $$F_0^n(\nu_1, \ldots, \nu_n) = F_0^{n+2}(\nu_1, \ldots, \nu_n, 0, F_0^{n+1}(\nu_1, \ldots, \nu_n, 0)).$$

At this stage we distinguish between the two cases when $\nu \equiv 0$ and when $\nu \not\equiv 0$.

If $\nu \equiv 0$, then we can proceed by the help of (RR) from the equations (1) and (2) to

$$F_0^n(\nu_1, \ldots, \nu_n) = F_0^{n+2}(\nu_1, \ldots, \nu_n, 0, 0).$$

On the other hand we obtain by substitutions into the fourth equation from the bottom of $\binom{***}{***}$

$$F_0^{n+2}(\nu_1, \ldots, \nu_n, 0, 0) = 0.$$

These last two equations give, using (RR),

$$F_0^n(\nu_1, \ldots, \nu_n) = 0, \quad \text{i.e.} \quad F_0^n(\nu_1, \ldots, \nu_n) = \nu, \quad \text{q. e. d.}$$

We proceed now to the case $\nu \not\equiv 0$ (in this case we have that $h(k_1, \ldots, k_n, 0) = 1$. We must keep this in mind for an application later on). Let $\tilde{\nu}$ be an arbitrary numeral. We show that

if $S(\tilde{\nu}) \not\equiv \nu$, *then*

$$(3) \qquad F_0^{n+2}(\nu_1, \ldots, \nu_n, \tilde{\nu}, S(0)) = F_0^{n+2}(\nu, \ldots, \nu_n, S(\tilde{\nu}), S(0))$$

is derivable from $\left(\begin{smallmatrix}***\\***\end{smallmatrix}\right)$; *if, on the other hand,* $S(\tilde{\nu}) \equiv \nu$, *then*

$$(4) \qquad F_0^{n+2}(\nu_1, \ldots, \nu_n, \tilde{\nu}, S(0)) = \nu$$

is derivable from $\left(\begin{smallmatrix}***\\***\end{smallmatrix}\right)$.

Proof. First we obtain by substitutions from the penultimate equation of $\left(\begin{smallmatrix}***\\***\end{smallmatrix}\right)$

$$(5) \quad F_0^{n+2}(\nu_1, \ldots, \nu_n, \tilde{\nu}, S(0)) = F_0^{n+2}(\nu_1, \ldots, \nu_n, S(\tilde{\nu}), F_0^{n+1}(\nu_1, \ldots, \nu_n, S(\tilde{\nu}))).$$

Let $\tilde{\nu}$ denote the number $\tilde{k}$.

If $S(\tilde{\nu}) \not\equiv \nu$, then $\tilde{k}' \neq k$ and so $h(k_1, \ldots, k_n, \tilde{k}') = 1$.

This gives, according to the hypothesis, that

$$(6) \qquad F_0^{n+1}(\nu_1, \ldots, \nu_n, S(\tilde{\nu})) = S(0)$$

is derivable from $\mathfrak{S}$ and so also from $\left(\begin{smallmatrix}***\\***\end{smallmatrix}\right)$. Now we obtain (3) from (6) and (5) by application of (RR).

If, on the other hand, $S(\tilde{\nu}) \equiv \nu$, then $\tilde{k}' = k$ and so $h(k_1, \ldots, k_n, \tilde{k}') = 0$. This gives that

$$(7) \qquad F_0^{n+1}(\nu_1, \ldots, \nu_n, S(\tilde{\nu})) = 0$$

is derivable from $\mathfrak{S}$ and so from $\left(\begin{smallmatrix}***\\***\end{smallmatrix}\right)$. Applying (RR) to (7) and (5) we obtain

$$F_0^{n+2}(\nu_1, \ldots, \nu_n, \tilde{\nu}, S(0)) = F_0^{n+2}(\nu_1, \ldots, \nu_n, S(\tilde{\nu}), 0).$$

On the other hand we get by substitutions into the fourth equation from the bottom in $\left(\begin{smallmatrix}***\\***\end{smallmatrix}\right)$

$$F_0^{n+2}(\nu_1, \ldots, \nu_n, S(\tilde{\nu}), 0) = S(\tilde{\nu}).$$

From the last two equations we obtain applying (RR)

$$F_0^{n+2}(\nu_1, \ldots, \nu_n, \tilde{\nu}, S(0)) = S(\tilde{\nu}),$$

i.e. (4), since $S(\tilde{\nu}) \equiv \nu$.

Now we consider the beginning $0, S(0), \ldots, \tilde{\tilde{\nu}}, \tilde{\nu}, \nu$ of the sequence of the numerals. We assert that the equations

$$(8) \quad \begin{cases} F_0^{n}(\nu_1, \ldots, \nu_n) = F_0^{n+2}(\nu_1, \ldots, \nu_n, 0, S(0)) \\ F_0^{n+2}(\nu_1, \ldots, \nu_n, 0, S(0)) = F_0^{n+2}(\nu_1, \ldots, \nu_n, S(0), S(0)) \\ F_0^{n+2}(\nu_1, \ldots, \nu_n, S(0), S(0)) = F_0^{n+2}(\nu_1, \ldots, \nu_n, S(S(0)), S(0)) \\ \cdots\cdots\cdots\cdots\cdots\cdots\cdots\cdots\cdots\cdots \\ F_0^{n+2}(\nu_1, \ldots, \nu_n, \tilde{\tilde{\nu}}, S(0)) = F_0^{n+2}(\nu_1, \ldots, \nu_n, \tilde{\nu}, S(0)) \\ F_0^{n+2}(\nu_1, \ldots, \nu_n, \tilde{\nu}, S(0)) = \nu \end{cases}$$

are derivable from $\binom{***}{***}$. This follows from (4) for the last equation, and from (3) for all the others apart from the first. For this first equation we obtain a derivation as follows. From the last equation of $\binom{***}{***}$ we obtain by substitutions

$$F_0^n(\nu_1, \ldots, \nu_n) = F_0^{n+2}(\nu_1, \ldots, \nu_n, 0, F_0^{n+1}(\nu_1, \ldots, \nu_n, 0)).$$

Since, according to our assumption, $h(k_1, \ldots, k_n, 0) = 1$, we can derive the equation

$$F_0^{n+1}(\nu_1, \ldots, \nu_n, 0) = S(0)$$

from $\mathfrak{S}$ and so from $\binom{***}{***}$. By an application of (RR) we obtain the first equation of (8) from the last two equations.

From the equations (8) we get, by applying (RR) repeatedly, starting from the last two equations,

$$F_0^{n+2}(\nu_1, \ldots, \nu_n, \tilde{\tilde{\nu}}, S(0)) = \nu,$$

etc., until finally

$$F_0^n(\nu_1, \ldots, \nu_n) = \nu,$$

q.e.d.[1]

9. Remarks. The proof given here is *constructive* in the sense that for every μ-recursive function which is traced back to the initial functions step by step by substitutions, inductive definitions and applications of the μ-operator we can give effectively a system of equations and a function variable with regard to which this function is standard defined.

We have carried out the proof of the recursiveness of the μ-recursive functions by the help of semantic considerations. This is an obvious procedure in view of the correctness of the rules. But, on the other hand, the concept of derivability is a purely formal, "syntactic" concept. Therefore we could omit the concept of interpretation.

---

[1] The procedure used in this section originates from KLEENE, although he does not show the existence of a *regular associated* system of equations. Therefore he does not use the equation which is the last but two in $\binom{***}{***}$. We use this equation (or a modified form of it) to show that the F-terms $F_0^{n+2}(\nu_1, \ldots, \nu_{n+2})$ can be evaluated. If we left out the last equation but two this would not be possible in the case when $F_0^{n+1}(\nu_1, \ldots, \nu_n, 0) = 0$ is derivable from the other equations.

KLEENE has also given an account of another procedure for dealing with the μ-operator. Let the system $\mathfrak{S}$ of equations define the function h^* (cf. text above) with regard to F_0^{n+1}. We take new function variables F_1^3 and F_0^n. Then the system

$$\binom{***}{***}' \begin{cases} \mathfrak{S} \\ F_1^3(S(x_1), 0, x_0) = x_0 \\ F_0^n(x_1, \ldots, x_n) = F_1^3(F_0^{n+1}(x_1, \ldots, x_n, x_0), F_0^{n+1}(x_1, \ldots, x_n, S(x_0)), S(x_0)) \end{cases}$$

of equations defines the function f with regard to F_0^n. Cf. Exercise.

It is possible to show *by purely formal considerations* that every μ-recursive function is definable by a system of equations. We can use for this purpose the systems given in this paragraph. (We can leave out the last equation but two of the system $\binom{***}{***}$ or use the system $\binom{***}{***}'$ given in the last note.) For the method of proof see KLEENE: Introduction to Metamathematics.

Exercise. Show for the system $\binom{***}{***}'$ given in the last note that

(a) we cannot find for every ν_1, ν_2, ν_3 a ν such that $F_1^3(\nu_1, \nu_2, \nu_3) = \nu$ is derivable;

(b) there exists in general no interpretation for the function variables occurring for which all equations are valid.

(Hint. In (a) show that an equation of the form $F_1^3(\tau_1, \tau_2, \tau_3) = \tau_4$ can only be derivable if τ_1 begins with S.)

Reference

KLEENE, S. C.: General Recursive Functions of Natural Numbers. Math. Ann. **112**, 727–742 (1936).

§ 21. The μ-Recursiveness of Recursive Functions

In this paragraph we shall show the

Theorem. Every recursive function is μ-recursive.

The proof is constructive in the sense that for any given system of equations which defines a function f we can trace back effectively the function f to the initial functions by the help of substitutions, inductive definitions and applications of the μ-operator to regular functions. Especially, we shall see that we need only apply the μ-operator once. The results of this and of the last paragraph provide us with the equipment to give a new proof of KLEENE's normal form theorem (cf. § 18.4). At the end of this paragraph (Section 7) we give an example for a system of equations which determines a function unambiguously although it is impossible to give an algorithm for the determination of the values of this function.

1. Gödel numbering of the terms. We shall characterize the terms τ unambiguously by numbers $\bar{\tau}$. We define $\bar{\tau}$ by induction and begin with the simple terms:[1]

$$\bar{0} = 1, \ \overline{F_j^0} = 4j + 3 \quad (j = 0, 1, 2, \ldots), \quad \bar{x}_j = 4j + 5 \quad (j = 0, 1, 2, \ldots).$$

[1] For typographical reasons we shall write $\bar{x}_j$ instead of $\overline{x_j}$ and $\bar{\tau}_1$ instead of $\overline{\tau_1}$, etc.

Thus, the Gödel numbers of the simple terms run through the set of odd numbers. The Gödel numbers of the compound terms must therefore be even. We put

$$\begin{aligned} \overline{S(\tau)} &= p_0 p_1^{\bar\tau} \\ \overline{F_j^r(\tau_1, \ldots, \tau_r)} &= p_0^{j+2} p_1^{\bar\tau_1} \ldots p_r^{\bar\tau_r} \qquad (r \geqq 1). \end{aligned}$$

It is clear that $\bar\tau_1 \neq \bar\tau_2$, if $\tau_1 \not\equiv \tau_2$.

We now introduce a few functions and predicates which are connected with this Gödel numbering. *All these functions and predicates are primitive recursive.*

Vx means that x is the Gödel number of a variable.[1] We obviously have

$$Vx \leftrightarrow x \geqq 5 \wedge 4/(x \dot- 5).$$

$d(k)$ is the Gödel number of the numeral which represents the number k. d is defined inductively by

$$\begin{aligned} d(0) &= 1 \\ d(k') &= p_0 \cdot p_1^{d(k)}. \end{aligned}$$

Nx means that x is the Gödel number of a numeral. We have

$$Nx \leftrightarrow \bigvee_{k=0}^{x} x = d(k).$$

The upper bound for k is chosen in view of the fact that $k < d(k)$, which can be shown by induction.

Fx means that x is the Gödel number of an F-term. We have

$$\begin{aligned} Fx \leftrightarrow (x \geqq 3 \wedge 4/(x \dot- 3)) \vee (&\exp(0, x) \geqq 2 \wedge \\ \wedge \bigvee_{r=0}^{x} \bigwedge_{k=0}^{x} &((k \neq 0 \wedge k \leqq r \rightarrow N \exp(k, x)) \wedge (k > r \rightarrow \exp(k, x) = 0))), \end{aligned}$$

where r is the number of argument places of the function variable occurring in the F-term. It is easily seen that $r \leqq x$.

Now we define a function C which, provided that x is the Gödel number of a numeral, has as value the number represented by this numeral.

$$C(x) = \mu_{k=0}^{x}\, x = d(k).$$

Further, we introduce an n-ary function B_n. $B_n(x_1, \ldots, x_n)$ gives the Gödel number of the term $F_0^n(v_1, \ldots, v_n)$, where $v_1, \ldots, v_n$ are the

[1] The predicate V should not be confused with the predecessor function introduced in § 10.4.

numerals which represent the numbers $x_1, \ldots, x_n$. For $n \geqq 1$ we have that

$$B_n(x_1, \ldots, x_n) = p_0^2 p_1^{d(x_1)} \ldots p_n^{d(x_n)}.$$

For $n = 0$ we define B_0 to be the Gödel number of the term F_0^0, i.e. 3.

Finally, we introduce the predicate T. Tt means that t is the Gödel number of a term. We must remember here that the concept of term was defined inductively in § 19.2. We can paraphrase this definition by the use of a secondary concept ("term of n-th order") in the following obvious way.

τ is a term of 0-th order *if and only if*
$\tau \equiv 0$ *or* τ is a variable *or* τ is a function variable of 0 arguments.

τ is a term of n'-th order *if and only if*
τ is a term of n-th order *or* τ is of the form $S(\tau_1)$, where τ_1 is a term of n-th order *or* τ is of the form $F_j^r(\tau_1, \ldots, \tau_r)$, where $\tau_1, \ldots, \tau_r$ are terms of the n-th order.

τ is a term *if and only if*
there exists an n such that τ is a term of n-th order.

Now we define a predicate $\mathring{T}$. $\mathring{T}tn$ means that t is the Gödel number of a term of n-th order. If we keep in mind that the set of odd numbers coincides with the set of Gödel numbers of terms of the 0-th order, then we have

$$\mathring{T}t0 \leftrightarrow Dt \text{ (i.e. } t \text{ is odd)}$$

$$\mathring{T}tn' \leftrightarrow \mathring{T}tn$$

$$\vee \left(\exp(0, t) = 1 \wedge \bigwedge_{k=0}^{t} (\exp(k, t) \neq 0 \rightarrow k = 0 \vee k = 1) \wedge \mathring{T} \exp(1, t)\, n\right)$$

$$\vee \Big(\exp(0, t) > 1 \wedge \exp(1, t) \neq 0 \wedge \bigwedge_{k=0}^{t} (\exp(k, t) > 0 \rightarrow \exp(k \dot{-} 1, t) > 0)$$

$$\wedge \bigwedge_{k=0}^{t} (k \geqq 1 \wedge \exp(k, t) > 0 \rightarrow \mathring{T} \exp(k, t)\, n)\Big).$$

Let f be the characteristic function of $\mathring{T}$. In view of the facts that the characteristic function of an alternative of predicates can be obtained by multiplication of the corresponding characteristic functions, and the characteristic function of a conjunction of predicates can be obtained by addition of the corresponding characteristic functions followed by sg-mapping (cf. the remark in § 11.3), we see that f is definable by two equations of the form § 12.5 (i) with primitive recursive functions g, h, H. Now the theorem of that section shows that f and with it $\mathring{T}$ are primitive recursive.

By induction on n we can easily show that $\mathring{T}tn \to \bigvee_{n_0} (\mathring{T}tn_0 \wedge t \geqq n_0)$. This proves that

$$Tt \leftrightarrow \bigvee_{n=0}^{t} \mathring{T}tn.$$

This representation shows that T is primitive recursive.

2. *The representation of $f(x_1, \ldots, x_n)$.* We presume that the n-ary function f is defined by a finite system $\mathfrak{S}$ of equations with regard to F_0^n. In Section 6 we shall introduce a predicate A. $At\tilde{t}m$ means that t and $\tilde{t}$ are the Gödel numbers of the terms τ and $\tilde{\tau}$ respectively and the equation $\tau = \tilde{\tau}$ can be derived *boundedly* in m steps from $\mathfrak{S}$, i.e. it can be derived so that if, for any number l, the substitution rule (SR) is applied in the l-th step, then no numeral will be substituted in this substitution which denotes a number greater than l. Furthermore, we also allow the passage from one equation to another which is identical with the first.

If the equation $\tau = \tilde{\tau}$ is derivable from $\mathfrak{S}$, then there exists an m such that $At\tilde{t}m$. This is easily seen. We only need to change the given derivation of $\tau = \tilde{\tau}$ into a bounded derivation. A derivation breaks the requirements for boundedness only if in the l-th step a numeral ν is substituted which denotes a number $k > l$. In this case we reproduce in the l-th, $(l+1)$-th, ..., $(k-1)$-th steps the equation which was obtained in the $(l-1)$-th step, then carry out the substitution in question in the k-th step. By such interpolations we obtain a bounded derivation of $\tau = \tilde{\tau}$. It will be shown in Section 6 that A is *primitive recursive*.

Now we introduce a predicate D_n by the definition

$$D_n s x_1 \ldots, x_n \leftrightarrow A\,\sigma_{31}(s)\,\sigma_{32}(s)\,\sigma_{33}(s) \wedge \sigma_{31}(s) = B_n(x_1, \ldots, x_n) \wedge N\sigma_{32}(s).$$

$D_n s x_1, \ldots, x_n$ means that s is the number of a triple which has the following properties. The first component $\sigma_{31}(s)$ is the Gödel number of the term $F_0^n(\nu_1, \ldots, \nu_n)$, where the numerals $\nu_1, \ldots, \nu_n$ represent the numbers $x_1, \ldots, x_n$. The second component $\sigma_{32}(s)$ is the Gödel number of a numeral ν. Finally, the equation $F_0^n(\nu_1, \ldots, \nu_n) = \nu$ is boundedly derivable from $\mathfrak{S}$ in $\sigma_{33}(s)$ steps. Since we have assumed that $\mathfrak{S}$ defines the function f with regard to F_0^n, one equation of the form $F_0^n(\nu_1, \ldots, \nu_n) = \nu$ must be derivable. Thus, there exists an s such that $D_n s x_1 \ldots x_n$. So in forming $\mu s D_n s x_1 \ldots x_n$ we are dealing with an application of the μ-operator to a *regular function*.

$\sigma_{32}(\mu s D_n s x_1 \ldots x_n)$ is the Gödel number of a numeral ν for which an equation of the form $F_0^n(\nu_1, \ldots, \nu_n) = \nu$ (where ν_i represents the number x_i)

is boundedly derivable from $\mathfrak{S}$ (in $\sigma_{33}(s)$ steps). Therefore ν represents the value $f(x_1, \ldots, x_n)$. Thus, we have

$$f(x_1, \ldots, x_n) = C(\sigma_{32}(\mu s D_n s x_1 \ldots x_n)).$$

This representation shows that f is μ-recursive.

The following Sections 3 to 6 serve to prove the primitive recursiveness of the predicate A. Naturally, A depends on the system $\mathfrak{S}$ of equations which we mentioned at the beginning of this section.

3. The substitution function $s(x, y, z)$. If x and z are Gödel numbers of the terms τ and τ_0 respectively and y is the Gödel number of a variable x_i, then $s(x, y, z)$ is the Gödel number of the term τ^{x_i}/τ_0. (We shall define s for all other cases as well.) It is immediately clear in view of the definition of substitution (cf. § 19.3) that the function defined by the following equations has the property we want (for δ and ε cf. § 10.4).

$$\begin{aligned} s(0, y, z) &= 0, \\ s(x, y, z) &= \delta(x, y) \cdot z + \varepsilon(x, y) \cdot x, \qquad \text{if } x \text{ is odd}, \\ s(p_0^{\nu_0} \ldots p_r^{\nu_r}, y, z) &= p_0^{\nu_0} p_1^{s(\nu_1, y, z)} \ldots p_r^{s(\nu_r, y, z)}, \text{ if } \nu_0 > 0. \end{aligned}$$

According to § 12.7 (see the formulae (13), (14), (15) there) we have that *s is a primitive recursive function.*

We derive now a few simple relations for the function s. These relations will be used later on for making estimates. (The reader should look into the problem of the meaning of these formulae in the case of substitution.)

(1) $s(s(x, y, z), y, z) = s(x, y, z)$ for odd z,

(2) $s(s(x, y, z), z, y) = s(x, z, y)$ for odd z,

(3) $s(s(x, u, v), y, z) = s(s(x, u, s(v, y, z)), y, z)$ for odd z,

(4) $s(x, y, z) \geqq x$, if $z \geqq y$,

(5) $s(x, 1, z) \geqq z$, if x is the Gödel number of a numeral,

(6) $s(x, 1, z) \geqq z$, if x is the Gödel number of an F-term which is not of the form F_j^0,

(7) $s(s(x, u, v), 1, z) \geqq x$ for odd z, if $z \geqq u$ and if v is the Gödel number of a numeral or of an F-term which is not of the form F_j^0.

Proof. (1)—(4) can be shown by induction corresponding to the definitions of $s(x, y, z)$ and $s(x, u, v)$. We shall carry out the proof completely for (1) only. For (2), (3) and (4) we only discuss the case when x is odd, since in all other cases the proof can be carried out similarly to that of (1).

About (1). For $x = 0$ we have on both sides 0.

If x is odd, then for $x \neq y$ we have on both sides $s(x, y, z)$. For $x = y$ we have $s(z, y, z)$ on the left hand side. This is equal to z, since z is odd, independently of whether $z = y$ or $z \neq y$. We have the value z also on the right hand side.

Finally, let $x = p_0^{\nu_0} \dots p_r^{\nu_r}$, where $\nu_0 > 0$. Then we have

$$\begin{aligned} s(s(p_0^{\nu_0} \dots p_r^{\nu_r}, y, z), y, z) &= s(p_0^{\nu_0} p_1^{s(\nu_1, y, z)} \dots p_r^{s(\nu_r, y, z)}, y, z) \\ &= p_0^{\nu_0} p_1^{s(s(\nu_1, y, z), y, z)} \dots p_r^{s(s(\nu_r, y, z), y, z)} \\ &= p_0^{\nu_0} p_1^{s(\nu_1, y, z)} \dots p_r^{s(\nu_r, y, z)} \qquad \text{(ind. hyp.)} \\ &= s(p_0^{\nu_0} \dots p_r^{\nu_r}, y, z). \end{aligned}$$

About (2). Let x be odd. For $x \neq y$ we have on both sides $s(x, z, y)$. For $x = y$ we have on the left hand side $s(z, z, y)$, i.e. y, and on the right hand side y.

About (3). Let x be odd. For $x \neq u$ we have on both sides $s(x, y, z)$. For $x = u$ we have on the left hand side $s(v, y, z)$ and on the right hand side $s(s(v, y, z), y, z) = s(v, y, z)$ according to (1).

About (4). Let x be odd. For $x \neq y$, $s(x, y, z) = x \geqq x$. For $x = y$ $s(x, y, z) = z$, and $z \geqq y = x$ according to hypothesis.

About (5). The numbers $d(m)$ run through the Gödel numbers of the numerals. It is therefore sufficient to show that $s(d(m), 1, z) \geqq z$. We show this by finite induction on m. $s(d(0), 1, z) = s(1, 1, z) = z \geqq z$. $s(d(m'), 1, z) = s(p_0 p_1^{d(m)}, 1, z) = p_0 p_1^{s(d(m), 1, z)} \geqq p_0 p_1^z \geqq z$.

About (6). In this case the term has the form $F_j^r(\nu_1, \dots, \nu_r)$ (with $r \geqq 1$), where $\nu_1, \dots, \nu_r$ are numerals. If $x_1, \dots, x_r$ are the Gödel numbers of $\nu_1, \dots, \nu_r$, then $x = p_0^{j+2} p_1^{x_1} \dots p_r^{x_r}$, and we have that $s(x, 1, z) = s(p_0^{j+2} p_1^{x_1} \dots p_r^{x_r}, 1, z) = p_0^{j+2} p_1^{s(x_1, 1, z)} \dots p_r^{s(x_r, 1, z)} \geqq p_0^{j+2} p_1^z \dots p_r^z$ (because of (5)) $\geqq z$.

About (7). According to (3) and (4) we have $s(s(x, u, v), 1, z) = s(s(x, u, s(v, 1, z)), 1, z) \geqq s(x, u, s(v, 1, z))$. According to (5) and (6) $s(v, 1, z) \geqq z$, and so by the hypothesis it is also $\geqq u$. This gives by (4) $s(x, u, s(v, 1, z)) \geqq x$.

4. The bound-variable x_k. In order to show that the predicate A, which we shall define later on, is primitive recursive, we need an estimate for the Gödel numbers of the variables which play a role in a derivation, and also for the Gödel numbers of the 0-ary function variables F_j^0 which occur in the derivation. We presume that the function f is defined by the system $\mathfrak{S}$ of equations with regard to F_0^n. *We now introduce k as the smallest number for which we have*

(1) k is larger than the index of any variable which occurs in $\mathfrak{S}$.

(2) $\bar{x}_k \geqq \overline{F_j^0}$ for all F_j^0 which occur in $\mathfrak{S}$.

In Sections (4), (5) and (6) k will always be used in this sense.

No equation which is derivable from $\mathfrak{S}$ contains a variable x_i with $i > k$. A substitution in a variable which does not appear in an equation leads to the same equation again, and so it is superfluous. *We therefore need only allow substitutions in which we substitute for a variable x_i with $i \leqq k$.*

Let us now discuss the *replacement*. Let us suppose that the equation $\tau = \tilde{\tau}$ is obtained from $\tau'' = \tilde{\tau}''$ by replacement of the F-term τ' by the numeral $\tilde{\tau}'$. Then, according to the definition of replacement there exist terms τ_0 and $\tilde{\tau}_0$ and a variable x_i such that

$$\tau'' \equiv {\tau_0}^{x_i}/\tau'; \qquad \tilde{\tau}'' \equiv {\tilde{\tau}_0}^{x_i}/\tau'; \qquad \tau \equiv {\tau_0}^{x_i}/\tilde{\tau}'; \qquad \tilde{\tau} \equiv {\tilde{\tau}_0}^{x_i}/\tilde{\tau}'.$$

Now we assume that the equations $\tau' = \tilde{\tau}'$ and $\tau'' = \tilde{\tau}''$ are derivable from $\mathfrak{S}$. We assert that there exist terms τ_1 and $\tilde{\tau}_1$ such that

$$\tau'' \equiv {\tau_1}^{x_k}/\tau'; \ \tilde{\tau}'' = {\tilde{\tau}_1}^{x_k}/\tau'; \qquad \tau \equiv {\tau_1}^{x_k}/\tilde{\tau}'; \qquad \tilde{\tau} \equiv {\tilde{\tau}_1}^{x_k}/\tilde{\tau}'.$$

We can assume that $x_i \not\equiv x_k$, since otherwise we have nothing to prove. *Further, we can assume that x_k occurs neither in τ_0 nor in $\tilde{\tau}_0$.* For, if x_k occurred in τ_0, say, then it would also occur in ${\tau_0}^{x_i}/\tau'$, and so in τ''. However, this cannot be the case, since the equation $\tau'' = \tilde{\tau}''$ is derivable from $\mathfrak{S}$ and x_k does not occur in any equation which is derivable from $\mathfrak{S}$.

Now we put

$$\tau_1 \equiv {\tau_0}^{x_i}/x_k, \qquad \tilde{\tau}_1 \equiv {\tilde{\tau}_0}^{x_i}/x_k.$$

Then ${\tau_1}^{x_k}/\tau' \equiv ({\tau_0}^{x_i}/x_k)^{x_k}/\tau' \equiv {\tau_0}^{x_i}/\tau'$ (since x_k does not occur in τ_0) $\equiv \tau''$. The other assertions can be shown in the same way.

From the result derived just now follows that *in connection with the present discussion we need only to consider such replacements for which the variable x_i occurring in the definition of replacement coincides with x_k.*

We put $K = \bar{x}_k$. According to the last remark *we can take K to be the upper bound of the Gödel numbers of all variables which play a role in a derivation from $\mathfrak{S}$.* Further we have $\bar{K} \geqq \overline{F_j^0}$ *for all F_j^0 which occur either in $\mathfrak{S}$ or in an equation derivable from $\mathfrak{S}$.*

5.[1] Now, we define *the predicates S_R and R_R.*

---

[1] The numbers t', t'', $\tilde{t}'$, etc., appearing here and in the next section correspond to the terms τ', τ'', $\tilde{\tau}'$, etc. *Therefore the ′ does not signify the "successor" here.*

$S_R t\tilde{t}t'\tilde{t}'m$ shall mean that $t, \tilde{t}, t', \tilde{t}'$ are the Gödel numbers of terms $\tau, \tilde{\tau}, \tau', \tilde{\tau}'$ and that the equation $\tau = \tilde{\tau}$ is obtained from the equation $\tau' = \tilde{\tau}'$ by the *substitution* of a numeral which represents a number $\leqq m$ for a variable x_i, where $i \leqq k$. Thus the predicate S_R, just like the predicate R_R introduced below, depends on $\mathfrak{S}$. As we have already seen $\bar{x}_i \leqq K$. The representation

$$S_R t\tilde{t}t'\tilde{t}'m \leftrightarrow Tt \wedge T\tilde{t} \wedge Tt' \wedge T\tilde{t}' \wedge \bigvee_{v=0}^{K} \bigvee_{n=0}^{d(m)} (Vv \wedge Nn \wedge t = s(t', v, z) \wedge \tilde{t} = s(\tilde{t}', v, z))$$

therefore shows that S_R is *primitive recursive.*

$R_R t\tilde{t}t'\tilde{t}'t''\tilde{t}''$ shall mean that $t, \tilde{t}, t', \tilde{t}', t'', \tilde{t}''$ are the Gödel numbers of terms $\tau, \tilde{\tau}, \tau', \tilde{\tau}', \tau'', \tilde{\tau}''$ (where especially τ' is an F-term and $\tilde{\tau}'$ is a numeral) and $\tau = \tilde{\tau}$ is obtained from $\tau'' = \tilde{\tau}''$ by *replacement* of τ' by $\tilde{\tau}'$, i.e. there exist terms τ_0 and $\tilde{\tau}_0$ such that $\tau'' \equiv \tau_0{}^{x_k}/\tau'$, $\tilde{\tau}'' \equiv \tilde{\tau}_0{}^{x_k}/\tau'$, $\tau \equiv \tau_0{}^{x_k}/\tilde{\tau}'$, $\tilde{\tau} \equiv \tilde{\tau}_0{}^{x_k}/\tilde{\tau}'$. (We should keep in mind here, what we have seen in the previous section, that we can confine ourselves when dealing with derivations from $\mathfrak{S}$ (and such derivations are being discussed here) to the case when the variable occurring in the definition of replacement is identical with x_k.) We have

$$\begin{aligned} R_R t\tilde{t}t'\tilde{t}'t''\tilde{t}'' \leftrightarrow\ & Tt \wedge T\tilde{t} \wedge Ft' \wedge N\tilde{t}' \wedge Tt'' \wedge T\tilde{t}'' \wedge \\ & \wedge \bigvee_{t_0=0}^{s(t,1,K)} \bigvee_{\tilde{t}_0=0}^{s(\tilde{t},1,K)} (Tt_0 \wedge T\tilde{t}_0 \wedge t'' = s(t_0, K, t') \wedge \tilde{t}'' = s(\tilde{t}_0, K, t') \wedge \\ & \wedge t = s(t_0, K, \tilde{t}') \wedge \tilde{t} = s(\tilde{t}_0, K, \tilde{t}')). \end{aligned}$$

Here we need only show how the upper bound for t_0 (and similarly for $\tilde{t}_0$) is obtained. $\tilde{t}'$ is the Gödel number of a numeral. Therefore, according to Section 3 (7), $s(s(t_0, K, \tilde{t}'), 1, K) \geqq t_0$, from which the required estimate follows because $s(t_0, K, \tilde{t}') = t$. This can only be done since K, the Gödel number of a variable, is odd. The representation above shows that R_R is *primitive recursive.*

6. The predicate A was in substance already introduced in Section 2. The equations which can be derived boundedly in 0 steps are the equations of $\mathfrak{S}$. The equations which are boundedly derivable in m' steps are those which are boundedly derivable in m steps and further those which we obtain by m-bounded substitution or by replacement from an equation boundedly derivable in m steps. $Gt\tilde{t}$ will mean that $t, \tilde{t}$ are Gödel numbers of terms $\tau, \tilde{\tau}$ which are such that $\tau = \tilde{\tau}$ is an equation of $\mathfrak{S}$. $Gt\tilde{t}$ is valid for finitely many t and $\tilde{t}$ only, and so it is primitive recursive according to § 11.6, Corollary 2. Further, there exists an M such that

$$Gt\tilde{t} \rightarrow \max(t, \tilde{t}) \leqq M.$$

Obviously we can write

$$A t\tilde{t} 0 \leftrightarrow G t\tilde{t}$$

$$A t\tilde{t} m' \leftrightarrow A t\tilde{t} m \vee \bigvee_{t'} \bigvee_{\tilde{t}'} \bigvee_{t''} \bigvee_{\tilde{t}''} [A t'\tilde{t}' m \wedge A t''\tilde{t}'' m \wedge (S_R t\tilde{t} t'\tilde{t}' m \vee R_R t\tilde{t} t'\tilde{t}' t''\tilde{t}'')].$$

We shall show that A is primitive recursive. We shall be able to make use of the result of § 12.6. We need for that an estimate of the form

$$(0) \qquad S_R t\tilde{t} t'\tilde{t}' m \vee R_R t\tilde{t} t'\tilde{t}' t''\tilde{t}'' \rightarrow \max(t, \tilde{t}) \leqq \varphi(t', \tilde{t}', t'', \tilde{t}'', m)$$

where φ is a primitive recursive function.

$S_R t\tilde{t} t'\tilde{t}' m$ gives $t = s(t', v, n)$ with $v \leqq K$ and $n \leqq d(m)$. Thus,

$$(1) \qquad S_R t\tilde{t} t'\tilde{t}' m \rightarrow t \leqq \max_{v=0}^{K} \max_{z=0}^{d(m)} s(t', v, z).$$

Now we assume that $R_R t\tilde{t} t'\tilde{t}' t''\tilde{t}''$. Then there exist a t_0 and a $\tilde{t}_0$ such that

$$t = s(t_0, K, \tilde{t}'), \qquad \tilde{t} = s(\tilde{t}_0, K, \tilde{t}'),$$

$$t'' = s(t_0, K, t'), \qquad \tilde{t}'' = s(\tilde{t}_0, K, t').$$

t' is the Gödel number of an F-term τ'. There are two possibilities:

(a) τ' has the form F_j^0. According to Section 3 (2) we have that $s(s(t_0, K, t'), t', K) = s(t_0, t', K)$, and so $s(t'', t', K) = s(t_0, t', K)$. In Section 4 K was chosen so that $K \geqq t'$. Therefore, by Section 3 (4), $s(t_0, t', K) \geqq t_0$. This gives $t_0 \leqq s(t'', t', K)$. From this we obtain, since $t = s(t_0, K, \tilde{t}')$, the estimate

$$(2\text{a}) \qquad R_R t\tilde{t} t'\tilde{t}' t''\tilde{t}'' \rightarrow t \leqq \max_{t_0=0}^{s(t'', t', K)} s(t_0, K, \tilde{t}'),$$

provided t' corresponds to an F-term of the form F_j^0.

(b) τ' is an F-term not of the form F_j^0. Then we can apply Section 3 (7) and obtain that $s(s(t_0, K, t'), 1, K) \geqq t_0$, i.e. $s(t'', 1, K) \geqq t_0$ and from this finally, since $t = s(t_0, K, \tilde{t}')$, we obtain the estimate

$$(2\text{b}) \qquad R_R t\tilde{t} t'\tilde{t}' t''\tilde{t}'' \rightarrow t \leqq \max_{t_0=0}^{s(t'', 1, K)} s(t_0, K, \tilde{t}'),$$

provided t' denotes an F-term not of the form F_j^0.

From the estimates (1), (2a), (2b) and from corresponding estimates for $\tilde{t}$ (which are given by symmetry) we can easily obtain an estimate of the form (0). This completes the proof of the primitive recursiveness of A.

7. Example for a system of equations which unambiguously determines a not μ-recursive function. In § 19 we mentioned that it is possible that a system $\mathfrak{S}$ of equations determines a function unambiguously without making it possible to compute arbitrary values of the function from $\mathfrak{S}$ effectively. We shall give here an *example* which originates from KALMÁR. We start from the primitive recursive predicate T_0, which we introduced in § 18.3. We shall call the characteristic function of this predicate g. g is primitive recursive, $g(x, y) = 0 \leftrightarrow T_0 xy$. (We write here x for t, cf. § 18.3.) We write $h(x)$ for $\mu y g(x, y) = 0$, i.e. $h(x) = \mu y T_0 xy$. (For the unbounded μ-operator cf. § 12.1.) *h is not μ-recursive,* as we have shown in § 18.5. We should notice for future reference that $h(x) = 0$ if there exists no y such that $g(x, y) = 0$. We can easily see that the following two assertions are valid.

(1) $$g(x, y) = 0 \wedge \bigwedge_z (z < y \rightarrow g(x, z) \neq 0) \rightarrow h(x) = y,$$

(2) $$\bigwedge_z (z < y \rightarrow g(x, z) \neq 0) \rightarrow h(x) \geqq y \vee h(x) = 0.$$

Furthermore h is unambiguously determined by these relations. We shall show that for every function h which satisfies (1) and (2) $h(x) = \mu y g(x, y) = 0$. There are two possibilities. (a) for every x there exists a y such that $g(x, y) = 0$. Let $y_0 = \mu y g(x, y) = 0$. We put in (1) $y = y_0$. Then the hypotheses are valid and therefore $h(x) = y_0 = \mu y g(x, y) = 0$. (b) $g(x, y) \neq 0$ for all y. Then, for all y, the hypothesis of (2) is satisfied. Therefore, for every y $h(x) \geqq y \vee h(x) = 0$. From this follows that $h(x) = 0$. On the other hand, in this case $\mu y g(x, y) = 0$ is 0 as well.

The hypotheses occurring in (1) and (2), i.e.

$$g(x, y) = 0 \wedge \bigwedge_z (z < y \rightarrow g(x, z) \neq 0) \quad \text{and} \quad \bigwedge_z (z < y \rightarrow g(x, z) \neq 0),$$

are primitive recursive relations. We shall call the corresponding primitive recursive characteristic functions f_1 and f_2 respectively. Then we can paraphrase (1) and (2):

(1′) $$f_1(x, y) = 0 \rightarrow h(x) = y,$$

(2′) $$f_2(x, y) = 0 \rightarrow h(x) \geqq y \vee h(x) = 0.$$

These are, according to a rule[1] of the sentential calculus equivalent to

(1″) $$f_1(x, y) \neq 0 \vee h(x) = y,$$

(2″) $$f_2(x, y) \neq 0 \vee h(x) \geqq y \vee h(x) = 0.$$

---

[1] Namely, $p \rightarrow q$ is equivalent to $\neg p \vee q$.

There exist primitive recursive functions f_3, f_4, f_5, f_6 such that

$$f_3(x, y) = 0 \leftrightarrow f_1(x, y) \neq 0, \quad f_4(x, y) = 0 \leftrightarrow f_2(x, y) \neq 0,$$
$$f_5(u, v) = 0 \leftrightarrow u = v, \qquad f_6(u, v) = 0 \leftrightarrow u \geqq v.$$

Using these, (1″) and (2″) are equivalent to

$$(1''') \qquad f_3(x, y) = 0 \vee f_5(h(x), y) = 0,$$
$$(2''') \qquad f_4(x, y) = 0 \vee f_6(h(x), y) = 0 \vee h(x) = 0$$

and finally to

$$(1^*) \qquad f_3(x, y) \cdot f_5(h(x), y) = 0,$$
$$(2^*) \qquad f_4(x, y) \cdot f_6(h(x), y) \cdot h(x) = 0.$$

According to the main result of the last paragraph there exist for the primitive recursive functions f_3, f_4, f_5, f_6 and for the product function $f_7(x, y)$ systems $\mathfrak{S}_3, \mathfrak{S}_4, \mathfrak{S}_5, \mathfrak{S}_6, \mathfrak{S}_7$ of equations (which have no function variables in common) and function variables $F_3^2, F_4^2, F_5^2, F_6^2, F_7^2$ such that $\mathfrak{S}_j$ defines f_j with regard to F_j^2. If we add to all these equations two further equations (with the new function variable F_0^1)

$$F_7^2(F_3^2(x_1, x_2), F_5^2(F_0^1(x_1), x_2)) = 0$$
$$F_7^2(F_7^2(F_4^2(x_1, x_2), F_6^2(F_0^1(x_1), x_2)), F_0^1(x_1)) = 0,$$

then we obtain a system $\mathfrak{S}$ of equations. According to the previous considerations this whole system $\mathfrak{S}$ of equations is valid under an interpretation $\mathfrak{J}$ if and only if $\mathfrak{J}(F_j^2) = f_j$ $(j = 3, \ldots, 7)$ and $\mathfrak{J}(F_0^1) = h$. Then $\mathfrak{S}$ characterizes the function h. On the other hand the values of h cannot be obtained from $\mathfrak{S}$ using the rules (SR) and (RR) only, since otherwise h would be recursive and so μ-recursive. Furthermore, *there cannot possibly exist a rule system* by the help of which we can calculate the values of h from $\mathfrak{S}$, because the existence of such a system would imply computability and so the μ-recursiveness of h.

References

Kleene, S. C.: General Recursive Functions of Natural Numbers. Math. Ann. **112**, 727–742 (1936).

Kalmár, L.: Über ein Problem, betreffend die Definition des Begriffes der allgemein-rekursiven Funktion. Z. math. Logik **1**, 93–96 (1955). (Here we find the example dealt with in Section 7.)

CHAPTER 6

UNDECIDABLE PREDICATES

After giving a precise definition of the concept of decidability it is possible to show for certain predicates (properties or relations) that they are undecidable. It is easy to show the undecidability of many predicates P which are definable by the help of concepts which are directly connected with the concept of algorithm. Typical of these proofs is that they operate using a diagonal procedure.

However, a mathematician is more interested in predicates which came to light during the historical development of mathematics and have so far remained undecided. Such is, in the group theory, the predicate of being a consequence of a finite set of defining relations. The *word problem of group theory* asks for a decision procedure for this predicate. Further we mention here *Hilbert's tenth problem* which asks for a procedure by the help of which we can decide for an arbitrary given diophantine equation whether or not it is solvable. As it happens, we can prove, starting from the above mentioned predicates P, which can directly be proved undecidable, the undecidability of many mathematically interesting predicates Q. This is done by a kind of reduction: We show that the decidability of Q would imply the decidability of P. In such a way it was possible to show the unsolvability of the word problem of group theory, while we still cannot say anything definite about Hilbert's tenth problem.

We shall show the unsolvability of the word problem for Thue systems (in § 23), which is related to the word problem for groups but is essentially easier to handle. For the word problem for groups the reader should consult the references given in § 23. Furthermore, we shall show (in § 25) the undecidability of the predicate calculus. As a consequence of this we shall show (in § 26) that the predicate calculus of the second order is incomplete. Finally we shall prove (in § 27) the undecidability and incompleteness of arithmetic.

Since a predicate is decidable if and only if the corresponding characteristic function is computable, every undecidable predicate provides us with a *non-computable function* as well.

§ 22. Simple Undecidable Predicates

We shall begin with a heuristic discussion, and show afterwards the undecidability of predicates which are directly connected with different precise replacements of the concept of algorithm.

1. We cannot decide whether an arbitrary finite set of words (in the everyday language) describes an algorithm which is suitable for the computa-

tion of the value of a singulary function for an arbitrary argument. We shall treat this from the intuitive point of view. We use *reductio ad absurdum* and assume that the problem stated is solvable. We can effectively order all possible finite sets of words in the language in question according to length and when of equal length lexicographically. Let the sequence be $\mathfrak{F}_0, \mathfrak{F}_1, \mathfrak{F}_2, \ldots$. According to our assumption we can select from this sequence those finite sets of words which provide us with singulary functions. Let the subsequence of these finite sets of words be $\mathfrak{F}'_0, \mathfrak{F}'_1, \mathfrak{F}'_2, \ldots$. We can produce this effectively as long as we like. Every finite set $\mathfrak{F}'_n$ of words describes an algorithm for the computation of a singulary function f_n, and every finite set of words which describes the computation of such a function occurs in the sequence $\mathfrak{F}'_0, \mathfrak{F}'_1, \mathfrak{F}'_2, \ldots$.

Now we define by diagonal procedure a function f as follows.

$$f(n) = f_n(n) + 1.$$

f is computable. In order to compute $f(n)$ we first find the finite set $\mathfrak{F}'_n$ of words. Then we compute the value of $f_n(n)$ by the help of the procedure given in $\mathfrak{F}'_n$. To this number we add 1. That provides us with the value of $f(n)$. If we write out completely the computing instruction, which is given here rather briefly, then we obtain a finite set of words which describes an algorithm for the computation of f. This finite set of words must be a member of the sequence $\mathfrak{F}'_0, \mathfrak{F}'_1, \mathfrak{F}'_2, \ldots$. Let it be identical with $\mathfrak{F}'_m$. $\mathfrak{F}'_m$ describes the computation of the function f_m. This gives a contradiction, because $f(m) = f_m(m) + 1 \neq f_m(m)$.[1]

2. *Undecidable predicates in connection with Turing machines.* We shall show that certain simple properties of Turing machines are undecidable. We begin with a few preliminary remarks.

(1) In this section we consider only such Turing machines whose alphabet is a finite initial part of the fixed infinite alphabet $\{a_1, a_2, a_3, \ldots\}$ (we loose nothing essential by our assumption; cf. § 1.2). We identify the symbol a_1 with the stroke |. So the stroke belongs to the alphabet of every Turing machine (considered here). (The empty symbol is again represented by a_0 or $*$.)

(2) To be able to speak meaningfully of the undecidability of a property of things we must be able to describe these things by words of a fixed finite alphabet (cf. § 2.3). Thus, we have the task of denoting

[1] In the proof a certain finite set of words (namely the one which serves to compute f) uses in its definition the collection of all finite sets of words. That is a quite usual procedure in *classical* mathematics, which is used for instance in defining a certain real number by a Dedekind cut using the collection of all real numbers. This Dedekind cut is used for instance in the proof of the intermediate value theorem of real analysis.

every Turing machine by a word of a fixed alphabet. In § 17.3 we characterized Turing machines by their Gödel numbers. These numbers can be represented by sequences of strokes. We shall call the sequence of strokes which is associated with M in this way (and which characterizes M unambiguously) the *machine word* W_{M}. According to (1) W_{M} is a word over the alphabet of every Turing machine, and so especially of M.

(3) If we say that a property P of Turing machines is decidable in the intuitive sense, then we obviously mean that it is decidable in the representation of Turing machines given in (2) whether or not an arbitrary sequence of strokes, *which is a machine word*, is a word for a machine which has the property P. Thus, we are dealing primarily with a relative decidability (cf. § 2.3, here we have to identify a property with the set of things which have this property). Now, it is certainly decidable whether or not a sequence of strokes is a machine word (see § 17.3). Therefore, according to a result of § 2.3, the relative decidability of P is equivalent to the absolute decidability of P. In other words, we have to consider the property which is valid for an *arbitrary* sequence of strokes if and only if it is a machine word of a machine with the property P. We shall call this associated property over the domain of arbitrary sequences of strokes $\tilde{P}$. Thus, P is decidable relative to the set of machine words if and only if $\tilde{P}$ is decidable.

(4) The reader should remember the definition of decidability in § 6.3 together with the remark in § 6.4. According to these a predicate P which is defined in the domain of all words W over an alphabet $\{a_1, \ldots, a_N\}$ is decidable if and only if there exists a Turing machine M over $\{a_1, \ldots, a_N\}$ such that

P is valid for W *if and only if* M, placed behind W, stops operating over I.

P is not valid for W *if and only if* M, placed behind W, stops operating over $*$.

We should keep in mind here the comments in § 6.3 on the arbitrariness of the "indicating" letters (here I and $*$).

We now consider the properties P_1, P_2, P_3 of machines which are defined by the following:

P_1 is valid for M *if and only if* M, placed behind W_{M}, stops operating over $*$.

P_2 is valid for M *if and only if* M, placed behind W_{M}, stops operating.

P_3 is valid for M *if and only if* M, placed onto the empty tape, stops operating.

We assert the

Theorem. There exists no general procedure to decide for any arbitrary machine M *whether* M *has the property* P_1. *The same applies for* P_2 *and* P_3.

In order to prove this theorem for P_1 it is sufficient, according to (3), to prove the undecidability of the corresponding property $\tilde{P}_1$ over the alphabet $\{|\}$. $\tilde{P}_1$ is valid for a sequence of strokes if and only if the sequence of strokes is a machine word for a machine which has the property P_1.

$\tilde{P}_1$ *is undecidable.* Proof by *reductio ad absurdum.* We assume that $\tilde{P}_1$ is decidable. Then there exists a machine M_1 over $\{|\}$ such that for every sequence W of strokes we have that

$\tilde{P}_1$ is valid for W *if and only if* M_1, placed behind W, stops operating over $|$;

$\tilde{P}_1$ is not valid for W *if and only if* M_1, placed behind W, stops operating over $*$.

We are only interested in the second of these assertions. What is valid for *every* sequence of strokes must especially also be valid for the machine word W_{M_1}. Thus, we have (diagonal procedure):

$\tilde{P}_1$ is not valid for W_{M_1} if and only if M_1, placed behind W_{M_1}, stops operating over $*$. Since W_{M_1} is a machine word and, indeed, for the machine M_1, $\tilde{P}_1$ is not valid for W_{M_1} if and only if M_1 has not the property P_1, i.e. if and only if M_1, placed behind W_{M_1}, does not stop operating over $*$. From this we have the *contradiction:*

M_1, placed behind W_{M_1}, does not stop operating over $*$ if and only if M_1, placed behind W_{M_1}, stops operating over $*$.

P_2 *is undecidable.* To prove this we show that a decision procedure for P_2 implies the following decision procedure for P_1. In order to determine whether a machine M has the property P_1 we first find out whether it has the property P_2. If this is *not* the case, then M, placed behind W_{M}, never stops operting, i.e. does not have the property P_1. However, if we find that M has the property P_2, then we place M behind W_{M}. After finitely many steps M stops operating. Then we can find out whether M stopped operating over $*$ or over another symbol. This decides whether or not M has the property P_1.

P_3 *is undecidable.* Again we use *reductio ad absurdum.* We assume that P_3 is decidable. Then we show that P_2 is also decidable. Let M be an arbitrary Turing machine. W_{M} is a finite word. We can easily describe a machine M' which prints onto the originally empty computing tape the word W_{M} and then stops operating behind this word (cf. § 6, Exercise 2). Now consider the machine $\mathsf{M}'' = \mathsf{M}'\mathsf{M}$. If we place M'' on the empty computing tape, then at first M'' operates like M' until the word W_{M} is printed on the tape and M' is at rest behind this word. Then M'' operates further like M placed behind W_M. Thus, M'', placed on the empty tape, stops operating after finitely many steps if and only if M, placed behind W_M, stops operating after finitely many steps. Now, if we could decide P_3, then we could decide whether M'', placed onto the empty tape, stops operating, and so whether M, placed behind W_M, stops operating, i.e. whether M has the property P_2.

3. Undecidable predicates which are defined by μ-recursive predicates. In § 18.3 we introduced for every $n \geqq 0$ a μ-recursive and thus decidable $(n+2)$-ary predicate T_n. We now show the

Theorem. For $n \geqq 1$ *the* $(n+1)$-ary *predicate*

$$S_n t\mathfrak{x} \leftrightarrow \bigvee_y T_n t\mathfrak{x} y$$

is undecidable.

Proof by *reductio ad absurdum.* We assume that S_n is decidable and so μ-recursive. We consider the following predicates (where x_1 is the first component of $\mathfrak{x}$).

$$S'_n\mathfrak{x} \leftrightarrow S_n x_1 \mathfrak{x}$$

$$S''_n\mathfrak{x} \leftrightarrow \neg S'_n\mathfrak{x}$$

$$S^*_n\mathfrak{x}y \leftrightarrow S''_n\mathfrak{x} \wedge y = y.$$

According to § 14.2 S'_n, S''_n and S^*_n are μ-recursive. If we apply Kleene's enumeration theorem (cf. § 18.6) to S^*_n, then we see that there exists a number t such that

$$\bigwedge_{\mathfrak{x}} (\bigvee_y S^*_n\mathfrak{x}y \leftrightarrow \bigvee_y T_n t\mathfrak{x}y).$$

Now, if we consider that $\bigvee_y S^*_n\mathfrak{x}y \leftrightarrow S''_n\mathfrak{x} \leftrightarrow \neg S_n x_1\mathfrak{x} \leftrightarrow \neg \bigvee_y T_n x_1 \mathfrak{x} y$, then we obtain

$$\bigwedge_{\mathfrak{x}} (\neg \bigvee_y T_n x_1 \mathfrak{x} y \leftrightarrow \bigvee_y T_n t\mathfrak{x}y).$$

If we take especially an n-tuple $\mathfrak{x}$ for which $x_1 = t$ (*diagonal procedure*), then we have the contradiction

$$\neg \bigvee_y T_n t\mathfrak{x}y \leftrightarrow \bigvee_y T_n t\mathfrak{x}y.$$

§ 23. The Unsolvability of the Word Problem for Semi-Thue Systems and Thue Systems

All finite and many denumerably infinite groups can be generated by finitely many generators whose interdependability is described by finitely many so-called defining relations. Let us for example consider the rotations of a cube about its centre which move the vertices into points which were previously also occupied by vertices. We can show that there exist special rotations A, B, C, D, E such that every rotation can be represented as a product of these rotations.[1] Every such product is a word in the generators, e.g. $BBCEAC$. The identical rotation, i.e. the identity element of the group, is represented by the empty word $\square$. The representation of an element of the group is not unambiguous. For example $AB = CCC$. For this reason we call the pair (AB, CCC) of words a "relation". It is possible to put down a finite system of such relations (so-called "defining relations") and to give rules (which are the same for every group given in this way) so that we are able to obtain every relation from the defining relations by the help of the rules. In the example under consideration the following relations for instance form a system of defining relations.

$$AA = \square,\quad BD = \square,\quad DB = \square,\quad CE = \square,\quad EC = \square,\quad D = BB,$$
$$\square = AA,\quad \square = BD,\quad \square = DB,\quad \square = CE,\quad \square = EC,\quad BB = D,$$
$$E = CCC,\quad ABC = \square,$$
$$CCC = E,\qquad \square = ABC.$$

(For a more precise definition of these concepts cf. Sections 1 and 2.)

If a group is given in this way by finitely many generators and defining relations, then we can ask whether there exists an algorithm by the help of which we can decide for arbitrary words W_1 and W_2 whether or not these represent the same element of the group. This is the "word problem" for *this* group. In general we can ask for an algorithm which solves the word problem for *arbitrary* groups given in this way. That is the *general word problem* for groups. It has been shown during the last few years that the general word problem for groups is unsolvable and furthermore that there exist special groups with unsolvable word problems.

The proof of the unsolvability of the *general* word problem for groups is based upon the fact that it is not decidable whether or not a Turing machine, placed behind its Gödel number (or on the empty

---

[1] A, B, C, D, E are chosen as follows. Let α, β, γ be the three pairwise orthogonal axes through the centre of the cube and the centres of the faces, which are so oriented that they form a righthanded system in the given sequence. This provides us for every axis with a positive sense of rotation for the rotations about the axis in question. Let A be a rotation of $-\pi/2$ degrees about γ followed by a rotation of π degrees about α. Let B be a rotation of $\pi/2$ degrees about α followed by a rotation of $\pi/2$ degrees about γ. Let C be a rotation of $\pi/2$ degrees about β. Let D be equal to B^{-1}, E to C^{-1}. — We adopt the convention that a product $R_1 R_2$ of rotations means that first the rotation R_1 is carried out, followed by the rotation R_2.

computing tape), stops operating after finitely many steps (§ 22.2). That there exists a *special* group with unsolvable word problem is proved in the same way by reduction to the fact that there exists a special Turing machine (universal machine) for which we cannot decide whether or not it stops operating when placed behind an arbitrary word (§ 30).

In this paragraph we shall carry out the first step to prove the unsolvability of the word problem for groups. This consists in the proof that the corresponding problems for "semi-Thue systems" and "Thue systems" are unsolvable. (We shall later refer to the result for semi-Thue systems in order to prove the unsolvability of the decision problems of the predicate calculus (§ 25).) The result of this paragraph was proved by Post and Markov independently. We follow the proof by Post.

1. Semi-Thue systems. Let a finite alphabet $\{S_1, \ldots, S_N\}$ $(N \geqq 1)$ be given. We consider *words* over this alphabet. In this paragraph we shall definitely *allow the empty word* $\square$ as well. We use $W_1 W_2$ to denote the word which is obtained by writing W_1 and W_2 behind each other in the given sequence. The associative law $(W_1 W_2) W_3 \equiv W_1 (W_2 W_3)$ is valid for this operation of *juxtaposition* and so we can write $W_1 W_2 W_3$ without parentheses. For every word W, $W\square \equiv \square W \equiv W$. The fact that in the domain of words we have a binary associative operation (juxtaposition), with respect to which there exists an element $\square$ such that $W\square \equiv \square W \equiv W$, can be expressed by saying that the words over the alphabet $\{S_1, \ldots, S_N\}$ form a *semi-group with identity element* with respect to juxtaposition.

A *semi-Thue system* $\mathfrak{S}$ over $\{S_1, \ldots, S_N\}$ is given by a finite non-empty set of ordered pairs

$$(D_i, D_i') \qquad (i = 1, \ldots, m).$$

of words over this alphabet. These pairs of words are called the *defining relations* of $\mathfrak{S}$. In the following only words which are over $\{S_1, \ldots, S_N\}$ will be considered. We shall say that the word W' is an *immediate consequence with respect to* $\mathfrak{S}$ of the word W (in symbols $W \Rightarrow_{\mathfrak{S}} W'$, or in short $W \Rightarrow W'$) if there exist an i $(1 \leqq i \leqq m)$ and words U and V such that

$$W \equiv U D_i V \qquad \text{and} \qquad W' \equiv U D_i' V,$$

in other words, if W' is obtained from W by replacing a part D_i of the word W by the corresponding D_i'. We should keep in mind here that U and V may be empty. We have for instance, if $(S_1S_2S_1, S_3)$ is a defining relation of $\mathfrak{S}$, that $S_1S_2S_1S_2S_1 \Rightarrow S_3S_2S_1$, but also $S_1S_2S_1S_2S_1 \Rightarrow S_1S_2S_3$.

W' is called a *consequence* of W with respect to $\mathfrak{S}$ (in symbols[1] $W \rightarrow_{\mathfrak{S}} W'$ or in short $W \rightarrow W'$), if there exists a finite chain of words $W_0, \ldots, W_p$ ($p \geqq 0$) such that

$$W \equiv W_0, W_0 \Rightarrow W_1, W_1 \Rightarrow W_2, \ldots, W_{p-1} \Rightarrow W_p, W_p \equiv W'.$$

The following laws are valid.

(a) $W \rightarrow W$.

(b) If $W \rightarrow W'$ and $W' \rightarrow W''$, then $W \rightarrow W''$.

(c) If $W \rightarrow W'$ and $\tilde{W} \rightarrow \tilde{W}'$, then $W\tilde{W} \rightarrow W'\tilde{W}'$.

(a) follows from the definition for $p = 0$. (b) is obtained by the combination of the chains which lead from W to W' and from W' to W'' respectively. (c) is shown as follows. According to hypothesis there exist chains $W_0, \ldots, W_p$ and $\tilde{W}_0, \ldots, \tilde{W}_q$ such that

$$W \equiv W_0 \Rightarrow W_1 \Rightarrow \cdots \Rightarrow W_{p-1} \Rightarrow W_p \equiv W'$$

and

$$\tilde{W} \equiv \tilde{W}_0 \Rightarrow \tilde{W}_1 \Rightarrow \cdots \Rightarrow \tilde{W}_{q-1} \Rightarrow \tilde{W}_q \equiv \tilde{W}'.$$

Then the chain

$$W\tilde{W} \equiv W_0\tilde{W} \Rightarrow W_1\tilde{W} \Rightarrow \cdots \Rightarrow W_{p-1}\tilde{W} \Rightarrow W_p\tilde{W} \equiv$$
$$\equiv W'\tilde{W}_0 \Rightarrow W'\tilde{W}_1 \Rightarrow \cdots \Rightarrow W'\tilde{W}_{q-1} \Rightarrow W'\tilde{W}_q \equiv W'\tilde{W}'.$$

shows that $W\tilde{W} \rightarrow W'\tilde{W}'$.

The *word problem for* $\mathfrak{S}$ is to find an algorithm which decides in finitely many steps for arbitrary words W and W' whether or not $W \rightarrow_{\mathfrak{S}} W'$. The general *word problem for semi-Thue systems* is to find an algorithm by the help of which we can decide in finitely many steps for an arbitrarily given semi-Thue system $\mathfrak{S}$ and words W and W' over the alphabet of $\mathfrak{S}$ whether or not $W \rightarrow_{\mathfrak{S}} W'$.

2. *Thue systems and group systems.* A *Thue system* is a semi-Thue system in which for every defining relation (D, D') the *inverse relation* (D', D) is also a defining relation. A Thue system is called a *group system* if there exists an involution σ of the alphabet $\{S_1, \ldots, S_N\}$ (i.e. a mapping of the alphabet onto itself such that the condition $\sigma(\sigma(S_i)) \equiv S_i$ is satisfied for every S_i) and if all the pairs

$$(*) \qquad (S_i\sigma(S_i), \square)$$

of words belong to the set of defining relations.[2]

---

[1] The reader ought not to confuse this arrow with the symbol introduced in § 11.1 to denote "if-then".

[2] The further requirement that, for every i, $\sigma(S_i) \not\equiv S_i$ is often presumed. In this case we must work with an alphabet with an even number of elements.

The *word problem* and the *general word problem* for *Thue systems* and for *group systems* respectively are defined similarly to the corresponding problems for semi-Thue systems.

For Thue systems (and a fortiori for group systems) we get from $W \Rightarrow W'$ that $W' \Rightarrow W$, and so from $W \rightarrow W'$ that $W' \rightarrow W$. Thus, the relation $\rightarrow$ is an *equivalence relation*. Furthermore, in view of Section 1 (c), it is a *congruence relation* in the semi-group of words. We can therefore build up an algebra of residue classes in the usual way. This algebra, in the case of a *Thue system*, is again *a semi-group with identity element*. In the case of a *group system* the algebra of residue classes is even a *group*.

Let a group element be given by the word

$$W \equiv S_{i_1} S_{i_2} \ldots S_{i_r}.$$

Then the group element which is inverse to this is given by the word

$$W' \equiv \sigma(S_{i_r}) \ldots \sigma(S_{i_2})\, \sigma(S_{i_1}).$$

(Thus, W' is obtained from W by the forming of the σ-images and reversing the order of the sequence.) As a matter of fact we have, because of $(*)$, that

$$\begin{aligned} W W' &\equiv S_{i_1} \ldots S_{i_{r-1}} S_{i_r} \sigma(S_{i_r})\, \sigma(S_{i_{r-1}}) \ldots \sigma(S_{i_1}) \\ &\Rightarrow S_{i_1} \ldots S_{i_{r-1}}\, \sigma(S_{i_{r-1}}) \ldots \sigma(S_{i_1}) \\ &\Rightarrow \cdots \Rightarrow S_{i_1}\, \sigma(S_{i_1}) \Rightarrow \square. \end{aligned}$$

3. Sketch of the proof. We start from a Turing machine M over an alphabet $\mathfrak{A}_0$. In Section 4 we shall introduce a further alphabet $\mathfrak{A}_1$ and show that every *configuration* K of M (for the concept of configuration and for other concepts which will occur in this paragraph in connection with configurations cf. § 5.3 and § 5.4) can be described by a word W_K over the alphabet $\mathfrak{A}_1$. If K is known, then we can determine W_K explicitly. These words W_K, which describe the configurations, will be called in short the *configuration words*. (Strictly speaking, W_K only describes a configuration up to a possible *shift*, cf. § 5.5.) In Sections 5 and 6 we shall introduce, by giving finitely many defining rules, a semi-Thue system $\mathfrak{S}(\mathsf{M})$ over an alphabet $\mathfrak{A}$, with $\mathfrak{A}_1 \subset \mathfrak{A}$. Further, in Section 6, we shall define a word W^* over $\mathfrak{A}$. We shall show the validity of the following assertions:

(1) If K' is a consecutive configuration of K, then $W_K \Rightarrow W_{K'}$.

(2) If K' is a consecutive configuration of K and $W_K \Rightarrow W$, then $W \equiv W_{K'}$.

(3) If K is a terminal configuration, then $W_K \rightarrow W^*$.

(4) W^* is not a configuration word.

From (1), (2), (3) and (4) follows the

Lemma. M, *placed on* B *over* A, *stops operating after finitely many steps if and only if* $W_K \to W^*$, *where* $K = (A, B, c_{\mathsf{M}})$.

Proof. (a) Let M, placed on B over A, stop operating after finitely many steps. Let K_n be the terminal configuration reached. Then, according to (1), $W_K \to W_{K_n}$ and, according to (3), $W_{K_n} \to W^*$ and so $W_K \to W^*$.

(b) Conversely, let $W_K \to W^*$. Then there exists a chain $W_K \equiv W_0 \Rightarrow W_1 \Rightarrow \cdots \Rightarrow W_p \equiv W^*$. We have to show that if M is placed on B over A, then it stops operating after finitely many steps. If this were not the case, then there would exist for every i an i-th configuration K_i. We shall show that $W_i \equiv W_{K_i}$ for $i \leqq p$. This would give that $W_p \equiv W^*$ is a configuration word, which contradicts (4).

Proof that $W_i \equiv W_{K_i}$ for all $i \leqq p$: First $K_0 = (A, B, c_{\mathsf{M}}) = K$, thus $W_0 \equiv W_K \equiv W_{K_0}$. Let us assume that it is already shown for an $i < n$ that $W_i \equiv W_{K_i} \cdot K_{i+1}$ is a consecutive configuration of K_i. We have that $W_{K_i} \equiv W_i \Rightarrow W_{i+1}$. Thus, according to (2), $W_{i+1} \equiv W_{K_{i+1}}$.

Now, if the general word problem for semi-Thue systems were solvable, then we could decide on the basis of the Lemma proved just now whether a machine M, placed on an arbitrary tape expression B over an arbitrary square A, stops operating after finitely many steps. To do this we make use of the following procedure:

(α) Form the semi-Thue system $\mathfrak{S}(\mathsf{M})$ which is associated with M. (It is possible to do this effectively on the basis of the definition of $\mathfrak{S}(\mathsf{M})$ given in Sections (5) and (6).)

(β) Produce for $K = (A, B, c_{\mathsf{M}})$ the word W_K. (This, too, can be done effectively.)

(γ) Decide whether or not $W_K \to_{\mathfrak{S}} W^*$. (This is possible according to our hypothesis.)

However, since according to § 22.2 it is impossible to decide by a general procedure whether a machine placed behind its machine word stops operating after finitely many steps, we have that the *general* word problem for semi-Thue systems is unsolvable.

If the word problem were solvable for a semi-Thue system $\mathfrak{S}(\mathsf{M})$ associated with a *special Turing machine* M, then we could decide, in the way discussed just now, whether M, placed behind an arbitrary word, stops operating after finitely many steps. However, this is not possible for *the universal machine* U_0, which we shall consider in § 30. Therefore the word problem for $\mathfrak{S}(\mathsf{U}_0)$ is unsolvable. In conclusion we have the

Theorem. The general word problem for semi-Thue systems is unsolvable. Furthermore, the word problem for the semi-Thue system $\mathfrak{S}(\mathsf{U}_0)$ *associated with the universal machine* U_0 *is unsolvable.*

4. Configuration words. Let M be a Turing machine over the alphabet $\mathfrak{A}_0 = \{a_1, \ldots, a_N\}$. The configuration words are words over the alphabet $\mathfrak{A}_1 = \{A_0, \ldots, A_N, Q_0, \ldots, Q_m, E\}$. A_0 corresponds to the empty symbol; $A_1, \ldots, A_N$ correspond to the symbols $a_1, \ldots, a_N$; $Q_0, \ldots, Q_m$ to the states $0, \ldots, m$; E is a letter which occurs at the beginning and at the end of a configuration word. We obtain the configuration word W_K corresponding to the configuration $K = (A, B, C)$ using the following instruction (cf. the examples given in the Figure!).

(a) Consider the smallest continuous section of the computing tape which contains all the marked squares of B and the square A.

(b) Represent the contents of this section by a word, more precisely write A_0 for an empty square and A_j for a square with the letter a_j $(j = 1, \ldots, N)$. (In this way the tape expression B will be represented up to a "shift" (cf. § 5.5).)

(c) Immediately on the left of the symbol for the square A insert a symbol Q_C. (In this way both the scanned square A and the state C are represented.)

(d) Attach both on the right and on the left a letter E. The word obtained in this way is the configuration word W.

The reader should notice that W_K contains one and only one of the letters $Q_0, \ldots, Q_m$.

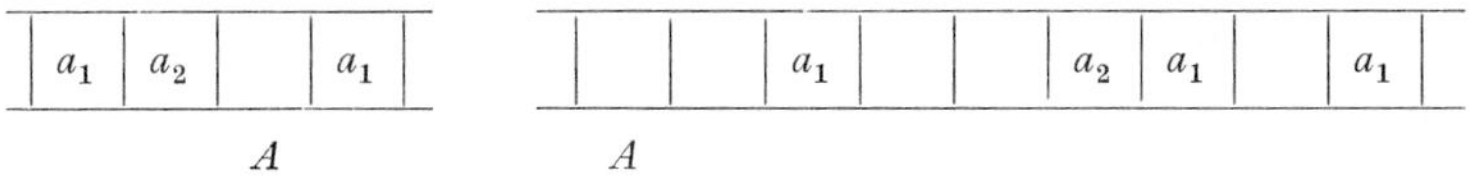

$EA_1A_2Q_4A_0A_1E$ $\qquad$ $EQ_4A_0A_0A_1A_0A_0A_2A_1A_0A_1E$

Fig. 23.1. Examples for configuration words (in both cases we let $C = 4$).

5. The rules of $\mathfrak{S}(\mathsf{M})$. First part. We shall now associate a semi-Thue system $\mathfrak{S}(\mathsf{M})$ with the machine M so that the assertions (1), (2), (3) and (4) in Section 3 are satisfied. The alphabet of $\mathfrak{S}(\mathsf{M})$ is

$$\mathfrak{A} = \{A_0, A_1, \ldots, A_N, Q_0, \ldots, Q_m, E, R, S\}.$$

We shall first give some of the defining rules of $\mathfrak{S}(\mathsf{M})$. These rules are chosen so that (1) is valid. That such a thing is possible is clear if we notice that if K' is a consecutive configuration of K, then only the symbols which are in the immediate neighbourhood of the symbol Q_C will be changed in moving from W_K to $W_{K'}$.

We shall associate with each line $i a_j b k$ of M one or more rules of $\mathfrak{S}(\mathsf{M})$. At first we only have to do this for lines for which $b \neq h$, since (1) makes sense only for such lines.

We associate with a line $i a_j a_l k$ of M the defining rule

(a) $$(Q_i A_j, Q_k A_l).$$

We associate with a line $i a_j r k$ of M ($j \neq 0$) the defining rules

(b$'_r$) $$\begin{cases} (Q_i A_j A_t, A_j Q_k A_t) & (t = 0, \ldots, N) \\ (Q_i A_j E, A_j Q_k A_0 E). \end{cases}$$

The last rule corresponds to the case when the scanned square is the last square of the section corresponding to K and so the section must be elongated by one square during the step to K'. The "end letter" E serves the purpose of indicating this case. — Corresponding boundary rules will also be given in the following cases.

We associate with a line $i a_0 r k$ of M the rules

(b$''_r$) $$\begin{cases} (A_u Q_i A_0 A_t, A_u A_0 Q_k A_t) & (u, t = 0, \ldots, N) \\ (E Q_i A_0 A_t, E Q_k A_t) & (t = 0, \ldots, N) \\ (A_u Q_i A_0 E, A_u A_0 Q_k A_0 E) & (u = 0, \ldots, N) \\ (E Q_i A_0 E, E Q_k A_0 E). \end{cases}$$

Corresponding rules for a line $i a_j l k$ (with $j \neq 0$) are

(b$'_l$) $$\begin{cases} (A_u Q_i A_j, Q_k A_u A_j) & (u = 0, \ldots, N) \\ (E Q_i A_j, E Q_k A_0 A_j). \end{cases}$$

Finally, for a line $i a_0 l k$ the rules are

(b$''_l$) $$\begin{cases} (A_u Q_i A_0 A_t, Q_k A_u A_0 A_t) & (u, t = 0, \ldots, N) \\ (A_u Q_i A_0 E, Q_k A_u E) & (u = 0, \ldots, N) \\ (E Q_i A_0 A_t, E Q_k A_0 A_0 A_t) & (t = 0, \ldots, N) \\ (E Q_i A_0 E, E Q_k A_0 E). \end{cases}$$

On the basis of the rules given so far assertion (1) of Section 3 is easily verified.

6. The rules of $\mathfrak{S}(\mathsf{M})$. *Second part.* According to Section 3 (3), the remaining rules of $\mathfrak{S}(\mathsf{M})$ serve to make the word W^* the consequence of any word which corresponds to a terminal configuration. We define

$$W^* \equiv S.$$

Since, according to Section 4, every configuration word contains a letter Q_k, W^* is not a configuration word, as it was asserted in Section 3 (4).

A configuration is a terminal configuration if and only if the corresponding configuration line is of the form $i a_j h k$. We associate with such a line of **M** the following defining rules of $\mathfrak{S}$(**M**).

$$\text{(c)} \quad \begin{cases} (Q_i A_j, R) & \\ (A_u R, R) & (u = 0, \ldots, N) \\ (ER, S) & \\ (SA_t, S) & (t = 0, \ldots, N) \\ (SE, S). & \end{cases}$$

It is clear that by the help of these rules we can derive the word W^* from any word W_K for which K is a terminal configuration, as it was asserted in Section 3(3). According to the first of the rules (c), $Q_i A_j$ can be replaced by R so that a word of the form $EA_{u_p} \ldots A_{u_2} A_{u_1} R A_{t_1} A_{t_2} \ldots A_{t_q} E$ is obtained. By the rules in the second line we move to $EA_{u_p} \ldots A_{u_2} R A_{t_1} A_{t_2} \ldots A_{t_q} E$, etc. until $ERA_{t_1} A_{t_2} \ldots A_{t_q} E$ is reached. The third rule provides us with $SA_{t_1} A_{t_2} \ldots A_{t_q} E$, and the rules in the fourth line with $SA_{t_2} \ldots A_{t_q} E$, etc. until SE is reached, and finally the last rule gives S, i.e. W^*.[1]

7. *Normal words.* Now we have to show Section 3(2). For this we give a definition. A word W over the alphabet belonging to $\mathfrak{S}$(**M**) is *a normal word* if it contains one and only one of the letters $Q_0, \ldots, Q_m, R, S$. We have the following:

(5) For every configuration K, W_K is a normal word according to the construction given in Section 4.

(6) If W is a normal word and $W \Rightarrow_{\mathfrak{S}(\mathsf{M})} W'$, then W' is uniquely determined.

This is true since we can apply at most one of the rules (a), (b'_r), (b''_r), (b'_l), (b''_l), (c).

From (5) and (6) we obtain Section 3(2). — For application later on we also note the following.

(7) If W is a normal word and $W \Rightarrow_{\mathfrak{S}(\mathsf{M})} W'$ or $W' \Rightarrow_{\mathfrak{S}(\mathsf{M})} W$, then W' is also a normal word. This follows from the fact that in every rule of $\mathfrak{S}$ one and only one of the letters $Q_0, \ldots, Q_m, R, S$ occurs in each of the two components.

---

[1] If we only wanted to give a proof of the Theorem of Section 3 without considering the application to Thue systems in Section 8, then we could put $W^* = ESE$ and choose instead of the rules (c) the following rules:

$$\text{(c')} \quad \begin{cases} (Q_i A_j, S) & \\ (A_u S, S) & (u = 0, \ldots, N) \\ (SA_t, S) & (t = 0, \ldots, N). \end{cases}$$

(8) There exists no W such that $W^* \Rightarrow_{\mathfrak{S}(\mathsf{M})} W$. This is true, since none of the rules of $\mathfrak{S}$ is applicable to W^*.

8. The word problem for Thue systems. We shall show that the Theorem in Section 3 on the unsolvability of the word problem for semi-Thue systems is valid for Thue systems as well. For this purpose we associate with the semi-Thue system $\mathfrak{S}(\mathsf{M})$ a *Thue system* $\mathfrak{T}(\mathsf{M})$. The alphabet of $\mathfrak{T}(\mathsf{M})$ is the same as the alphabet of $\mathfrak{S}(\mathsf{M})$. A pair (W, W') of words is a defining rule of $\mathfrak{T}(\mathsf{M})$ if and only if either (W, W') or (W', W) is a defining rule of $\mathfrak{S}(\mathsf{M})$. Thus, $\mathfrak{T}(\mathsf{M})$ is obtained from $\mathfrak{S}(\mathsf{M})$ by taking the inverse rules as additional defining rules. For $\mathfrak{T}(\mathsf{M})$ we assert the

Lemma. For every configuration word W_K

$$W_K \rightarrow_{\mathfrak{T}(\mathsf{M})} W^* \quad \textit{if and only if} \quad W_K \rightarrow_{\mathfrak{S}(\mathsf{M})} W^*.$$

To prove this we only have to show that if $W_K \rightarrow_{\mathfrak{T}(\mathsf{M})} W^*$, then $W_K \rightarrow_{\mathfrak{S}(\mathsf{M})} W^*$. The proof is by *reductio ad absurdum.* We assume that there exists a configuration K for which $W_K \rightarrow_{\mathfrak{T}(\mathsf{M})} W^*$, but not $W_K \rightarrow_{\mathfrak{S}(\mathsf{M})} W^*$. Because $W_K \rightarrow_{\mathfrak{T}(\mathsf{M})} W^*$, there exists a chain of words such that

$$W_K \equiv W_0 \Rightarrow_{\mathfrak{T}(\mathsf{M})} W_1 \Rightarrow_{\mathfrak{T}(\mathsf{M})} \cdots \Rightarrow_{\mathfrak{T}(\mathsf{M})} W_p \equiv W^*.$$

From (5) and (7) we have that all words $W_0, \ldots, W_p$ are normal words. Since, according to our assumption, not $W_0 \rightarrow_{\mathfrak{S}(\mathsf{M})} W^*$, but certainly $W_p \rightarrow_{\mathfrak{S}(\mathsf{M})} W^*$, there must be an index $i < p$ such that not $W_i \rightarrow_{\mathfrak{S}(\mathsf{M})} W^*$, but $W_{i+1} \rightarrow_{\mathfrak{S}(\mathsf{M})} W^*$. Let $\mathfrak{R}$ be a defining rule of $\mathfrak{T}(\mathsf{M})$ which provides the passage $W_i \Rightarrow_{\mathfrak{T}(\mathsf{M})} W_{i+1}$. $\mathfrak{R}$ is not a rule of $\mathfrak{S}(\mathsf{M})$, since otherwise $W_i \rightarrow_{\mathfrak{S}(\mathsf{M})} W^*$. Thus, the inverse $\overline{\mathfrak{R}}$ is a rule of $\mathfrak{S}(\mathsf{M})$. By the application of $\overline{\mathfrak{R}}$ we have that $W_{i+1} \Rightarrow_{\mathfrak{S}(\mathsf{M})} W_i$. Further, because $W_{i+1} \rightarrow_{\mathfrak{S}(\mathsf{M})} W^*$, there exists a chain

$$W_{i+1} \equiv \overline{W}_0 \Rightarrow_{\mathfrak{S}(\mathsf{M})} \overline{W}_1 \Rightarrow_{\mathfrak{S}(\mathsf{M})} \cdots \Rightarrow_{\mathfrak{S}(\mathsf{M})} \overline{W}_q \equiv W^*.$$

Here $q \geqq 1$, since otherwise $W_{i+1} \equiv W^*$. But this is impossible, since it would imply $W^* \Rightarrow_{\mathfrak{S}(\mathsf{M})} W_i$ in contradiction to (8). From $W_{i+1} \Rightarrow_{\mathfrak{S}(\mathsf{M})} W_i$ and $W_{i+1} \Rightarrow_{\mathfrak{S}(\mathsf{M})} \overline{W}_1$ we obtain by the help of (6) that $\overline{W}_1 \equiv W_i$, But then the second chain shows that $W_i \rightarrow_{\mathfrak{S}(\mathsf{M})} W^*$, by which our assumption has led to a contradiction.

We have found out in the proof of the Theorem of Section 3 that there exists no algorithm by the help of which we can decide for arbitrary K and M whether $W_K \rightarrow_{\mathfrak{S}(\mathsf{M})} W^*$, and that there exists no algorithm by the help of which we can decide for the *universal Turing machine* U_0 and arbitrary K whether $W_K \rightarrow_{\mathfrak{S}(\mathsf{U}_0)} W^*$. From this follows, due to the equivalence stated in the Lemma proved just now, the

Theorem. The general word problem for Thue systems is unsolvable. Furthermore, the word problem of the Thue system $\mathfrak{T}(\mathsf{U}_0)$ associated with the machine U_0 is unsolvable.

References

THUE, A.: Probleme über Veränderungen von Zeichenreihen nach gegebenen Regeln. Skr. Vidensk. Selsk. I, **10**, 34 pp. (1914).

POST, E. L.: Recursive Unsolvability of a Problem of Thue. J. symbolic Logic **12**, 1–11 (1947).

MARKOV, A. A.: The impossibility of certain algorithms in the theory of associative systems [Russ.]. Dokl. Akad. Nauk SSSR. **55**, 587–590 (1947).

TURING, A. M.: The Word Problem in Semi-Groups with Cancellation. Ann. Math., Princeton **52**, 491–505 (1950).

KALMÁR, L.: Another Proof of the Markov-Post Theorem. Acta math. Hungaricae **3**, 1–25; 26–27 [Russ.]; (1952).

On the unsolvability of the word problem, compare

NOVIKOV, P. S.: On the algorithmic unsolvability of the word problem in group theory [Russ.]. Akad. Nauk SSSR., Matém. Inst. Trudy **44**, Moscow 1955. Engl. transl. by K. A. Hirsch, Amer. Math. Soc. Translations **9** (1958), 122 pp.

BOONE, W. W.: Certain Simple, Unsolvable Problems of Group Theory. V. Proc. Kon. Nederl. Akad. (A) **60**, 22–27 (1957).

— Certain Simple, Unsolvable Problems of Group Theory. VI. Ibid., pp. 227–232.

— The Word Problem. Ann. Math. **70**, 207–265 (1959).

BRITTON, J. L.: The Word Problem for Groups. Proc. London math. Soc. **8**, 493 to 506 (1958).

§ 24. The Predicate Calculus

Among the formal languages known today the language of the *predicate calculus* (predicate calculus of the first order, restricted predicate calculus) must be considered to be the most important. For this language certain lines of signs are selected to be *formulae* (expressions). It is decidable whether or not a line of signs is a formula. Formulae can be *interpreted* and it is defined what it means that a formula is *valid* under an interpretation. A formula is called *valid* if it is valid under *every* interpretation. This kind of definition of valid formulae belongs to the *semantics* of the predicate calculus. The valid formulae are the logical consequences which belong to the domain of the predicate calculus.

We can give an account (in different ways) of a calculus for which the class of derivable formulae coincides with the class of valid formulae of the predicate calculus. This fact is called the *completeness theorem* of the predicate calculus (GÖDEL 1930).[1] The completeness theorem says

[1] The name "completeness theorem" can be explained by saying that this theorem asserts the existence of a calculus which completely comprehends the

that the set $\mathfrak{M}_0$ of the valid formulae is an *enumerable set.* CHURCH showed in 1936 that the set of the valid formulae is *not decidable.* This, together with Gödel's completeness theorem, shows that the set of *not* valid formulae cannot be enumerable, as we shall see in Theorem 4 of § 28.2.

In the next paragraph we shall give a proof of the undecidability of the set of valid formulae of the predicate calculus. In preparation for this we shall in this paragraph build up the predicate calculus as far as it is necessary for the proof. Later on we shall only refer back to Theorems (T 1), ..., (T 8) of Section 3, (T 9) of Section 4, and (T 10) of Section 5.

The predicate calculus is formed in the literature concerned in very different ways. If we start from a different structure than that given here, then we only need to verify the analogues of the Theorems (T 1), ..., (T 10), and by that we have an access to the proof of undecidability carried out in this book (cf. especially the remark in Section 3).

1. Formulae of the predicate calculus. The formulae are special words over the alphabet {o, O, I, ∗, ¬, ∧, ⋀, (,)}. The words oI, oII, oIII, ... are called *individual variables*, the words O∗I, O∗II, O∗III, ..., O∗∗I, O∗∗II, O∗∗III, ..., O∗∗∗I, ..., ... are called *predicate variables*[1]. The number of stars occurring in a predicate variable Π is called the *number of places* of Π. Thus, the number of places of O∗∗III is 2.

A word α is called an *atomic formula* if it starts with a predicate variable Π followed by as many individual variables as determined by the number of places of Π. We shall say that these individual variables *occur* in α. O∗IIoI and O∗∗∗IoIoIIIoI are examples of atomic formulae. oI and oIII (but not oII) occur in the second example.

Formulae are all the atomic formulae and all such words which can be obtained from the atomic formulae by one or more applications of the following processes.

(i) Proceed from α to $\neg\alpha$.

(ii) Proceed from α and β to $(\alpha\wedge\beta)$.

(iii) Proceed from α to $\bigwedge\xi\alpha$, where ξ is an arbitrary individual variable.

Examples of formulae are

¬¬O∗∗IoIIIoI, ⋀oIO∗∗IoIIoI and ⋀oI¬⋀oII(O∗IoII∧¬O∗∗IoIIIoI).

---

valid formulae. — In relation to the coming remarks cf. also the references at the end of the next paragraph.

[1] The latter are sometimes called predicate *constants*. Similarly we sometimes speak of individual *constants* when speaking of individual variables which, according to the terminology introduced in Section 4, occur *free* in a formula.

Instead of $\bigwedge \xi \alpha$ we write also $\bigwedge\limits_{\xi} \alpha$, instead of $\neg \bigwedge\limits_{\xi} \neg \alpha$ we write in short $\bigvee\limits_{\xi} \alpha$, and instead of $\neg(\alpha \wedge \neg \beta)$ we write in short $(\alpha \rightarrow \beta)$.

2. *Semantics of the predicate calculus.* A *domain* ω *of individuals* is an arbitrary non-empty set. The elements of ω are called *individuals.* An (empty or non-empty) set of ordered n-tuples of elements of ω is called an *n-ary predicate* over ω. An interpretation $\mathfrak{J}$ over ω is a mapping of the individual variables ξ onto the individuals $\mathfrak{J}(\xi)$ of ω and the predicate variables Π onto the predicates $\mathfrak{J}(\Pi)$ over ω, where we require that $\mathfrak{J}(\Pi)$ is a set of n-tuples if the number of places of Π is n.

We shall explain what it means that a formula α is *valid* under an interpretation $\mathfrak{J}$. For this we give a definition by *induction on the structure of the formulae.* In the definition below $\mathfrak{J} \underset{\xi}{=} \mathfrak{J}'$ (read $\mathfrak{J} = \mathfrak{J}'$ up to ξ) shall mean that $\mathfrak{J}$ and $\mathfrak{J}'$ are interpretations over the same domain of individuals and that $\mathfrak{J}$ and $\mathfrak{J}'$ differ at most for the argument ξ.

Definition 1.

(a) An atomic formula $\Pi \xi_1 \ldots \xi_n$ *is valid under* $\mathfrak{J}$ if and only if the n-tuple $\langle \mathfrak{J}(\xi_1), \ldots, \mathfrak{J}(\xi_2) \rangle \in \mathfrak{J}(\Pi)$.

(b) $\neg \alpha$ *is valid under* $\mathfrak{J}$ if and only if α is not valid under $\mathfrak{J}$.

(c) $(\alpha \wedge \beta)$ *is valid under* $\mathfrak{J}$ if and only if both α and β are valid under $\mathfrak{J}$.

(d) $\bigwedge\limits_{\xi} \alpha$ *is valid under* $\mathfrak{J}$ if and only if α is valid under every $\mathfrak{J}'$ for which $\mathfrak{J} \underset{\xi}{=} \mathfrak{J}'$.

$\neg$ corresponds according to (b) to the negation, $\wedge$ according to (c) to the conjunction, and $\bigwedge$ according to (d) to the generalization, where in the passage to the interpretation $\mathfrak{J}'$ we realize the phrase "for all x".

With respect to the abbreviations $\rightarrow$ and $\bigvee$, which were introduced by definition, we have (e) and (f):

(e) $(\alpha \rightarrow \beta)$ *is valid under* $\mathfrak{J}$ if and only if *whenever* α *is valid under* $\mathfrak{J}$, *then* β *is also valid under* $\mathfrak{J}$.

To *prove* this, we first assume that $(\alpha \rightarrow \beta)$ and α are valid (under $\mathfrak{J}$). We must show that so is β. If this were not the case, then $\neg \beta$ would be valid under $\mathfrak{J}$ and with it also $(\alpha \wedge \neg \beta)$. This would contradict the assumption that $\neg(\alpha \wedge \neg \beta)$ (i.e. $(\alpha \rightarrow \beta)$) is valid under $\mathfrak{J}$. Conversely, let us assume that β is valid under $\mathfrak{J}$ whenever α is valid under $\mathfrak{J}$. We must show that $(\alpha \rightarrow \beta)$ is also valid under $\mathfrak{J}$. If this were not the case, then $\neg(\alpha \wedge \neg \beta)$ would not be valid under $\mathfrak{J}$. Then $(\alpha \wedge \neg \beta)$ would have to be valid under $\mathfrak{J}$ and with it α and $\neg \beta$ as well. This would mean that β is not valid under $\mathfrak{J}$, in contradiction to our assumption that the validity of α under $\mathfrak{J}$ implies the validity of β under $\mathfrak{J}$.

(f) $\bigvee_{\xi} \alpha$ *is valid under* $\mathfrak{J}$ if and only if *there exists an* $\mathfrak{J}'$, *such that* $\mathfrak{J}' \underset{\xi}{=} \mathfrak{J}$, *under which* α *is valid.*

Proof. $\neg \bigwedge_{\xi} \neg \alpha$ is valid under $\mathfrak{J}$ if and only if $\bigwedge_{\xi} \neg \alpha$ is not valid under $\mathfrak{J}$, i.e. if and only if it is not true that $\neg \alpha$ is valid under all $\mathfrak{J}'$ such that $\mathfrak{J}' \underset{\xi}{=} \mathfrak{J}$, in other words, if and only if it is not true that α is not valid under all $\mathfrak{J}'$ such that $\mathfrak{J}' \underset{\xi}{=} \mathfrak{J}$. This says that there exists an $\mathfrak{J}'$, such that $\mathfrak{J}' \underset{\xi}{=} \mathfrak{J}$, under which α is valid.

Definition 1'. A set $\mathfrak{M}$ *of formulae is valid under* $\mathfrak{J}$ if and only if every element of $\mathfrak{M}$ is valid under $\mathfrak{J}$.

Examples. Let ω be the set of natural numbers, M the set of prime numbers and K the set of all ordered pairs $\langle \mathfrak{x}, \mathfrak{y} \rangle$ of natural numbers such that $\mathfrak{x} < \mathfrak{y}$. Let $x = \text{o|}$, $y = \text{o||}$, $P = \text{O}{*}\text{|}$, $Q = \text{O}{*}{*}\text{|}$. Let $\mathfrak{J}$ be an interpretation over ω for which we have, among others, that $\mathfrak{J}(x) = 3$, $\mathfrak{J}(y) = 4$, $\mathfrak{J}(P) = M$, $\mathfrak{J}(Q) = K$. For such an $\mathfrak{J}$ all the formulae Px, $\neg Py$ and $\bigwedge_{x} (Px \to \bigvee_{y} (Py \wedge Qxy))$ are valid, since 3 is a prime number, 4 is not a prime number, and for every prime number there exists another which is larger.

Definition 2. A formula α is *valid,* in symbols $\Vdash \alpha$, if it is valid under every interpretation over an arbitrary domain of individuals.

Definition 3. A formula α *follows* from a *finite* set $\mathfrak{M}$ of formulae, in symbols $\mathfrak{M} \Vdash \alpha$, if for every domain ω of individuals α is valid under every interpretation $\mathfrak{J}$ over ω under which every formula of $\mathfrak{M}$ is valid.

3. Simple consequences[1]. (T 1), ..., (T 7) of this Section, just like (T 8) of Section 4 and (T 9) of Section 5 have been chosen in a way to make applications later on as easy as possible. It was not considered important to obtain the simplest possible system of such theorems.

(T 1) $0 \Vdash \alpha$ *if and only if* $\Vdash \alpha$ (where 0 is the empty set of formulae).

(T 2) *If* $\alpha \in \mathfrak{M}$, *then* $\mathfrak{M} \Vdash \alpha$.

(T 3) *If* $\mathfrak{M} \Vdash \alpha$, *then* $\mathfrak{M} \cup \mathfrak{N} \Vdash \alpha$.

(T 4) *If* $\alpha \Vdash \beta$ *and* $\beta \Vdash \gamma$, *then* $\alpha \Vdash \gamma$.

(T 5) *If* $\mathfrak{M} \Vdash \alpha$, and $\mathfrak{N} \Vdash \beta$, then $\mathfrak{M} \cup \mathfrak{N} \Vdash (\alpha \wedge \beta)$.

---

[1] We can forgo the requirement that $\mathfrak{M}$ is finite in Definition 3. The theorems (T 1), ..., (T 10) are also valid (with the same proofs) in the case when $\mathfrak{M}$ is infinite. However, we do not need these generalizations. The relation $\mathfrak{M} \Vdash \alpha$ can also be defined by $\Vdash (\alpha_1 \to (\alpha_2 \to \dots (\alpha_n \to \alpha) \dots))$, if $\mathfrak{M} = \{\alpha_1, \dots, \alpha_n\}$, and by $\Vdash \alpha$, if $\mathfrak{M}$ is empty. This remark is of interest for instance in the case when the valid formulae are introduced *syntactically* as the formulae which can be obtained by a suitable calculus.

(T 6) *If* $\mathfrak{M} \Vdash (\alpha \to \beta)$ *and* $\mathfrak{N} \Vdash \alpha$, *then* $\mathfrak{M} \cup \mathfrak{N} \Vdash \beta$.

(T 7) $\mathfrak{M} \Vdash (\alpha \to \beta)$ *if and only if* $\mathfrak{M} \cup \{\alpha\} \Vdash \beta$.

(T 8) $\bigwedge_{\xi} \alpha \Vdash \alpha$.

We write in short $\alpha \Vdash \beta$ for $\{\alpha\} \Vdash \beta$.

(T 1), ..., (T 4) are immediate consequences of the definitions. To prove (T 5) we have to show that $(\alpha \wedge \beta)$ is valid under every interpretation $\mathfrak{J}$ under which $\mathfrak{M} \cup \mathfrak{N}$ is valid. Because $\mathfrak{M} \Vdash \alpha$, α is valid under $\mathfrak{J}$; because $\mathfrak{N} \Vdash \beta$, β is also valid under $\mathfrak{J}$. Thus, by Definition 1 (c) $(\alpha \wedge \beta)$ is valid under $\mathfrak{J}$.

For (T 8) we have to prove that α is valid under every interpretation $\mathfrak{J}$ under which $\bigwedge_{\xi} \alpha$ is valid. According to Definition 1 (d) α is valid under every $\mathfrak{J}'$ such that $\mathfrak{J}' \underset{\xi}{=} \mathfrak{J}$. Thus especially, α is valid under $\mathfrak{J}$.

To prove (T 6) we assume that $\mathfrak{M} \cup \mathfrak{N}$ is valid under $\mathfrak{J}$. Then because $\mathfrak{N} \Vdash \alpha$, α is also valid under $\mathfrak{J}$. Also $(\alpha \to \beta)$, because $\mathfrak{M} \Vdash (\alpha \to \beta)$. This gives in view of (e) that β is valid under $\mathfrak{J}$.

The proof from left to right of (T 7) is obtained from (T 6) for $\mathfrak{N} = \{\alpha\}$, since $\{\alpha\} \Vdash \alpha$ by (T 2). — Conversely, let us assume that $\mathfrak{M} \cup \{\alpha\} \Vdash \beta$ and that $\mathfrak{M}$ is valid under $\mathfrak{J}$. Now, if α is valid under $\mathfrak{J}$, then so is β. But according to (e) this means that $(\alpha \to \beta)$ is valid under $\mathfrak{J}$.

4. *Free occurrence of an individual variable.* We shall explain what it means that an individual variable η *occurs free in a formula* α.

Definition 4.

(a) In an atomic formula α those and only those individual variables occur free which occur in α according to Section 1.

(b) η occurs free in $\neg\alpha$ if and only if η occurs free in α.

(c) η occurs free in $(\alpha \wedge \beta)$ if and only if η occurs free either in α or in β or in both.

(d) η occurs free in $\bigwedge_{\xi} \alpha$ if and only if η occurs free in α and is different from ξ.

Definition 4'. ξ *occurs free in a set* $\mathfrak{M}$ *of formulae if* ξ occurs free in at least one element of $\mathfrak{M}$.

Lemma 1. If the individual variable η does not occur free in the formula α and if $\mathfrak{J}_1 \underset{\eta}{=} \mathfrak{J}_2$, then α is valid under $\mathfrak{J}_1$ if and only if α is valid under $\mathfrak{J}_2$. In other words, the interpretation of the individual variable η does not play an essential role in this context.

Proof by induction on the structure of α. Let $\mathfrak{J}_1$ and $\mathfrak{J}_2$ be arbitrary interpretations.

(a) An atomic formula $\Pi \xi_1 \ldots \xi_n$ is valid under $\mathfrak{J}_1$ if and only if $\langle \mathfrak{J}_1(\xi_1), \ldots, \mathfrak{J}_1(\xi_n) \rangle \in \mathfrak{J}_1(\Pi)$. Because $\mathfrak{J}_1 \underset{\eta}{=} \mathfrak{J}_2$ and $\eta \neq \xi_1, \ldots, \xi_n$ we have that $\mathfrak{J}_1(\xi_1) = \mathfrak{J}_2(\xi_1), \ldots$ and $\mathfrak{J}_1(\Pi) = \mathfrak{J}_2(\Pi)$. Thus we can say that $\langle \mathfrak{J}_2(\xi_1), \ldots, \mathfrak{J}_2(\xi_n)\rangle \in \mathfrak{J}_2(\Pi)$, i.e. $\Pi \xi_1 \ldots \xi_n$ is valid under $\mathfrak{J}$.

(b) We shall show the Lemma for $\neg\alpha$ under the assumption that it is already proved for α. If η does not occur free in $\neg\alpha$, then η does not occur free in α either. Thus, we have the following assertions, each equivalent to the one preceding it:

$\neg\alpha$ is valid under $\mathfrak{J}_1$
α is not valid under $\mathfrak{J}_1$
α is not valid under $\mathfrak{J}_2$ (induction hypothesis)
$\neg\alpha$ is valid under $\mathfrak{J}_2$.

(c) The proof for the conjunction is carried out similarly to that for the negation.

(d) Finally, we have to show that the assertion is valid for $\bigwedge_\xi \alpha$ if it is valid for α. By symmetry it is sufficient to show that $\bigwedge_\xi \alpha$ is valid under $\mathfrak{J}_2$ if it is valid under $\mathfrak{J}_1$. For this we must prove that if $\bigwedge_\xi \alpha$ is valid under $\mathfrak{J}_1$ and if $\mathfrak{J}_2' \underset{\xi}{=} \mathfrak{J}_2$, then α is valid under $\mathfrak{J}_2'$. We introduce an interpretation $\mathfrak{J}_1'$ by the following definition.

$$\mathfrak{J}_1' \underset{\xi}{=} \mathfrak{J}_1$$
$$\mathfrak{J}_1'(\xi) = \mathfrak{J}_2'(\xi).$$

α is valid under $\mathfrak{J}_1'$ since $\bigwedge_\xi \alpha$ is valid under $\mathfrak{J}_1$. We compare $\mathfrak{J}_1'$ and $\mathfrak{J}_2'$. Because $\mathfrak{J}_1' \underset{\xi}{=} \mathfrak{J}_1$, $\mathfrak{J}_1 \underset{\eta}{=} \mathfrak{J}_2$ and $\mathfrak{J}_2 \underset{\xi}{=} \mathfrak{J}_2'$, $\mathfrak{J}_1'$ and $\mathfrak{J}_2'$ can differ at most for the arguments ξ and η. Now, according to definition, $\mathfrak{J}_1'(\xi) = \mathfrak{J}_2'(\xi)$ and so $\mathfrak{J}_1' \underset{\eta}{=} \mathfrak{J}_2'$.

Now we make use of the fact that η does not occur free in $\bigwedge_\xi \alpha$. According to Definition 4(d) we must distinguish between two cases.

Case 1. η does not occur free in α. Then we can apply the induction hypothesis because $\mathfrak{J}_1' \underset{\eta}{=} \mathfrak{J}_2'$ (the reader should keep in mind here that we are showing by induction that the Lemma is valid *for all interpretations* $\mathfrak{J}_1$ *and* $\mathfrak{J}_2$). Thus, we have that α is valid under $\mathfrak{J}_1'$ if and only if it is valid under $\mathfrak{J}_2'$. We know that α is valid under $\mathfrak{J}_1'$. Thus, α is valid under $\mathfrak{J}_2'$, q.e.d.

Case 2. η is identical with ξ. Then $\mathfrak{J}_1'$ and $\mathfrak{J}_2'$ are completely identical because $\mathfrak{J}_1' \underset{\eta}{=} \mathfrak{J}_2'$ and $\mathfrak{J}_1'(\xi) = \mathfrak{J}_2'(\xi)$. Thus, α is valid under $\mathfrak{J}_2'$ because α is valid under $\mathfrak{J}_1'$.

(T 9) *If* ξ *does not occur free in* $\mathfrak{M} \cup \{\beta\}$ *and if* $\mathfrak{M} \cup \{\alpha\} \Vdash \beta$, *then* $\mathfrak{M} \cup \{\bigvee_\xi \alpha\} \Vdash \beta$.

Proof. Let $\mathfrak{I}$ be an interpretation under which $\mathfrak{M} \cup \{\bigvee_\xi \alpha\}$ is valid. We have to show that β is also valid under $\mathfrak{I}$. Because of Section 2(f) there exists an $\mathfrak{I}'$, such that $\mathfrak{I}' \underset{\xi}{=} \mathfrak{I}$, under which α is valid. By Lemma 1 we have that $\mathfrak{M}$ is also valid under $\mathfrak{I}'$. Now, we can see from $\mathfrak{M} \cup \{\alpha\} \Vdash \beta$ that β is valid under $\mathfrak{I}'$ and from this, again by Lemma 1, that β is valid under $\mathfrak{I}$.

5. *Substitution.* If we replace in a formula α an individual variable η by an individual variable η' in all places in which η is not "in the scope of $\bigwedge_\eta$", and if none of the variables η' introduced in this way is "in the scope of $\bigwedge_{\eta'}$", then we shall say that the resulting formula α' is obtained from α by substitution of η' for η, in symbols *Sub* $\alpha\eta\eta'\alpha'$. We give a precise definition of substitution by induction on the structure of α.

Definition 5.

(a) If α is atomic, then *Sub* $\alpha\eta\eta'\alpha'$ if and only if α' is obtained from α by replacing η by η' at all places where the variable η occurs in α.

(b) *Sub* $\neg\alpha\eta\eta'\beta$ if and only if there exists a formula α' such that *Sub* $\alpha\eta\eta'\alpha'$ and $\beta = \neg\alpha'$.

(c) *Sub* $(\alpha\wedge\beta)\eta\eta'\gamma$ if and only if there exist formulae α' and β' such that *Sub* $\alpha\eta\eta'\alpha'$, *Sub* $\beta\eta\eta'\beta'$ and $\gamma = (\alpha'\wedge\beta')$.

(d) *Sub* $\bigwedge_\xi\alpha\eta\eta'\beta$ if and only if one of the following requirements is satisfied.

(1) η does not occur free in $\bigwedge_\xi\alpha$ and $\beta = \bigwedge_\xi\alpha$.

(2) η occurs free in $\bigwedge_\xi\alpha$, $\eta' \neq \xi$ and there exists a formula α' for which *Sub* $\alpha\eta\eta'\alpha'$ and $\beta = \bigwedge_\xi\alpha'$.

Definition 5'. Let $\mathfrak{M}$ and $\mathfrak{M}'$ be sets of formulae. *Sub* $\mathfrak{M}\eta\eta'\mathfrak{M}'$ means that for every element α of $\mathfrak{M}$ there exists an element α' of $\mathfrak{M}'$ such that *Sub* $\alpha\eta\eta'\alpha'$ and every element of $\mathfrak{M}'$ can be obtained in this way.

Lemma 2. Under the hypotheses

1) $$Sub\ \alpha_1\eta_1\eta_2\alpha_2$$

2) $$\mathfrak{I}_1 \underset{\eta_1}{=} \mathfrak{I}_2$$

3) $$\mathfrak{I}_1(\eta_1) = \mathfrak{I}_2(\eta_2)$$

α_1 is valid under $\mathfrak{I}_1$ if and only if α_2 is valid under $\mathfrak{I}_2$.

Proof. We show by induction on the structure of α_1 that Lemma 2 is valid for arbitrary interpretations $\mathfrak{I}_1$ and $\mathfrak{I}_2$. We leave the simple cases when α_1 is atomic or a negation or a conjunction to the reader and turn at once to the case when $\alpha_1 = \bigwedge_\xi\alpha_1'$. According to Definition 5 (d) we have to distinguish between two cases.

Case 1. η_1 does not occur free in $\bigwedge_\xi \alpha_1'$. In this case Lemma 2 follows directly from Lemma 1.

Case 2. η_1 occurs free in $\bigwedge_\xi \alpha_1'$, $\eta_2 \neq \xi$ and there exists a formula α_2' for which *Sub* $\alpha_1'\eta_1\eta_2\alpha_2'$ and $\alpha_2 = \bigwedge_\xi \alpha_2'$. We show the required equivalence in both directions.

(1) Let α_1 be valid under $\mathfrak{J}_1$. Let $\mathfrak{J}_2' \underset{\xi}{=} \mathfrak{J}_2$. We have to show that α_2' is valid under $\mathfrak{J}_2'$. For this purpose we define an interpretation $\mathfrak{J}_1'$ as follows.

$$\mathfrak{J}_1' \underset{\xi}{=} \mathfrak{J}_1$$
$$\mathfrak{J}_1'(\xi) = \mathfrak{J}_2'(\xi).$$

We want to compare $\mathfrak{J}_1'$ and $\mathfrak{J}_2'$. We have that $\mathfrak{J}_1' \underset{\xi}{=} \mathfrak{J}_1$, $\mathfrak{J}_1 \underset{\eta_1}{=} \mathfrak{J}_2$ and $\mathfrak{J}_2 \underset{\xi}{=} \mathfrak{J}_2'$. Thus, $\mathfrak{J}_1'$ and $\mathfrak{J}_2$ coincide up to the arguments ξ and η_1. Because $\mathfrak{J}_1'(\xi) = \mathfrak{J}_2'(\xi)$ we even have that $\mathfrak{J}_1' \underset{\eta_1}{=} \mathfrak{J}_2'$. $\mathfrak{J}_1'(\eta_1) = \mathfrak{J}_1(\eta_1)$, since $\mathfrak{J}_1' \underset{\xi}{=} \mathfrak{J}_1$ and $\xi \neq \eta_1$ (because η_1 occurs free in $\bigwedge_\xi \alpha_1'$). Furthermore $\mathfrak{J}_2'(\eta_2) = \mathfrak{J}_2(\eta_2)$ because $\mathfrak{J}_2' \underset{\xi}{=} \mathfrak{J}_2$ and $\xi \neq \eta_2$ (according to hypothesis). Because $\mathfrak{J}_1(\eta_1) = \mathfrak{J}_2(\eta_2)$ we have that $\mathfrak{J}_1'(\eta_1) = \mathfrak{J}_2'(\eta_2)$. This shows that the induction hypothesis is applicable to $\mathfrak{J}_1', \mathfrak{J}_2', \alpha_1', \eta_1, \eta_2, \alpha_2'$. Thus, α_1' is valid under $\mathfrak{J}_1'$ if and only if α_2' is valid under $\mathfrak{J}_2'$. Since we obviously have from $\mathfrak{J}_1' \underset{\xi}{=} \mathfrak{J}_1$ that α_1' is valid under $\mathfrak{J}_1'$, α_2' must be valid under $\mathfrak{J}_2'$, q.e.d.

(2) Let α_2 be valid under $\mathfrak{J}_2$. Let $\mathfrak{J}_1' \underset{\xi}{=} \mathfrak{J}_1$. We have to show that α_1' is valid under $\mathfrak{J}_1'$. For this purpose we introduce the interpretation $\mathfrak{J}_2'$ by the following stipulations:

$$\mathfrak{J}_2' \underset{\xi}{=} \mathfrak{J}_2$$
$$\mathfrak{J}_2'(\xi) = \mathfrak{J}_1'(\xi).$$

Now we have the same relations between $\mathfrak{J}_1, \mathfrak{J}_1', \mathfrak{J}_2, \mathfrak{J}_2'$ as in (1). Thus, we can again apply the induction hypothesis to $\mathfrak{J}_1', \mathfrak{J}_2', \alpha_1', \eta_1, \eta_2, \alpha_2'$, according to which α_1' is valid under $\mathfrak{J}_1'$ if and only if α_2' is valid under $\mathfrak{J}_2'$. α_2' is valid under $\mathfrak{J}_2'$. Thus, α_1' is valid under $\mathfrak{J}_1'$, q.e.d.

(T 10) *If Sub* $\mathfrak{M}_1\eta_1\eta_2\mathfrak{M}_2$, *Sub* $\alpha_1\eta_1\eta_2\alpha_2$ *and* $\mathfrak{M}_1 \Vdash \alpha_1$, *then* $\mathfrak{M}_2 \Vdash \alpha_2$.

To *prove* this we consider an arbitrary interpretation $\mathfrak{J}_2$ under which $\mathfrak{M}_2$ is valid. We have to show that α_2 is also valid under $\mathfrak{J}_2$. For this purpose we define an interpretation $\mathfrak{J}_1$ by the stipulations

$$\mathfrak{J}_1 \underset{\eta_1}{=} \mathfrak{J}_2$$
$$\mathfrak{J}_1(\eta_1) = \mathfrak{J}_2(\eta_2).$$

On the basis of the Lemma proved just now $\mathfrak{M}_1$ and α_1 are valid under $\mathfrak{J}_1$ if and only if $\mathfrak{M}_2$ and α_2 respectively are valid under $\mathfrak{J}_2$. Therefore we have step by step that $\mathfrak{M}_2$ is valid under $\mathfrak{J}_2$, $\mathfrak{M}_1$ is valid under $\mathfrak{J}_1$, α_1 is valid under $\mathfrak{J}_1$ (because $\mathfrak{M}_1 \Vdash \alpha_1$) and α_2 is valid under $\mathfrak{J}_2$, q.e.d.

References

GÖDEL, K.: Die Vollständigkeit der Axiome des logischen Funktionenkalküls. Mh. Math. Phys. **37**, 349–360 (1930).

TARSKI, A.: Der Wahrheitsbegriff in den formalisierten Sprachen. Studia Philosophica **1**, 261–405 (1935). Cf. also: Logic, Semantics, Metamathematics. Papers from 1923 to 1938 by A. TARSKI, Translated by J. H. WOODGER, Oxford: Clarendon Press 1956. pp. 152–278.

HERMES, H., and H. SCHOLZ: Mathematische Logik. Enzyklopädie math. Wiss. I. 1 Heft 1, I. Leipzig: B. G. Teubner 1952.

CHURCH, A.: Introduction to Mathematical Logic I. Princeton, N. J.: Princeton University Press 1956.

SCHOLZ, H., and G. HASENJAEGER: Grundzüge der mathematischen Logik. Berlin-Göttingen-Heidelberg: Springer 1961.

Compare other text-books on logic (they do not however always treat the subject from a semantical point of view).

§ 25. The Undecidability of the Predicate Calculus

We shall show in this paragraph that there exists no algorithm by the help of which we can decide whether or not an arbitrarily given formula of the predicate calculus is valid. From this follows a fortiori, in view of § 24, Section 3 (T 1), that there exists no algorithm by the help of which we can decide, for an arbitrarily given finite set $\mathfrak{M}$ of formulae and an arbitrary formula α, whether or not α follows from $\mathfrak{M}$.

1. Sketch of the proof. We produce a relation between semi-Thue systems and the predicate calculus. We consider only those semi-Thue systems whose alphabet is contained in the denumerable alphabet $\{S_1, S_2, S_3, \ldots\}$, which is fixed once and for all. With every such semi-Thue system $\mathfrak{T}$ together with an ordered pair of words W', W'' over the alphabet of $\mathfrak{T}$ we associate a formula $\varphi(\mathfrak{T}, W', W'')$ of the predicate calculus so that the following stipulations are valid:

(1) For given $\mathfrak{T}, W'$ and W'', $\varphi(\mathfrak{T}, W', W'')$ can be produced by an effective procedure.

(2) $W' \rightarrow_{\mathfrak{T}} W''$ if and only if $\Vdash \varphi(\mathfrak{T}, W', W'')$.

If the predicate calculus were decidable, then we could determine in finitely many steps on the basis of (1) and (2) for an arbitrary semi-Thue system $\mathfrak{T}_0$ and arbitrary words W_0' and W_0'' over the alphabet of $\mathfrak{T}_0$

whether or not $W_0' \to_{\mathfrak{T}_0} W_0''$. This can be shown as follows. We associate the letters $A_1, \ldots, A_n$ of the alphabet $\{A_1, \ldots, A_n\}$ of $\mathfrak{T}_0$ with the letters $S_1, \ldots, S_n$ respectively. This provides us with an obvious translation of every word over the alphabet of $\mathfrak{T}_0$ into a word over $\{S_1, \ldots, S_n\}$. By translating the defining relations of $\mathfrak{T}_0$ we obtain a semi-Thue system $\mathfrak{T}$ over the alphabet $\{S_1, \ldots, S_N\}$. Let W' and W'' be the translations of W_0' and W_0'' respectively. Obviously, $W_0' \to_{\mathfrak{T}_0} W_0''$ if and only if $W' \to_{\mathfrak{T}} W''$. But this is equivalent, according to the assertion above, to $\Vdash \varphi(\mathfrak{T}, W', W'')$, which is decidable if the predicate calculus is decidable. Thus, the decidability of the predicate calculus would imply the solvability of the general word problem for semi-Thue systems. Since we have seen in § 23 that the general word problem for semi-Thue systems is unsolvable we have the

Theorem (*undecidability of the predicate calculus*). *There exists no general procedure by the help of which we can determine in finitely many steps, for any given formula of the predicate calculus, whether or not the formula is valid.*

2. *Definitions. A Lemma.* Our task is to give an account of a formula $\varphi(\mathfrak{T}, W', W'')$ which satisfies the requirements (1) and (2) of Section 1. For this purpose we first associate with every word W over $\{S_1, S_2, S_3, \ldots\}$ a Gödel number $g(W)$ by the following instruction.

$$g(\square) = 1 \quad \text{for the empty word } \square$$
$$g(S_{i_0} \ldots S_{i_r}) = p_0^{i_0} \ldots p_r^{i_r}.$$

If we know W, then we can calculate $g(W)$. Conversely, we can find W if $g(W)$ is known.

We associate with every word W the individual variable which contains $g(W)$ strokes. (This is a one-one mapping; cf. § 24.1.) We shall denote this individual variable in short by x_W.

Not every individual variable is associated in this way with a word. Examples are the individual variables with 3^j strokes ($j = 1, 2, 3, \ldots$) and the individual variables with a prime number 5, 7, 11, ... of strokes. We shall denote the individual variables with 3^j strokes by s_j ($j = 1, 2, 3, \ldots$) and the individual variables with 5, 7, 11, 13, 17, 19 strokes by x, y, z, u, v, w respectively.

Furthermore we use the abbreviations T for the binary predicate variable $\bigcirc{*}{*}|$ and C for the ternary predicate variable $\bigcirc{*}{*}{*}|$.

Now we associate with every word W a finite set $\mathfrak{D}_W$ of formulae by the following inductive definition.

$$\mathfrak{D}_\square = 0 \quad \text{(empty set of formulae)}$$
$$\mathfrak{D}_{WS_j} = \mathfrak{D}_W \cup \{C x_W s_j x_{WS_j}\}.$$

Remark. From the construction of $\mathfrak{D}_W$ directly follows that if a variable x_{US_j} occurs in an element of $\mathfrak{D}_W$, then $Cx_U s_j x_{US_j}$ is an element of $\mathfrak{D}_W$.

Let $\mathfrak{T}$ be a semi-Thue system over an alphabet contained in $\{S_1, S_2, S_3, \ldots\}$ with the relations

$$(L_k, R_k) \qquad (k = 1, \ldots, m).$$

We associate with this semi-Thue system $\mathfrak{T}$ a *finite set* $\mathfrak{A}_\mathfrak{T}$ *of formulae which consists of the following formulae:*

A 1: $\bigwedge_x \bigwedge_y \bigvee_z Cxyz$

A 2: $\bigwedge_x \bigwedge_y \bigwedge_z \bigwedge_u \bigwedge_v \bigwedge_w (((Cxyz \wedge Cyuv) \wedge Czuw) \rightarrow Cxvw)$

A 3: $\bigwedge_x Cxx_\square x$

A 4: $\bigwedge_x Txx$

A 5: $\bigwedge_x \bigwedge_y \bigwedge_z ((Txy \wedge Tyz) \rightarrow Txz)$

A 6_k: $\bigwedge_x \bigwedge_y \bigwedge_z \bigwedge_u \bigwedge_v \bigwedge_w ((((Cux_{L_k}w \wedge Cwvx) \wedge Cux_{R_k}z) \wedge Czvy) \rightarrow Txy)$ $(k = 1, \ldots, m)$

A 7_k: *all elements of* $\mathfrak{D}_{L_k}$ $(k = 1, \ldots, m)$

A 8_k: *all elements of* $\mathfrak{D}_{R_k}$ $(k = 1, \ldots, m)$.

The given formulae describe in a certain sense the semi-Thue system $\mathfrak{T}$. This follows from the fact that these formulae are valid under the interpretation $\mathfrak{I}$ described in Section 3 as we shall show in Section 4.

We have to keep in mind for applications later on that the variables x, y, z, u, v, w, *do not occur free in* $\mathfrak{A}_\mathfrak{T}$.

In Sections 3 and 4 we shall give a proof of the following

Lemma. For any semi-Thue system $\mathfrak{T}$ *and for any two words* W' *and* W'' *over the alphabet of* $\mathfrak{T}$.

$$\mathfrak{A}_\mathfrak{T} \cup \mathfrak{D}_{W'} \cup \mathfrak{D}_{W''} \Vdash Tx_{W'}x_{W''} \quad \textit{if and only if} \quad W' \rightarrow_\mathfrak{T} W''.$$

We can easily show that (1) and (2) of Section 1 follow from this Lemma.

For, looking at (T 7) of § 24.2 we see that, starting from $\mathfrak{A}_\mathfrak{T} \cup \mathfrak{D}_{W'} \cup \mathfrak{D}_{W''} \Vdash Tx_{W'}x_{W''}$, we can successively carry over all elements of the *finite* set $\mathfrak{A}_\mathfrak{T} \cup \mathfrak{D}_{W'} \cup \mathfrak{D}_{W''}$ to the right until we have the empty set 0 on the left of $\Vdash$. If we fix the order in which these elements are carried over, then, at the end of the procedure, we shall obtain on the right of $\Vdash$ an unambiguously determined formula which we shall call $\varphi(\mathfrak{T}, W', W'')$. We have given an account how $\varphi(\mathfrak{T}, W', W'')$ is produced. Therefore, (1) is valid.

$\mathfrak{A}_{\mathfrak{T}} \cup \mathfrak{D}_{W'} \cup \mathfrak{D}_{W''} \Vdash Tx_{W'}x_{W''}$ is, according to (T 7), equivalent to $0 \Vdash \varphi(\mathfrak{T}, W', W'')$. This, is equivalent to $\Vdash \varphi(\mathfrak{T}, W', W'')$, according to (T 1). Thus, (2) follows from the Lemma.

3. Proof of the Lemma, part 1. We shall show here that $\mathfrak{A}_{\mathfrak{T}} \cup \mathfrak{D}_{W'} \cup \mathfrak{D}_{W''} \Vdash Tx_{W'}x_{W''}$ implies that $W' \to_{\mathfrak{T}} W''$, provided that W' and W'' are words over the semi-Thue system $\mathfrak{T}$. For this purpose we shall give an account of an interpretation $\mathfrak{J}$ under which every formula of $\mathfrak{A}_{\mathfrak{T}} \cup \mathfrak{D}_{W'} \cup \mathfrak{D}_{W''}$ is valid, and so $Tx_{W'}x_{W''}$ is also valid, for $\mathfrak{A}_{\mathfrak{T}} \cup \mathfrak{D}_{W'} \cup \mathfrak{D}_{W''} \Vdash Tx_{W'}x_{W''}$. $\mathfrak{J}$ is chosen so that the validity of $Tx_{W'}x_{W''}$ is equivalent to $W' \to_{\mathfrak{T}} W''$.

We take the non-empty set of all words over the alphabet of $\mathfrak{T}$ to be the domain ω of individuals. We choose $\mathfrak{J}$ such that

$\mathfrak{J}(x_W) = W$ for every word over the alphabet of $\mathfrak{T}$,

$\mathfrak{J}(s_j) = S_j$ for every symbol S_j of the alphabet of $\mathfrak{T}$,

$\mathfrak{J}(C) =$ the set of all ordered triplets $\langle W_1, W_2, W_3\rangle$ of words with $W_1W_2 = W_3$[1],

$\mathfrak{J}(T) =$ the set of all ordered pairs $\langle W_1, W_2\rangle$ of words with $W_1 \to_{\mathfrak{T}} W_2$.

The interpretations of the variables not given here are not relevant. They can be defined in any way whatsoever. $\mathfrak{J}(C)$ is the juxtaposition relation and $\mathfrak{J}(T)$ is the transformation relation (consequence relation) in $\mathfrak{T}$.

If W' and W'' are words over the alphabet of $\mathfrak{T}$, then the validity of $Tx_{W'}x_{W''}$ under $\mathfrak{J}$ obviously means that $W' \to_{\mathfrak{T}} W''$. Thus, we only need to show that every element of $\mathfrak{A}_{\mathfrak{T}} \cup \mathfrak{D}_{W'} \cup \mathfrak{D}_{W''}$ is valid under $\mathfrak{J}$, provided W' and W'' are words over the alphabet of $\mathfrak{T}$.

A 1 is valid under $\mathfrak{J}$. To prove this we have to show, according to § 24, Section 2(d), that $\bigvee_z Cxyz$ is valid under every $\mathfrak{J}^*$ which differs from $\mathfrak{J}$ at most for the variables x and y. According to § 24 Section 2(f) this means that $Cxyz$ is valid under at least one interpretation $\mathfrak{J}^{*\prime}$ which may differ from $\mathfrak{J}^*$ for z. Let $\mathfrak{J}^*(x) = X$ and $\mathfrak{J}^*(y) = Y$. If we put $\mathfrak{J}^{*\prime} \underset{z}{=} \mathfrak{J}^*$ and $\mathfrak{J}^{*\prime}(z) = XY$, then we immediately see that $Cxyz$ is valid under $\mathfrak{J}^{*\prime}$.

In order to show that A 2 is valid under $\mathfrak{J}$ we must prove, in view of § 24 Section 2(e), that for every interpretation $\mathfrak{J}^*$, which differs from $\mathfrak{J}$

[1] Here it does not matter whether we confine ourselves to words over the alphabet $\mathfrak{T}$ or whether we allow all words over $\{S_1, S_2, S_3, \ldots\}$.

at most for the arguments x, y, z, u, v, w, $Cxvw$ is valid provided that $((Cxyz \wedge Cyuv) \wedge Czuw)$ is valid. If we temporarily put $\mathfrak{J}^*(x) = X$, $\mathfrak{J}^*(y) = Y$, $\mathfrak{J}^*(z) = Z$, $\mathfrak{J}^*(u) = U$, $\mathfrak{J}^*(v) = V$, $\mathfrak{J}^*(w) = W$, then we have to show that if $XY = Z$, $YU = V$ and $ZU = W$, then $XV = W$. In fact, we have in view of the *associative law* for the juxtaposition relation (§ 23.1) that $XV = X(YU) = (XY)U = ZU = W$.

A 3 is valid under $\mathfrak{J}$. In order to show this we must prove that $Cxx_\square x$ is valid under every interpretation $\mathfrak{J}'$ for which $\mathfrak{J}' \underset{x}{=} \mathfrak{J}$ is valid. This means that $\mathfrak{J}'(x)\,\mathfrak{J}'(x_\square) = \mathfrak{J}'(x)$. This is indeed the case, since $\mathfrak{J}'(x_\square) = \mathfrak{J}(x_\square) = \square$.

That A 4 is valid under $\mathfrak{J}$ means that Txx is valid under every $\mathfrak{J}'$ for which $\mathfrak{J}' \underset{x}{=} \mathfrak{J}$ is valid. $\mathfrak{J}'(x)$ is a word over the alphabet of $\mathfrak{T}$ and every such word is transformable into itself.

Similarly we can show that A 5 is valid under $\mathfrak{J}$ in view of the transitive law for the relation $\rightarrow_{\mathfrak{T}}$.

Further A 6_k is valid under $\mathfrak{J}$ for every k. In order to prove this we must show that for every $\mathfrak{J}^*$ which differs from $\mathfrak{J}$ at most for the arguments x, y, z, u, v, w, Txy is valid provided that $(((Cux_{L_k}w \wedge Cwvx) \wedge Cux_{R_k}z) \wedge Czvy)$ is valid. If we again put temporarily $\mathfrak{J}^*(x) = X, \ldots, \mathfrak{J}^*(w) = W$, then, because $\mathfrak{J}^*(x_{L_k}) = \mathfrak{J}(x_{L_k}) = L_k$ and $\mathfrak{J}^*(x_{R_k}) = \mathfrak{J}(x_{R_k}) = R_k$, we must show that $X \rightarrow_{\mathfrak{T}} Y$, provided $UL_k = W$, $WV = X$, $UR_k = Z$ and $ZV = Y$. But this is true, since $X = UL_kV$ and $Y = UR_kV$ and, therefore, $X \Rightarrow_{\mathfrak{T}} Y$, since (L_k, R_k) is a defining relation of $\mathfrak{T}$.

Finally, we have to show that every element of $\mathfrak{D}_{L_k} \cup \mathfrak{D}_{R_k} \cup \mathfrak{D}_{W'} \cup \mathfrak{D}_{W''}$ is valid under $\mathfrak{J}$. To do this it is sufficient to show that $\mathfrak{D}_W$ is valid under $\mathfrak{J}$ for every word W over the alphabet of $\mathfrak{T}$. Looking at the definition of $\mathfrak{D}_W$ we see that it is sufficient to show the validity of all formulae $Cx_W s_j x_{WS_j}$, where W is a word and S_j is an element of the alphabet of $\mathfrak{T}$. In fact, $\mathfrak{J}(x_W)\,\mathfrak{J}(s_j) = WS_j = \mathfrak{J}(x_{WS_j})$, which, in view of the definition of C, is equivalent to the validity of $Cx_W s_j x_{WS_j}$.

4. Proof of the Lemma, part 2. It remains to be shown that

$$\mathfrak{A}_{\mathfrak{T}} \cup \mathfrak{D}_{W'} \cup \mathfrak{D}_{W''} \Vdash Tx_{W'}x_{W''}, \quad \text{if } W' \rightarrow_{\mathfrak{T}} W''.$$

We shall use here the Theorems (T 1), ..., (T 10) of § 24.

We begin with a

Proposition. $\mathfrak{A}_{\mathfrak{T}} \cup \mathfrak{D}_V \cup \mathfrak{D}_{UV} \Vdash Cx_U x_V x_{UV}$ for arbitrary words U and V.

Proof by induction on the structure of V. For $V = \square$ $UV = U$, so that we have to show that

$$(*) \qquad \mathfrak{A}_{\mathfrak{T}} \cup \mathfrak{D}_\square \cup \mathfrak{D}_U \Vdash Cx_U x_\square x_U.$$

According to (T 8) we have that $\bigwedge_x Cxx_\square x \Vdash Cxx_\square x$, from which follows, according to (T 10), that $\bigwedge_x Cxx_\square x \Vdash Cx_U x_\square x_U$.[1]

This gives (*) because of (T 3), since $\{A\,3\} < \mathfrak{A}_\mathfrak{T} \cup \mathfrak{D}_\square \cup \mathfrak{D}_U$.

Now, we must show under the induction hypothesis

1) $$\mathfrak{A}_\mathfrak{T} \cup \mathfrak{D}_V \cup \mathfrak{D}_{UV} \Vdash Cx_U x_V x_{UV}$$

that

(**) $$\mathfrak{A}_\mathfrak{T} \cup \mathfrak{D}_{Vs_j} \cup \mathfrak{D}_{UVs_j} \Vdash Cx_U x_{Vs_j} x_{UVs_j} \qquad \text{for } j = 1, 2, 3, \ldots.$$

If we observe that $Cx_V s_j x_{Vs_j} \in \mathfrak{D}_{Vs_j}$ and $Cx_{UV} s_j x_{UVs_j} \in \mathfrak{D}_{UVs_j}$, then we have, because of (T 2), that

$$\mathfrak{D}_{Vs_j} \Vdash Cx_V s_j x_{Vs_j} \qquad \text{and} \qquad \mathfrak{D}_{UVs_j} \Vdash Cx_{UV} s_j x_{UVs_j}.$$

These, together with 1), give according to (T 5) and (T 3) (if we note that $\mathfrak{D}_V < \mathfrak{D}_{Vs_j}$ and $\mathfrak{D}_{UV} < \mathfrak{D}_{UVs_j}$) that

2) $$\mathfrak{A}_\mathfrak{T} \cup \mathfrak{D}_{Vs_j} \cup \mathfrak{D}_{UVs_j} \Vdash ((Cx_U x_V x_{UV} \wedge Cx_V s_j x_{Vs_j}) \wedge Cx_{UV} s_j x_{UVs_j}).$$

According to (T 8) and (T 4) $A\,2 \Vdash (((Cxyz \wedge Cyuv) \wedge Czuw) \to Cxvw)$, from which follows, using (T 10), that

3) $$A\,2 \Vdash (((Cx_U x_V x_{UV} \wedge Cx_V s_j x_{Vs_j}) \wedge Cx_{UV} s_j x_{UVs_j}) \to Cx_U x_{Vs_j} x_{UVs_j}).$$

From 3) and 2) we obtain (**) by the help of (T 6), which completes the proof of the proposition.

Now, we give the proof which shows that the assertion $\mathfrak{A}_\mathfrak{T} \cup \mathfrak{D}_{W'} \cup \mathfrak{D}_{W''} \Vdash Tx_{W'} x_{W''}$ follows from $W' \to_\mathfrak{T} W''$. It is obviously sufficient to show by induction on n that

(†) *If* $W_1 \Rightarrow_\mathfrak{T} W_2 \Rightarrow_\mathfrak{T} W_3 \Rightarrow_\mathfrak{T} \cdots \Rightarrow_\mathfrak{T} W_n$, *then* $\mathfrak{A}_\mathfrak{T} \cup \mathfrak{D}_{W_1} \cup \mathfrak{D}_{W_n} \Vdash Tx_{W_1} x_{W_n}$.

For $n = 1$ we must show that $\mathfrak{A}_\mathfrak{T} \cup \mathfrak{D}_{W_1} \Vdash Tx_{W_1} x_{W_1}$. According to (T 8) $\bigwedge_x Txx \Vdash Txx$. From this follows by an application of (T 10) that $\bigwedge_x Txx \Vdash Tx_{W_1} x_{W_1}$. In view of (T 3) this is sufficient, since $\bigwedge_x Txx$ is an element of $\mathfrak{A}_\mathfrak{T}$.

For the induction step let us assume that (†) is proved. Further we assume that $W_1 \Rightarrow_\mathfrak{T} \cdots \Rightarrow_\mathfrak{T} W_n \Rightarrow_\mathfrak{T} W_{n+1}$, and we have to show that

(††) $$\mathfrak{A}_\mathfrak{T} \cup \mathfrak{D}_{W_1} \cup \mathfrak{D}_{W_{n+1}} \Vdash Tx_{W_1} x_{W_{n+1}}$$

[1] We must keep in mind here that

$$\mathit{Sub} \bigwedge_x Cxx_\square x \; x \; x_U \; \bigwedge_x Cxx_\square x \qquad \text{and} \qquad \mathit{Sub}\; Cxx_\square x \; x \; x_U \;\; Cx_U x_\square x_U.$$

Similar considerations must be made in all later applications of (T 10).

According to the induction hypothesis we have that

4) $$\mathfrak{A}_{\mathfrak{T}} \cup \mathfrak{D}_{W_1} \cup \mathfrak{D}_{W_n} \Vdash T x_{W_1} x_{W_n}.$$

$W_n \Rightarrow_{\mathfrak{T}} W_{n+1}$ means that there exists a k, a U and a V such that

5) $$W_n = U L_k V \qquad \text{and} \qquad W_{n+1} = U R_k V.$$

By the help of (T 8), (T 4) and (T 10) we obtain that

6) $$A\,6_k \Vdash ((((C x_U x_{L_k} x_{UL_k} \wedge C x_{UL_k} x_V x_{UL_kV}) \wedge C x_U x_{R_k} x_{UR_k}) \wedge C x_{UR_k} x_V x_{UR_kV}) \rightarrow T x_{UL_kV} x_{UR_kV}).$$

Now we apply the Proposition, which we have just proved, four times:

$$\mathfrak{A}_{\mathfrak{T}} \cup \mathfrak{D}_{L_k} \cup \mathfrak{D}_{UL_k} \Vdash C x_U x_{L_k} x_{UL_k}$$
$$\mathfrak{A}_{\mathfrak{T}} \cup \mathfrak{D}_{V} \cup \mathfrak{D}_{UL_kV} \Vdash C x_{UL_k} x_V x_{UL_kV}$$
$$\mathfrak{A}_{\mathfrak{T}} \cup \mathfrak{D}_{R_k} \cup \mathfrak{D}_{UR_k} \Vdash C x_U x_{R_k} x_{UR_k}$$
$$\mathfrak{A}_{\mathfrak{T}} \cup \mathfrak{D}_{V} \cup \mathfrak{D}_{UR_kV} \Vdash C x_{UR_k} x_V x_{UR_kV}.$$

Now, we observe that $\mathfrak{D}_{UL_k} < \mathfrak{D}_{UL_kV}$, $\mathfrak{D}_{UR_k} < \mathfrak{D}_{UR_kV}$, $\mathfrak{D}_{L_k} < \mathfrak{A}_{\mathfrak{T}}$ and also $\mathfrak{D}_{R_k} < \mathfrak{A}_{\mathfrak{T}}$. From this we obtain, using (T 5), that

$$\mathfrak{A}_{\mathfrak{T}} \cup \mathfrak{D}_{UL_kV} \cup \mathfrak{D}_{UR_kV} \cup \mathfrak{D}_V$$
$$\Vdash (((C x_U x_{L_k} x_{UL_k} \wedge C x_{UL_k} x_V x_{UL_kV}) \wedge C x_U x_{R_k} x_{UR_k}) \wedge C x_{UR_k} x_V x_{UR_kV}).$$

Now, we have (if we introduce W_n and W_{n+1} according to 5)) in view of 6) and using (T 6) that

$$\mathfrak{A}_{\mathfrak{T}} \cup \mathfrak{D}_{W_n} \cup \mathfrak{D}_{W_{n+1}} \cup \mathfrak{D}_V \Vdash T x_{W_n} x_{W_{n+1}}.$$

This, together with 4) gives (because of (T 5)), that

7) $$\mathfrak{A}_{\mathfrak{T}} \cup \mathfrak{D}_{W_1} \cup \mathfrak{D}_{W_{n+1}} \cup \mathfrak{D}_{W_n} \cup \mathfrak{D}_V \Vdash (T x_{W_1} x_{W_n} \wedge T x_{W_n} x_{W_{n+1}}).$$

Furthermore, using (T 8) and (T 4) and (T 10) we obtain that

$$A\,5 \Vdash ((T x_{W_1} x_{W_n} \wedge T x_{W_n} x_{W_{n+1}}) \rightarrow T x_{W_1} x_{W_{n+1}}),$$

Now, this gives together with 7) (using (T 6)) that

$$\mathfrak{A}_{\mathfrak{T}} \cup \mathfrak{D}_{W_1} \cup \mathfrak{D}_{W_{n+1}} \cup \mathfrak{D}_{W_n} \cup \mathfrak{D}_V \Vdash T x_{W_1} x_{W_{n+1}}.$$

We shall comprehend the elements $\mathfrak{D}_{W_n} \cup \mathfrak{D}_V$ which are not alredy elements of $\mathfrak{A}_{\mathfrak{T}} \cup \mathfrak{D}_{W_1} \cup \mathfrak{D}_{W_{n+1}}$ in a set $\mathfrak{D}$. Then we can write that

8) $$\mathfrak{A}_{\mathfrak{T}} \cup \mathfrak{D}_{W_1} \cup \mathfrak{D}_{W_{n+1}} \cup \mathfrak{D} \Vdash T x_{W_1} x_{W_{n+1}}.$$

Now, we have almost reached our goal, i.e. (††). We only have to do away with $\mathfrak{D}$. According to Section 2 every element of $\mathfrak{D}$ is of the form

$C x_U s_j x_{US_j}$. We consider an element δ of $\mathfrak{D}$ such that the length of U for δ is greater than the lenght of U for any other element of $\mathfrak{D}$. Let $\mathfrak{D}'$ be the set of all elements of $\mathfrak{D}$ which are different from $\delta = C x_U s_j x_{US_j}$. Thus, we have that

$$9) \qquad \mathfrak{A}_{\mathfrak{T}} \cup \mathfrak{D}_{W_1} \cup \mathfrak{D}_{W_{n+1}} \cup \mathfrak{D}' \cup \{C x_U s_j x_{US_j}\} \Vdash T x_{W_1} x_{W_{n+1}}.$$

We shall show that the variable x_{US_j} does not occur free in any of the formulae which appear in 9), except of course for $\delta = C x_U s_j x_{US_j}$. Obviously, it is sufficient to show that x_{US_j} does not at all occur in any of the formulae in 9) which are different from δ.

First of all, this is obvious for the elements of $\mathfrak{D}'$, because U was chosen to be as long as possible.

If x_{US_j} occured in an element of $\mathfrak{D}_{W_1}$, then according to the *Remark* of Section 2, the formula δ would be an element of $\mathfrak{D}_{W_1}$ in contradiction to the construction of $\mathfrak{D}$. Similarly we see that x_{US_j} does not occur in any element of $\mathfrak{D}_{W_{n+1}}$ or in any element of $\mathfrak{D}_{L_k}$ or $\mathfrak{D}_{R_k}$ from $\mathfrak{A}_{\mathfrak{T}}$. Let us consider the other elements of $\mathfrak{A}_{\mathfrak{T}}$. A 1, A 2, A 4 and A 5 contain no element of the form x_W at all, A 3 contains only $x_{\square}$, which is certainly different from x_{US_j}. A 6_k contains the variables x_{L_k} and x_{R_k}. But these variables also occur in $\mathfrak{D}_{I_k}$ and $\mathfrak{D}_{R_k}$ and so they are different from x_{US_j}, as we have just seen. Finally the variables x_{W_1} and $x_{W_{n+1}}$ occur in $T x_{W_1} x_{W_{n+1}}$. But these variables occur also in $\mathfrak{D}_{W_1}$ and $\mathfrak{D}_{W_{n+1}}$ and so they are different from x_{US_j}.

Now we shall substitute z for the variable x_{US_j} in all formulae which occur in 9). Then all formulae which are different from δ remain unaltered, and we obtain by the help of (T 10) that

$$\mathfrak{A}_{\mathfrak{T}} \cup \mathfrak{D}_{W_1} \cup \mathfrak{D}_{W_{n+1}} \cup \mathfrak{D}' \cup \{C x_U s_j z\} \Vdash T x_{W_1} x_{W_{n+1}}.$$

It is immediately obvious that z occurs free only in the formula $C x_U s_j z$. Thus we can apply (T 9) and proceed to

$$\mathfrak{A}_{\mathfrak{T}} \cup \mathfrak{D}_{W_1} \cup \mathfrak{D}_{W_{n+1}} \cup \mathfrak{D}' \cup \{\bigvee_z C x_U s_j z\} \Vdash T x_{W_1} x_{W_{n+1}}.$$

And from here by (T 7) to

$$10) \qquad \mathfrak{A}_{\mathfrak{T}} \cup \mathfrak{D}_{W_1} \cup \mathfrak{D}_{W_{n+1}} \cup \mathfrak{D}' \Vdash (\bigvee_z C x_U s_j z \to T x_{W_1} x_{W_{n+1}}).$$

Finally, using (T 8), (T 4) and (T 10) we obtain that A 1 $\Vdash \bigvee_z C x_U s_j z$, and from this together with 10) we obtain (using (T 6)) that

$$11) \qquad \mathfrak{A}_{\mathfrak{T}} \cup \mathfrak{D}_{W_1} \cup \mathfrak{D}_{W_{n+1}} \cup \mathfrak{D}' \Vdash T x_{W_1} x_{W_{n+1}}.$$

By the same method by which we have eliminated *one* element of $\mathfrak{D}$ we can now obviously eliminate a suitable element of $\mathfrak{D}'$ as well, etc., until we finally arrive at (††).

References

CHURCH, A.: A Note on the Entscheidungsproblem. J. symbolic Logic 1, 40–41 (1936); Correction ibid. pp. 101–102.

KALMÁR, L.: Ein direkter Beweis für die allgemein-rekursive Unlösbarkeit des Entscheidungsproblems des Prädikatenkalküls der ersten Stufe mit Identität. Z. math. Logik 2, 1–14 (1956).

TRACHTÉNBROT, B. A.: Impossibility of an algorithm for the decision problem in finite classes [Russ.]. Dokl. Akad. Nauk SSSR, 70, 569–572 (1950). (This shows that there exists no algorithm by the help of which we can decide whether or not an arbitrarily given formula of the predicate calculus is valid over a finite domain of individuals.)

For special classes of formulae of the predicate calculus there exist algorithms by the help of which we can decide whether or not any formula of this class is valid. For this consult

ACKERMANN, W.: Solvable Cases of the Decision Problem. Amsterdam: North-Holland Publishing Company 1954.

However, for other special classes of formulae of the predicate calculus we can show that there exists no such algorithm by reducing the (unsolvable) decision problem of the whole predicate calculus to the decision problem of this class. For this compare

SURÁNYI, J.: Reduktionstheorie des Entscheidungsproblems im Prädikatenkalkül der ersten Stufe. Budapest—Berlin: Ungarische Akademie der Wissenschaften—VEB Deutscher Verlag der Wissenschaften 1959.

KAHR, A. S., E. F. MOORE, and H. WANG: Entscheidungsproblem Reduced to the AEA Case. Proc. Nat. Acad. Sci., USA. 48, 365–377 (1962).

§ 26. The Incompleteness of the Predicate Calculus of the Second Order

We want to show in this paragraph that there exists no algorithm for the predicate calculus of the second order (the structure of which will be sketched in Section 1) such that this algorithm provides us precisely with the class of the valid formulae of this calculus. For the ordinary predicate calculus, which is sometimes called predicate calculus of the first order, there exists such an algorithm, i.e. the predicate calculus of the first order is *complete* (cf. the introduction to §24). We therefore say that the predicate calculus of the second order is *incomplete*. (The same applies to the predicate calculi of higher orders. These calculi can be defined similarly to the predicate calculus of the second order.)

The incompleteness of the predicate calculus of the second order was first proved by GÖDEL in 1931. In this book we give a proof by reducing the problem to the problem of the undecidability of the ordinary predicate calculus, which has been dealt with in the last paragraph. (This method was suggested by HASENJAEGER.)

1. The structure of the predicate calculus of the second order. In the formation of the predicate calculus, which we have developed at the beginning of § 24, the individual variables (*variables of the first order*) ξ were allowed to be quantified by means of $\bigwedge_{\xi}$. On the other hand we did not allow the quantification of predicate variables (*variables of the second order*). This is why we sometimes call the predicate calculus *the predicate calculus of the first order* (first order calculus) in contrast to the *predicate calculus of the second order* (second order calculus) in which predicate variables π may also be quantified by means of $\bigwedge_{\pi}$. We could furthermore introduce into the predicate calculus of the second order a new kind of variables, the predicate variables of the second order, which however may not be quantified. However, for the sake of simplicity we shall consider a formalism in which such predicate variables of the second order do not occur. The main result of this paragraph is also valid (by virtue of the same proof) for the complete second order calculus.

We shall briefly sketch the structure of that part of the second order calculus which is considered here. We refer to the structure of the ordinary predicate calculus as it was given in § 24. In the construction of the formulae we allow as a further possibility to proceed from α to $\bigwedge \pi \alpha$, where π is a predicate variable. Instead of $\bigwedge \pi \alpha$ we write $\bigwedge_{\pi} \alpha$, instead of $\neg \bigwedge_{\pi} \neg \alpha$ we write in short $\bigvee_{\pi} \alpha$. The semantic relation α *is valid under* $\mathfrak{J}$ of the ordinary predicate calculus (§ 24.2) shall be extended to the predicate calculus of the second order by the following additional definition.

(d′) $\bigwedge_{\pi} \alpha$ *is valid under* $\mathfrak{J}$ if and only if α is valid under every $\mathfrak{J}'$, where $\mathfrak{J}$ and $\mathfrak{J}'$ are interpretations over the same domain of individuals and $\mathfrak{J}'$ differs from $\mathfrak{J}$ at most for the argument π.

The assertions § 24.2 (e) and (f) can obviously be carried over also to the predicate calculus of the second order. The Definitions 2 and 3 of § 24.2 are taken over without alteration. We shall however use the symbol $\Vdash_{\text{II}}$ in order to show that we are dealing with a formula of the second order calculus. $\Vdash_{\text{I}}$ is equivalent to $\Vdash$. For a formula α of the first order calculus (which is naturally also a formula of the second order calculus), we have that $\Vdash_{\text{I}} \alpha$ if and only if $\Vdash_{\text{II}} \alpha$.

2. Remarks on the interpretations of the predicate calculi of the first and the second order. We can, in analogy to Definition 4 of § 24.4, speak of the *free occurrence* of a *predicate variable* π in a formula α. Lemma 1 of § 24.4 can also be proved in case of a predicate variable; the procedure applied even shows that if two interpretations $\mathfrak{J}_1$ and $\mathfrak{J}_2$ are over the same domain of individuals and coincide for all individual and predicate variables which occur free in a formula α, then α is either valid under both or under neither of the interpretations. In the second order calculus (in contrast to the first order calculus) there exist formulae without free variables. According to the previous remarks the following assertion is valid for such formulae.

(1) If α is a formula of the second order calculus without free variables and if ω is an arbitrary (non-empty) domain of individuals, then either α is valid under *all* interpretations over ω or α is valid under *no* interpretation over ω.

We introduce υ (Greek Ypsilon) as an abbreviation for

$$\bigvee_R \left(\bigwedge_x \bigvee_y Rxy \wedge \bigwedge_x \bigwedge_y \bigwedge_z ((Rxy \wedge Ryz) \rightarrow Rxz) \wedge \bigwedge_x \neg Rxx\right).$$

The validity of υ in a domain ω of individuals implies that ω is infinite, because we have (as can easily be shown) that

(2) υ is valid under no interpretation over a finite domain of individuals. υ is valid under every interpretation over an infinite domain of individuals, especially over the domain of natural numbers.

In the following considerations we shall make use of two more remarks about the first order calculus. These are given here without proofs[1].

(3) Let α be a formula of the first order calculus. If α is valid under an interpretation over a domain ω_1 of individuals and if ω_2 is a domain of individuals whose cardinal number is not smaller than that of ω_1, then there also exists an interpretation over ω_2 under which α is valid.

(4) Let α be a formula of the first order calculus. If α is valid under an interpretation over any domain of individuals, then there exists an interpretation over the domain of the natural numbers under which α is valid (*Theorem of Löwenheim and Skolem*).

Now let α be a formula of the first order calculus, $\xi_1, \ldots, \xi_r, \pi_1, \ldots, \pi_s$ be the variables which occur free in α. The sequence of these variables may be determined by any method which makes certain that the predicate variables come at the end. It can happen that no individual variable occurs free in α, but certainly at least one predicate variable occurs free in α. Then we call the expression

$$\bigvee_{\xi_1} \ldots \bigvee_{\xi_r} \bigvee_{\pi_1} \ldots \bigvee_{\pi_s} \alpha$$

the particularization of α. *We abbreviate this by* $\bigvee \alpha$. $\bigvee \alpha$ is a formula of the second order calculus without free variables. We should note that the predicate variable quantified last stands immediately before α. From this we can easily verify the following assertion.

---

[1] Cf. for instance SCHOLZ-HASENJAEGER (see references after § 24), pp. 199 and 207.

(5) For every formula β of the *second order calculus* we can decide whether β is a particularization $\bigvee\alpha$ of a formula α of the first order. If this is the case, then α is unambiguously determined and is effectively constructible from β.

3. The relationship between $\Vdash_{\mathrm{I}}$ *and* $\Vdash_{\mathrm{II}}$.

Theorem. If α is a formula of the first order calculus, then we have that

$$\textit{not } \Vdash_{\mathrm{I}}\alpha \textit{ if and only if } \Vdash_{\mathrm{II}}(v \to \bigvee\neg\alpha).$$

Proof. We prove the assertions (*) and (**).

(*) $\Vdash_{\mathrm{I}}\alpha$ and $\Vdash_{\mathrm{II}}(v \to \bigvee\neg\alpha)$ contradict each other. Proof: Let $\mathfrak{J}$ be an arbitrary interpretation over the domain of the natural numbers. If $\Vdash_{\mathrm{II}}(v \to \bigvee\neg\alpha)$, then $(v \to \bigvee\neg\alpha)$, i.e. $\neg(v\wedge\neg\bigvee\neg\alpha)$ is valid under $\mathfrak{J}$. Therefore $(v\wedge\neg\bigvee\neg\alpha)$ is not valid under $\mathfrak{J}$. Since according to (2) v is valid under $\mathfrak{J}$, $\neg\bigvee\neg\alpha$ cannot be valid under $\mathfrak{J}$. Thus, $\bigvee\neg\alpha$ is valid under $\mathfrak{J}$. From this follows (cf. § 24.2 (f); analogous results are also valid for predicate variables) that there exists an interpretation $\mathfrak{J}'$ over the domain of the natural numbers under which $\neg\alpha$ is valid and, therefore, α is not valid. This contradicts $\Vdash_{\mathrm{I}}\alpha$.

(**) *Not* $\Vdash_{\mathrm{I}}\alpha$ and *not* $\Vdash_{\mathrm{II}}(v \to \bigvee\neg\alpha)$ contradict each other. Proof: We assume *not* $\Vdash_{\mathrm{II}}(v \to \bigvee\neg\alpha)$. Thus, there exists an interpretation $\mathfrak{J}_2$ over a domain ω_2 of individuals under which $(v \to \bigvee\neg\alpha)$, i.e. $\neg(v\wedge\neg\bigvee\neg\alpha)$ is not valid, i.e. under which $(v\wedge\neg\bigvee\neg\alpha)$ is valid. Thus, v and $\neg\bigvee\neg\alpha$ are valid under $\mathfrak{J}_2$. Since v is valid under $\mathfrak{J}_2$, ω_2 must be infinite according to (2). Since $\neg\bigvee\neg\alpha$ contains no free variables, $\neg\bigvee\neg\alpha$ is valid under *every* interpretation over ω_2.

Furthermore we conclude from *not* $\Vdash_{\mathrm{I}}\alpha$ that there exists an interpretation $\mathfrak{J}_1$ over a domain ω_1 of individuals under which α is not valid and, therefore, under which $\neg\alpha$ is valid. From this follows by (4) that there exists an interpretation over the domain of natural numbers under which $\neg\alpha$ is valid. The cardinal number of the domain of the natural numbers is the smallest infinite cardinal number. ω_2 is infinite. Now, we can deduce by (3) that there exists an interpretation $\mathfrak{J}_2^*$ over the domain ω_2 of individuals under which $\neg\alpha$ is valid. Then $\bigvee\neg\alpha$ is also valid under $\mathfrak{J}_2^*$ (cf. § 24.2 (f)). From this follows that $\neg\bigvee\neg\alpha$ is not valid under $\mathfrak{J}_2^*$, in contradiction to the validity of $\neg\bigvee\neg\alpha$ under every interpretation over ω_2.

4. Proof of the incompleteness. Now we can easily prove the incompleteness of the second order calculus. This incompleteness means that there exists no algorithm by the help of which we can obtain precisely the set of valid formulae of the second order calculus; in other words, that the set of the valid formulae of the second order calculus is not enumerable.

Proof by *reductio ad absurdum*. We assume that the set of valid formulae of the second order calculus is enumerable. Thus there exists a computable sequence $\beta_0, \beta_1, \beta_2, \beta_3, \ldots$ of valid formulae of the second order calculus which contains all the valid formulae. By an application of (5) we can determine effectively for any formula β_i of this sequence whether it has the form $(v \rightarrow \gamma)$, where γ is a particularization of a formula β of the first order, and if this is the case, we can produce β effectively. We only consider those cases in which there exists an α such that $\beta \equiv \neg\alpha$. In this manner we obtain a sequence $\alpha_0, \alpha_1, \alpha_2, \ldots$ of formulae of the first order. Now, it follows from the last theorem that the sequence $\alpha_0, \alpha_1, \alpha_2, \ldots$ runs through the set of all *not* valid formulae of the first order. Thus, this set is enumerable. Since on the other hand the set of the valid formulae of the first order is enumerable on the basis of Gödel's completeness theorem (cf. the introduction to § 24), this set is decidable according to § 2.4 (f). But this contradicts the undecidability of the first order calculus, which we have shown in § 25.1. From this we have the

Theorem (*Gödel* 1931). *The second order calculus is incomplete, i.e. there exists no algorithm by the help of which we can obtain precisely the class of valid formulae of the second order calculus.*

Reference

Gödel, K.: Über formal unentscheidbare Sätze der Principia Mathematica und verwandter Systeme. I. Mh. Math. Phys. **38**, 173—198 (1931).

§ 27. The Undecidability and Incompleteness of Arithmetic

By the help of the symbols $+$ for addition and $\cdot$ for multiplication we can introduce the *arithmetical formulae* (Section 1). The arithmetical formulae (we shall examine them more closely in § 29) contain in general free variables for natural numbers. Arithmetical formulae *without* free variables are called *arithmetical sentences*. Arithmetical sentences are either true or false. We show in this paragraph the

Theorem (Incompleteness of arithmetic). There exists no algorithm by the help of which we can derive precisely the set of all true arithmetical sentences[1].

From this we obtain directly the

Corollary (Undecidability of arithmetic). There exists no algorithm by the help of which we can decide for every arithmetical sentence in finitely many steps whether it is true or false.

The proof of the above theorem is carried out according to the following outline. We shall give (in § 28, Section 3 where we shall deal

[1] For the notion of "incompleteness" cf. the corresponding remark for the second order calculus (§ 26).

collectively with the *enumerable* predicates) a binary primitive recursive predicate P_1 such that the singulary predicate P which is defined for all x by

$$Px \leftrightarrow \bigwedge_y P_1 xy$$

is not enumerable.

We show in Lemma 3 (Sections 2 and 3) that for every primitive recursive function f which is traced back to the initial functions by means of substitutions and inductive definitions we can find effectively an arithmetical sentence α_f, which "defines" the predicate given by the relation $f(x_1, \ldots, x_n) = y$ (cf. Section 1, Definition 2). Using this we prove in Lemma 4 (Section 2) that for the predicate P and every number r we can find effectively an arithmetical sentence α_r such that Pr *if and only if* α_r is true. Now, if there existed an algorithm by the help of which we could derive precisely the set of true arithmetical sentences, then we could use this algorithm to enumerate all true sentences among the α_r's. By this we also have an enumeration of all r's for which the predicate P is valid, in contradiction to the non-enumerability of P.

In this book we define the arithmetical sentences with respect to the natural numbers, i.e. semantically. Semantics in its modern form was built up by Tarski (1935). In his paper (cf. references given below) Tarski shows that the concept of true arithmetical sentence is not definable in arithmetic, from which follows the undecidability of arithmetic. Instead of starting with the semantic conception we could also start with a system A_0 of axioms for the natural numbers, e.g. the Peano axioms, to which we add as further axioms the equations which are usually used to define addition and multiplication (see § 10.4). Peano's induction axiom, which deals with arbitrary properties is treated as a *schema*, which allows for every property which is definable by an arithmetical formula the application of the induction process. Now we can consider the set A of those sentences which can be derived from A_0 by the help of the rules of inference of the predicate calculus (increased by obvious rules dealing with function symbols). The set A is enumerable on the basis of its definition. In 1931 Gödel showed, by using an argument similar to the antinomy of the liar, that for every such system A_0 of axioms (assuming the so-called ω-consistency) there exists an arithmetical sentence α such that neither α nor $\neg\alpha$ is derivable from A_0. In 1936 Rosser proved that the assumption of ω-consistency can be replaced by the simple requirement that A should be consistent, i.e. that A does not coincide with the set of all arithmetical sentences. At least one of the sentences α and $\neg\alpha$ is true when considered semantically, and hence this true sentence cannot be derived from A. In 1936 Church proved for the above mentioned system A_0 of axioms that the (enumerable) set A of the sentences derivable from A_0 is not decidable. This assertion of the non-decidability of arithmetic should be distinguished from the assertion of the Corollary given above, since A does not coincide with the set of all true arithmetical sentences. ROSSER furthermore proved that not only A, but every consistent "supertheory" of A is undecidable, a property which TARSKI called *essential undecidability*. In the meantime the property of essential undecidability has been proved for appreciably weaker systems of axioms. We shall not go into these results in this book. We refer the reader to the book by TARSKI-MOSTOWSKI-ROBINSON, in which the essential undecidability of other theories is also dealt with.

1. Arithmetical formulae, sentences and predicates. Arithmetical sentences are special arithmetical formulae. The arithmetical formulae are formed from (individual) variables $x_1, x_2, x_3, \ldots$ for natural numbers, the symbols $+$ and $\cdot$ for addition and multiplication, the equality symbol $=$, the symbols $\neg$, $\wedge$ of the propositional calculus, the operator $\bigwedge$ of the predicate calculus and the parentheses (,). The structure has a lot in common with the structure of the predicate calculus (§ 24). For the omission of the parentheses and the application of the abbreviations $\rightarrow$ and $\bigvee$ the reader should consult § 24.1 and the note of § 11.1. Furthermore we use the abbreviation $\alpha_1 \wedge \ldots \wedge \alpha_n$ for $(\ldots(\alpha_1 \wedge \alpha_2) \ldots \wedge \alpha_n)$.

First we introduce the *arithmetical terms*. Every variable is a term. If t_1 and t_2 are terms, then $(t_1 + t_2)$ and $(t_1 \cdot t_2)$ are also terms.

The parentheses containing the whole expression will often be omitted.

The *arithmetical formulae* are defined inductively as follows. If t_1 and t_2 are arithmetical terms, then $t_1 = t_2$ is an arithmetical formula. If α and β are arithmetical formulae and ξ is a variable, then $\neg\alpha$, $(\alpha \wedge \beta)$ and $\bigwedge_\xi \alpha$ are arithmetical formulae. We write, as usual, $t_1 t_2$ instead of $t_1 \cdot t_2$ and use the well-known abbreviations $\xi \neq \eta$ for $\neg \xi = \eta$, $\xi \leqq \eta$ for $\bigvee_\zeta \xi + \zeta = \eta$ (here, for the sake of unambiguity, ζ is the first variable after ξ and η) and $\xi < \eta$ for $\xi \leqq \eta \wedge \xi \neq \eta$.

That a variable x occurs *free* in an arithmetical formula is defined similarly to the Definition in § 24.4. An arithmetical formula is called an *arithmetical sentence* if it contains no free variable.

The variables have a *natural sequence* which is determined by the sequence of their indices. We use the expression $\alpha(\xi_1, \ldots, \xi_n)$ to denote that α is a formula in which the variables $\xi_1, \ldots, \xi_n$ and no other occur free, where $\xi_1, \ldots, \xi_n$ are written down according to their order in the natural sequence of variables. If $\eta_1, \ldots, \eta_n$ are variables, written down according to their order in the natural sequence of variables, for which $\eta_i \equiv \xi_i$ or η_i does not at all occur in α $(i = 1, \ldots, n)$, then there exists one and only one formula β for which the following conditions are valid.

(1) β is obtained from α by successive substitutions of η_1 for ξ_1, η_2 for $\xi_2, \ldots, \eta_n$ for ξ_n. (2) Conversely α is obtained from β by successive substitutions of ξ_1 for $\eta_1, \ldots, \xi_n$ for η_n. We shall denote this formula β by $\alpha(\eta_1, \ldots, \eta_n)$.

By an *interpretation* $\mathfrak{J}$ we mean a mapping of the *variables* onto the natural numbers. This mapping can be extended in a natural manner to *terms* by the stipulations that $\mathfrak{J}(t_1 + t_2)$ and $\mathfrak{J}(t_1 t_2)$ are equal to the sum and product of $\mathfrak{J}(t_1)$ and $\mathfrak{J}(t_2)$ respectively. We shall say that an *equation* $t_1 = t_2$ *is valid under the interpretation* $\mathfrak{J}$ if $\mathfrak{J}(t_1)$ is the same number as $\mathfrak{J}(t_2)$. Now we can define for an arbitrary arithmetical formula α, what it means that α is valid under $\mathfrak{J}$. This definition coincides completely with § 24.2, Definition 1, (b), (c), (d)[1]. The theorems derived in § 24 on the basis of Definition 1, (b), (c), (d) are also valid for arithmetical formulae. We shall make use of them without always referring to them explicitly. Especially Lemma 1 of § 24.4 is also valid. From that follows that an arithmetical *sentence* α, which indeed does not contain any free individual variables, is either valid under every interpretation or under none. In the first case we shall say that α is *true*, in the second case that α is *false*. For example the assertion $\bigwedge_{x_1} \bigwedge_{x_2} x_1 + x_2 = x_2 + x_1$ is true, whereas $\bigwedge_{x_1} \bigwedge_{x_2} x_1 + x_2 = x_1$ is false.

For every natural number n we shall use $\xi = n$ as an abbreviation for an arithmetical formula, namely

$$\xi = 0 \quad \text{for} \quad \xi + \xi = \xi$$

$$\xi = n' \quad \text{for} \quad \bigvee_{\eta} \bigvee_{\zeta} (\eta = n \wedge \neg \zeta = 0 \wedge \zeta\zeta = \zeta \wedge \xi = \eta + \zeta).$$

Here (for the sake of unambiguity) η is defined to be the variable following ξ, ζ the variable following η. Finally, we put

$$\xi + 1 = \eta \quad \text{for} \quad \bigvee_{\zeta} (\zeta = 1 \wedge \xi + \zeta = \eta).$$

Here ζ is the first variable which comes after both ξ and η.

Zero is the only number n which added to itself gives n again. Thus, $\xi = 0$ is valid if and only if ξ is interpreted by 0. Zero and one are the only numbers n which multiplied by themselves give n again. Now it follows generally that $\xi = n$ is valid if and only if ξ is interpreted by n. $\xi + 1 = \eta$ is valid if and only if η is interpreted by the successor of ξ.

---

[1] The reader should note that arithmetical formulae contain no predicate variables. Thus, an interpretation of arithmetical formulae (in contrast to an interpretation of formulae of the predicate calculus) maps only the individual variables. A further difference to the predicate calculus is that $\mathfrak{J}(x)$ for the predicate calculus can be an element of an *arbitrary* domain ω of individuals, whereas here $\mathfrak{J}(x)$ is always a *natural number*.

We shall associate with a formula $\alpha(\xi_1, \ldots, \xi_n)$ an n-ary predicate P_α (for the concept of predicate cf. § 11.1) by the following

Definition 1. Let $r_1, \ldots, r_n$ be arbitrary natural numbers. $P_\alpha r_1 \ldots r_n$ if and only if there exists an interpretation $\mathfrak{J}$ under which α is valid and for which $\mathfrak{J}(\xi_1) = r_1$ and ... and $\mathfrak{J}(\xi_n) = r_n$.

Definition 2. α defines the predicate P if and only if $P = P_\alpha$.

Definition 3. A predicate P is called arithmetical if there exists an arithmetical formula α such that $P = P_\alpha$.

We show

Lemma 1. $\alpha(\xi_1, \ldots, \xi_n)$ defines the predicate P if and only if for every interpretation $\mathfrak{J}$ we have that
α is valid under $\mathfrak{J}$ if and only if $P\mathfrak{J}(\xi_1) \ldots \mathfrak{J}(\xi_n)$.

Proof. We start by assuming that α defines the predicate P. This means according to Definition 2 that $P = P_\alpha$, i.e. that for all $r_1, \ldots, r_n$ $Pr_1 \ldots r_n$ if and only if $P_\alpha r_1 \ldots r_n$. This is equivalent to saying that for any interpretation $\mathfrak{J}$: $P\mathfrak{J}(\xi_1) \ldots \mathfrak{J}(\xi_n)$ if and only if $P_\alpha \mathfrak{J}(\xi_1) \ldots \mathfrak{J}(\xi_n)$. According to Definition 1 $P_\alpha \mathfrak{J}(\xi_1) \ldots \mathfrak{J}(\xi_n)$ means that there exists an interpretation $\mathfrak{J}'$ under which α is valid and for which $\mathfrak{J}'(\xi_1) = \mathfrak{J}(\xi_1)$ and ... and $\mathfrak{J}'(\xi_n) = \mathfrak{J}(\xi_n)$. But such an interpretation $\mathfrak{J}'$ exists if and only if α is valid under $\mathfrak{J}$. This can be shown as follows. If there exists such an $\mathfrak{J}'$, then α is valid under $\mathfrak{J}$ according to Lemma 1 of § 24.4. Conversely, if α is valid under $\mathfrak{J}$, then we can put $\mathfrak{J}' = \mathfrak{J}$. We have shown that α defines the predicate P if and only if for every interpretation $\mathfrak{J}$: $P\mathfrak{J}(\xi_1) \ldots \mathfrak{J}(\xi_n)$ if and only if α is valid under $\mathfrak{J}$, q.e.d.

Lemma 2. $\alpha(\xi_1, \ldots, \xi_n)$ and $\alpha(\eta_1, \ldots, \eta_n)$ define the same predicate.

Proof. We assume that $P_{\alpha(\xi_1, \ldots, \xi_n)} r_1 \ldots r_n$, with arbitrary $r_1, \ldots, r_n$. According to Definition 1 there exists an interpretation $\mathfrak{J}$ under which α is valid and for which $\mathfrak{J}(\xi_1) = r_1, \ldots, \mathfrak{J}(\xi_n) = r_n$. Now we define an interpretation $\mathfrak{J}'$ by the following stipulations[1]:

$$\mathfrak{J}' \underset{\eta_1, \ldots, \eta_n}{=} \mathfrak{J}, \qquad \mathfrak{J}'(\eta_1) = \mathfrak{J}(\xi_1), \ldots, \mathfrak{J}'(\eta_n) = \mathfrak{J}(\xi_n).$$

Then, according to Lemma 2 (§ 24.5), $\alpha(\eta_1, \ldots, \eta_n)$ is valid under $\mathfrak{J}'$ if and only if $\alpha(\xi_1, \ldots, \xi_n)$ is valid under $\mathfrak{J}$. Thus, $\alpha(\eta_1, \ldots, \eta_n)$ is valid under $\mathfrak{J}'$. Further we have that $\mathfrak{J}'(\eta_1) = \mathfrak{J}(\xi_1) = r_1$, etc. This shows

[1] We introduced the notation $\mathfrak{J}_1 \underset{\xi}{=} \mathfrak{J}_2$ in § 24.2. We extend this notation by the stipulation that $\mathfrak{J}_1 \underset{\xi_1, \ldots, \xi_r}{=} \mathfrak{J}_2$ shall mean that $\mathfrak{J}_1$ and $\mathfrak{J}_2$ are interpretations over the same domain of individuals and that $\mathfrak{J}_1$ and $\mathfrak{J}_2$ differ at most for the arguments $\xi_1, \ldots, \xi_r$.

that $P_{\alpha(\eta_1,\ldots,\eta_n)} r_1 \ldots r_n$. By symmetry we also have the converse; $P_{\alpha(\xi_1,\ldots,\xi_n)} r_1 \ldots r_n$ follows from $P_{\alpha(\eta_1,\ldots,\eta_n)} r_1 \ldots r_n$. Thus we have that $P_{\alpha(\xi_1,\ldots,\xi_n)} = P_{\alpha(\eta_1,\ldots,\eta_n)}$.

2. *Lemma 4* is stated at the end of this section. In its proof we make use of Lemma 3 which is proved in the next section.

Lemma 3. For every n-ary primitive recursive function f which is given effectively in the sense that it is traced back to the initial functions by means of substitutions and inductive definitions we can give effectively an arithmetical formula $\alpha_f(\xi_1, \ldots, \xi_n, \xi)$ such that for every interpretation $\mathfrak{J}$:

$\alpha_f(\xi_1, \ldots, \xi_n, \xi)$ is valid under $\mathfrak{J}$ if and only if $f(\mathfrak{J}(\xi_1), \ldots, \mathfrak{J}(\xi_n)) = \mathfrak{J}(\xi)$.

We make use of Example 1 of § 28.3. Let f be the characteristic function of the binary predicate $\neg T_1 r_1 r_1 r_2$. f is primitive recursive. We shall show in § 28.3 that the singulary predicate P, which is defined by the stipulation that for all r_1

$$P r_1 \text{ if and only if } f(r_1, r_2) = 0 \text{ for all } r_2$$

is not enumerable. We determine an arithmetical formula $\alpha_f(\xi_1, \xi_2, \xi_3)$ for the primitive recursive function f according to Lemma 3. Then we form the arithmetical sentence

$$\alpha_{r_1} \equiv \bigvee_{\xi_1} \big(\bigwedge_{\xi_2} \bigvee_{\xi} (\alpha_f(\xi_1, \xi_2, \xi) \wedge \xi = 0) \wedge \xi_1 = r_1\big).$$

If this sentence is true, then there exists an interpretation under which it is valid. Then there also exists (cf. § 22.2 Definition 1 (f)) an interpretation $\mathfrak{J}$ under which $\bigwedge_{\xi_2} \bigvee_{\xi} (\alpha_f(\xi_1, \xi_2, \xi) \wedge \xi = 0) \wedge \xi_1 = r_1$ is valid. It follows that $\mathfrak{J}(\xi_1) = r_1$. Furthermore, $\bigvee_{\xi} (\alpha_f(\xi_1, \xi_2, \xi) \wedge \xi = 0)$ is valid under *every* $\mathfrak{J}'$ with $\mathfrak{J}' \underset{\xi_2}{=} \mathfrak{J}$. Thus, $\mathfrak{J}'(\xi_2)$ can be an arbitrary number r_2. For every such $\mathfrak{J}'$ there exists an $\mathfrak{J}''$ with $\mathfrak{J}'' \underset{\xi}{=} \mathfrak{J}$ such that $\alpha_f(\xi_1, \xi_2, \xi) \wedge \xi = 0$ is valid under $\mathfrak{J}''$. From the validity of $\alpha_f(\xi_1, \xi_2, \xi)$ under $\mathfrak{J}''$ follows that $f(\mathfrak{J}''(\xi_1), \mathfrak{J}''(\xi_2)) = \mathfrak{J}''(\xi)$. From the validity of $\xi = 0$ under $\mathfrak{J}''$ follows that $\mathfrak{J}''(\xi) = 0$. Thus, $f(\mathfrak{J}''(\xi_1), \mathfrak{J}''(\xi_2)) = 0$. Because $\mathfrak{J}''(\xi_1) = \mathfrak{J}'(\xi_1) = \mathfrak{J}(\xi_1) = r_1$ and since $\mathfrak{J}''(\xi_2) = \mathfrak{J}'(\xi_2) = r_2$ is an arbitrary number it follows that $f(r_1, r_2) = 0$ for all r_2, i.e. that $P r_1$.

Now we start conversely from $P r_1$. Then we have that $f(r_1, r_2) = 0$ for all r_2. If we choose an interpretation $\mathfrak{J}$ such that $\mathfrak{J}(\xi_1) = r_1$, $\mathfrak{J}(\xi) = 0$, whereas $\mathfrak{J}(\xi_2)$ can be chosen arbitrarily, then we obtain $f(\mathfrak{J}(\xi_1), \mathfrak{J}(\xi_2)) = \mathfrak{J}(\xi)$. Thus, for such an $\mathfrak{J}$ the following formulae are valid:

$\alpha_f(\xi_1, \xi_2, \xi)$	according to Lemma 3
$\alpha_f(\xi_1, \xi_2, \xi) \wedge \xi = 0$	because $\mathfrak{I}(\xi) = 0$
$\bigvee_{\xi} (\alpha_f(\xi_1, \xi_2, \xi) \wedge \xi = 0)$	since the formula above is valid
$\bigwedge_{\xi_2}\bigvee_{\xi} (\alpha_f(\xi_1, \xi_2, \xi) \wedge \xi = 0)$	since $\mathfrak{I}(\xi_2)$ was arbitrary
$\bigwedge_{\xi_2}\bigvee_{\xi} (\alpha_f(\xi_1, \xi_2, \xi) \wedge \xi = 0) \wedge \xi_1 = r_1$	because $\mathfrak{I}(\xi_1) = r_1$
$\bigvee_{\xi_1}(\bigwedge_{\xi_2}\bigvee_{\xi} (\alpha_f(\xi_1, \xi_2, \xi) \wedge \xi = 0) \wedge \xi_1 = r_1)$	since the formula above is valid.

This shows that α_{r_1} is valid. — In conclusion we have

Lemma 4. We can give an account of a not-enumerable predicate P such that for every number r_1 we can construct an arithmetical formula α_{r_1} which is true if and only if Pr_1.

3. *Proof of Lemma 3 of Section 2.* First we prove the assertion for the initial functions and then we consider the processes of substitution and inductive definition.

(a) *The initial functions.*

(a_1) Let f be the *successor function*. We put

$$\alpha_f \equiv x_1 + 1 = x_2 .$$

According to Section 1 α_f is valid under $\mathfrak{I}$ if and only if $\mathfrak{I}(x_1) + 1 = \mathfrak{I}(x_2)$, i.e. if and only if $f(\mathfrak{I}(x_1)) = \mathfrak{I}(x_2)$.

(a_2) Let f be the *identity function* U_n^i. We put

$$\alpha_f \equiv x_1 = x_1 \wedge \ldots \wedge x_n = x_n \wedge x_i = x_{n+1} .$$[1]

α_f is valid under $\mathfrak{I}$ if and only if $\mathfrak{I}(x_i) = \mathfrak{I}(x_{n+1})$, i.e. if and only if $U_n^i(\mathfrak{I}(x_1), \ldots, \mathfrak{I}(x_n)) = \mathfrak{I}(x_{n+1})$.

(a_3) Let f be the *constant* C_0^0. We put

$$\alpha_f \equiv x_1 = 0 .$$

α_f is valid under $\mathfrak{I}$ if and only if $\mathfrak{I}(x_i) = 0$, i.e. if and only if $C_0^0 = \mathfrak{I}(x_1)$.

(b) *The substitution process.* Let for all $\mathfrak{x}$

$$f(\mathfrak{x}) = g(h_1(\mathfrak{x}), \ldots, h_m(\mathfrak{x})) .$$

Let us assume that for the functions $h_1, \ldots, h_m, g$ Lemma 3 is already proved. In view of Lemma 2 we can find pairwise different variables

---

[1] The first n members of the conjunction serve to make sure that $x_1, \ldots, x_n$ occur free in α_f.

$\xi_1, \ldots, \xi_n, \eta_1, \ldots, \eta_m, \zeta$ such that the arithmetical formulae corresponding to $h_1, \ldots, h_m, g$ can be written more precisely in the form

$$\alpha_{h_1}(\xi_1, \ldots, \xi_n, \eta_1), \ldots, \alpha_{h_m}(\xi_1, \ldots, \xi_n, \eta_m), \alpha_g(\eta_1, \ldots, \eta_m, \zeta).$$

Now, we put

$$\alpha_f(\xi_1, \ldots, \xi_n, \zeta) \equiv$$
$$\bigvee_{\eta_1} \ldots \bigvee_{\eta_m} (\alpha_{h_1}(\xi_1, \ldots, \xi_n, \eta_1) \wedge \ldots \wedge \alpha_{h_m}(\xi_1, \ldots, \xi_n, \eta_m) \wedge \alpha_g(\eta_1, \ldots, \eta_m, \zeta)).$$

Let this formula be valid under an interpretation $\mathfrak{J}$. Then there exists an $\mathfrak{J}' \underset{\eta_1, \ldots, \eta_m}{=} \mathfrak{J}$ such that $\alpha_{h_1} \wedge \ldots \alpha_{h_m} \wedge \alpha_g$ is valid under $\mathfrak{J}'$. According to the hypotheses about $\alpha_{h_1}, \ldots, \alpha_{h_m}, \alpha_g$ we have that

$$h_1(\mathfrak{J}'(\xi_1), \ldots, \mathfrak{J}'(\xi_n)) = \mathfrak{J}'(\eta_1), \ldots, h_m(\mathfrak{J}'(\xi_1), \ldots, \mathfrak{J}'(\xi_n)) = \mathfrak{J}'(\eta_m),$$
$$g(\mathfrak{J}'(\eta_1), \ldots, \mathfrak{J}'(\eta_m)) = \mathfrak{J}'(\zeta),$$

from which follows

$$f(\mathfrak{J}'(\xi_1), \ldots, \mathfrak{J}'(\xi_n)) = \mathfrak{J}'(\zeta), \text{ i.e. } f(\mathfrak{J}(\xi_1), \ldots, \mathfrak{J}(\xi_n)) = \mathfrak{J}(\zeta).$$

Conversely, let $f(\mathfrak{J}(\xi_1), \ldots, \mathfrak{J}(\xi_n)) = \mathfrak{J}(\zeta)$. We define an interpretation $\mathfrak{J}'$ by the stipulations that

$$\mathfrak{J}' \underset{\eta_1, \ldots, \eta_m}{=} \mathfrak{J}$$

and

$$\mathfrak{J}'(\eta_1) = h_1(\mathfrak{J}(\xi_1), \ldots, \mathfrak{J}(\xi_n)), \ldots, \mathfrak{J}'(\eta_m) = h_m(\mathfrak{J}(\xi_1), \ldots, \mathfrak{J}(\xi_n)).$$

This gives that $\alpha_{h_1} \wedge \ldots \wedge \alpha_{h_m} \wedge \alpha_g$ is valid under $\mathfrak{J}'$ and consequently α_f is valid under $\mathfrak{J}$. This completes the proof of Lemma 3 for f.

(c) *The induction process.* To deal with the induction process we need a 4-*ary predicate* G (introduced by GÖDEL) with the following properties:

(I) For every a, b, i there exists one and only one k such that $Gabik$.

(II) For every finite sequence $k_0, \ldots, k_n$ we can find an a and a b such that $Gabik_i$ for every $i \leqq n$.

We shall define G in Section 4. We shall also show there that G is an *arithmetical predicate.*

Let for all $\mathfrak{x}$ and r

$$f(\mathfrak{x}, 0) = g(\mathfrak{x})$$
$$f(\mathfrak{x}, r+1) = h(\mathfrak{x}, r, f(\mathfrak{x}, r)).$$

We assume that Lemma 3 is valid for the functions g and h. Let the corresponding formulae be α_g and α_h respectively. Further, there exists for the arithmetical predicate G a defining arithmetical formula γ. In view of Lemma 2 we can find pairwise different variables $\xi_1, \ldots, \xi_n, \vartheta_1, \vartheta_2, \eta, \zeta, \eta_1, \zeta_1, \eta_2, \zeta_2$ (which are written here according to their order in the natural sequence of variables) such that the above mentioned formulae can be written more precisely in the form

$$\alpha_g(\xi_1, \ldots, \xi_n, \zeta), \alpha_h(\xi_1, \ldots, \xi_n, \eta, \zeta, \zeta_1), \gamma(\vartheta_1, \vartheta_2, \eta, \zeta),$$

and that we may assume that the formulae

$$\gamma(\vartheta_1, \vartheta_2, \eta_1, \zeta_1) \qquad \text{and} \qquad \gamma(\vartheta_1, \vartheta_2, \eta_2, \zeta_2)$$

are defined (cf. Section 1). Then we form the arithmetical formula

$$\begin{aligned}\alpha_f(\xi_1, \ldots, \xi_n, \eta_2, \zeta_2) \equiv \\ \bigvee_{\vartheta_1}\bigvee_{\vartheta_2}(\bigwedge_{\eta}\bigwedge_{\zeta}(\gamma(\vartheta_1, \vartheta_2, \eta, \zeta) \wedge \eta = 0 \rightarrow \alpha_g(\xi_1, \ldots, \xi_n, \zeta)) \\ \wedge \bigwedge_{\eta}\bigwedge_{\zeta}\bigwedge_{\eta_1}\bigwedge_{\zeta_1}(\eta < \eta_2 \wedge \gamma(\vartheta_1, \vartheta_2, \eta, \zeta) \wedge \gamma(\vartheta_1, \vartheta_2, \eta_1, \zeta_1) \wedge \eta + 1 = \eta_1 \\ \rightarrow \alpha_h(\xi_1, \ldots, \xi_n, \eta, \zeta, \zeta_1)) \\ \wedge \gamma(\vartheta_1, \vartheta_2, \eta_2, \zeta_2)).\end{aligned}$$

We shall show that for this formula α_f the assertion of Lemma 3 is valid

1) *We assume first that* $f(\mathfrak{J}(\xi_1), \ldots, \mathfrak{J}(\xi_n), \mathfrak{J}(\eta_2)) = \mathfrak{J}(\zeta_2)$ for an interpretation $\mathfrak{J}$. We use the following abbreviations

$$r_1 = \mathfrak{J}(\xi_1), \ldots, r_n = \mathfrak{J}(\xi_n), \qquad r = \mathfrak{J}(\eta_2), \qquad s = \mathfrak{J}(\zeta_2).$$

Then $f(r_1, \ldots, r_n, r) = s$. Now we form the finite sequence

$$\begin{aligned} k_0 &= f(r_1, \ldots, r_n, 0) \\ k_1 &= f(r_1, \ldots, r_n, 1) \\ &\ldots\ldots\ldots \\ k_r &= f(r_1, \ldots, r_n, r). \end{aligned}$$

For this sequence we choose an a and a b according to (II). Thus, we have

(0) $\qquad Gabif(r_1, \ldots, r_n, i) \qquad$ for every $i \leqq r$.

We put

$$\mathfrak{J}^* \underset{\vartheta_1, \vartheta_2}{=} \mathfrak{J}, \ \mathfrak{J}^*(\vartheta_1) = a, \ \mathfrak{J}^*(\vartheta_2) = b.$$

In order to prove that α_f is valid under $\mathfrak{J}$ it is sufficient to show that the following formulae are valid under $\mathfrak{J}^*$.

(1) $\bigwedge_{\eta}\bigwedge_{\zeta}(\gamma(\vartheta_1, \vartheta_2, \eta, \zeta) \wedge \eta = 0 \rightarrow \alpha_g(\xi_1, \ldots, \xi_n, \zeta))$,

(2) $\bigwedge_{\eta}\bigwedge_{\zeta}\bigwedge_{\eta_1}\bigwedge_{\zeta_1}(\eta < \eta_2 \wedge \gamma(\vartheta_1, \vartheta_2, \eta, \zeta) \wedge \gamma(\vartheta_1, \vartheta_2, \eta_1, \zeta_1) \wedge \eta + 1 = \eta_1$
$$\rightarrow \alpha_h(\xi_1, \ldots, \xi_n, \eta, \zeta, \zeta_1)),$$

(3) $\gamma(\vartheta_1, \vartheta_2, \eta_2, \zeta_2)$.

About (1). Let $\mathfrak{J}_1^* \underset{\eta,\zeta}{=} \mathfrak{J}^*$. We have to show that α_g is valid under $\mathfrak{J}_1^*$ if $\gamma(\vartheta_1, \vartheta_2, \eta, \zeta) \wedge \eta = 0$ is valid under $\mathfrak{J}_1^*$. The latter means that $Gab\,\mathfrak{J}_1^*(\eta)\,\mathfrak{J}_1^*(\zeta)$ and $\mathfrak{J}_1^*(\eta) = 0$, i.e. that $Gab0\,\mathfrak{J}_1^*(\zeta)$. Then, because of (0) and (I) we have that $\mathfrak{J}_1^*(\zeta) = f(r_1, \ldots, r_n, 0) = g(r_1, \ldots, r_n) = g(\mathfrak{J}_1(\xi_1), \ldots, \mathfrak{J}_1(\xi_n)) = g(\mathfrak{J}_1^*(\xi_1), \ldots, \mathfrak{J}_1^*(\xi_n))$, from which follows, because of the hypothesis about α_g, that α_g is valid under $\mathfrak{J}_1^*$.

About (2). Let $\mathfrak{J}_1^* \underset{\eta,\zeta,\eta_1,\zeta_1}{=} \mathfrak{J}^*$. We have to show that α_h is valid under $\mathfrak{J}_1^*$ if $\eta < \eta_2$, $\gamma(\vartheta_1, \vartheta_2, \eta, \zeta)$, $\gamma(\vartheta_1, \vartheta_2, \eta_1, \zeta_1)$ and $\eta + 1 = \eta_1$ are valid under $\mathfrak{J}_1^*$. But that means that $\mathfrak{J}_1^*(\eta) < r$, and

$$Gab\,\mathfrak{J}_1^*(\eta)\,\mathfrak{J}_1^*(\zeta),\quad Gab\,\mathfrak{J}_1^*(\eta_1)\,\mathfrak{J}_1^*(\zeta_1),\quad \mathfrak{J}_1^*(\eta) + 1 = \mathfrak{J}_1^*(\eta_1).$$

From this follows in view of (0) and (I) that

$$\mathfrak{J}_1^*(\zeta) = f(r_1, \ldots, r_n, \mathfrak{J}_1^*(\eta)),\quad \mathfrak{J}_1^*(\zeta_1) = f(r_1, \ldots, r_n, \mathfrak{J}_1^*(\eta) + 1),$$

and from this, because of the inductive definition of f, that

$$\mathfrak{J}_1^*(\zeta_1) = h(r_1, \ldots, r_n, \mathfrak{J}_1^*(\eta), \mathfrak{J}_1^*(\zeta)),$$

i.e. $\mathfrak{J}_1^*(\zeta_1) = h(\mathfrak{J}_1^*(\xi_1), \ldots, \mathfrak{J}_1^*(\xi_n), \mathfrak{J}_1^*(\eta), \mathfrak{J}_1^*(\zeta))$. Thus, α_h is valid under $\mathfrak{J}_1^*$ because of the hypothesis about h.

About (3). We have assumed that $f(r_1, \ldots, r_n, r) = s$. From this we obtain by (0) that $Gabrs$, i.e. $G\mathfrak{J}^*(\vartheta_1)\,\mathfrak{J}^*(\vartheta_2)\,\mathfrak{J}^*(\eta_2)\,\mathfrak{J}^*(\zeta_2)$. Thus, $\gamma(\vartheta_1, \vartheta_2, \eta_2, \zeta_2)$ is valid under $\mathfrak{J}^*$ since this formula defines G.

2) *Now we assume conversely that α_f is valid under an interpretation $\mathfrak{J}$.* Then there exists an $\mathfrak{J}^*$ with $\mathfrak{J}^* \underset{\vartheta_1,\vartheta_2}{=} \mathfrak{J}$ such that the formulae (1), (2) and (3) are valid under $\mathfrak{J}^*$.

We have to show that $f(\mathfrak{J}(\xi_1), \ldots, \mathfrak{J}(\xi_n), \mathfrak{J}(\eta_2)) = \mathfrak{J}(\zeta_2)$, i.e. that $f(\mathfrak{J}^*(\xi_1), \ldots, \mathfrak{J}^*(\xi_n), \mathfrak{J}^*(\eta_2)) = \mathfrak{J}^*(\zeta_2)$. We put $a = \mathfrak{J}^*(\vartheta_1)$, $b = \mathfrak{J}^*(\vartheta_2)$, $r_1 = \mathfrak{J}^*(\xi_1)$, $\ldots$, $r_n = \mathfrak{J}^*(\xi_n)$, $r = \mathfrak{J}^*(\eta_2)$, $s = \mathfrak{J}^*(\zeta_2)$ and assert

(') $Gab0f(r_1, \ldots, r_n, 0)$.

Proof: According to (I) there exists a number k such that $Gab0k$. We put $\mathfrak{J}_1^* \underset{\eta,\zeta}{=} \mathfrak{J}^*$ and $\mathfrak{J}_1^*(\eta) = 0$, $\mathfrak{J}_1^*(\zeta) = k$. Then the following formulae are valid under $\mathfrak{J}_1^*$:

$\gamma(\vartheta_1, \vartheta_2, \eta, \zeta) \wedge \eta = 0 \rightarrow \alpha_g(\xi_1, \ldots, \xi_n, \zeta)$ (because (1) is valid under $\mathfrak{J}^*$)

$\gamma(\vartheta_1, \vartheta_2, \eta, \zeta)$ (because $Gab0k$)

$\eta = 0$ (because $\mathfrak{J}_1^*(\eta) = 0$)

and thus we get that $\alpha_g(\xi_1, \ldots, \xi_n, \zeta)$ is also valid under $\mathfrak{J}_1^*$. From this follows that $g(r_1, \ldots, r_n) = k$, i.e. that $k = f(r_1, \ldots, r_n, 0)$. Now the assertion follows, because $Gab0k$.

('') *If* $Gabif(r_1, \ldots, r_n, i)$ *for* $i < r$, *then* $Gab(i+1)\, f(r_1, \ldots, r_n, i+1)$.

Proof: According to (I) there exists a k such that $Gab(i+1)k$. We put $\mathfrak{J}_1^* \underset{\eta,\zeta,\eta_1,\zeta_1}{=} \mathfrak{J}^*$ and $\mathfrak{J}_1^*(\eta) = i$, $\mathfrak{J}_1^*(\zeta) = f(r_1, \ldots, r_n, i)$, $\mathfrak{J}_1^*(\eta_1) = i+1$, $\mathfrak{J}_1^*(\zeta_1) = k$. Then the following formulae are valid under $\mathfrak{J}_1^*$.

$$\eta < \eta_2 \wedge \gamma(\vartheta_1, \vartheta_2, \eta, \zeta) \wedge \gamma(\vartheta_1, \vartheta_2, \eta_1, \zeta_1) \wedge \eta + 1 = \eta_1 \rightarrow \alpha_h(\xi_1, \ldots, \xi_n, \eta, \zeta, \zeta_1)$$
(because (2) is valid under $\mathfrak{J}^*$)

$\eta < \eta_2$ (because $i < r$)

$\gamma(\vartheta_1, \vartheta_2, \eta, \zeta)$ (because $Gabif(r_1, \ldots, r_n, i)$)

$\gamma(\vartheta_1, \vartheta_2, \eta_1, \zeta_1)$ (because $Gab(i+1)k$)

$\eta + 1 = \eta_1$ (because $\mathfrak{J}_1^*(\eta) + 1 = \mathfrak{J}_1^*(\eta_1)$).

Thus, $\alpha_h(\xi_1, \ldots, \xi_n, \eta, \zeta, \zeta_1)$ is also valid under $\mathfrak{J}_1^*$. From this follows that

$$h(r_1, \ldots, r_n, i, f(r_1, \ldots, r_n, i)) = k,$$

i.e. that $k = f(r_1, \ldots, r_n, i+1)$. Now the assertion follows because $Gab(i+1)k$.

Now we can complete the proof as follows.

(3) is valid under $\mathfrak{J}^*$. That implies $Gabrs$. From (') and ('') follows that $Gabrf(r_1, \ldots, r_n, r)$. Because of (I) we have $f(r_1, \ldots, r_n, r) = s$, i.e. $f(\mathfrak{J}(\xi_1), \ldots, \mathfrak{J}(\xi_n), \mathfrak{J}(\eta_2)) = \mathfrak{J}(\zeta_2)$, q.e.d.

4. GÖDEL's *predicate* G. We put

$$\gamma^*(\vartheta_1, \vartheta_2, \eta, \zeta) \equiv \zeta < 1 + (\eta+1)\,\vartheta_2 \wedge \bigvee_{\xi} (1 + (\eta+1)\,\vartheta_2)\,\xi + \zeta = \vartheta_1.$$

We can turn γ^* into an arithmetical formula γ by eliminating the digit 1 by means of the arithmetical predicates $\zeta = 1$ and $\zeta_1 + 1 = \zeta_2$ (cf. Section 1). Thus γ^* (or γ) defines an arithmetical predicate G. It follows from the definition of G that $Gabik$ *if and only if* k *is the remainder which we obtain when we divide* a *by the number*

$$b_i = 1 + (i+1)\,b.$$

(This number is different from zero.) Then, there exists for a, b, i one and only one k such that $Gabik$, as it was asserted in Section 3 (I).

In order to show Section 3 (II) we start from an arbitrary sequence $k_0, \ldots, k_n$ of numbers. Let $m = \max(n, k_0, \ldots, k_n)$. We put $b = m!$.

First we show *that for* $0 \leq i \leq n$, $0 \leq j \leq n$, $i \neq j$ *the numbers* b_i *and* b_j *are relatively prime.* Proof by *reductio ad absurdum.* We assume that there exists a prime number p such that p/b_i and p/b_j. Then we have that $p/b_i - b_j = (i-j)\,b$ and therefore that $p/i-j$ or p/b. $|i-j| \leqq n \leqq m$ and $b = m!$. Therefore p/b follows from $p/i-j$. Thus, we only need to refute p/b. From p/b and p/b_i follows that $p/b_i - (i+1)\,b$ and therefore that $p/1$, which cannot possibly happen.

Now we show that (II) is valid if a is suitably chosen from the numbers $0, 1, 2, \ldots, b_0 b_1 \ldots b_n - 1$. (This assertion is also known under the name "Chinese remainder theorem".) Let

$$0 \leqq \bar{a} < b_0 b_1 \ldots b_n, \qquad 0 \leqq \bar{\bar{a}} < b_0 b_1 \ldots b_n, \qquad \bar{a} \neq \bar{\bar{a}}.$$

We consider the remainder systems

$$\bar{k}_i = \text{remainder of } \bar{a} \text{ modulo } b_i \qquad (i = 0, \ldots, n)$$

$$\bar{\bar{k}}_i = \text{remainder of } \bar{\bar{a}} \text{ modulo } b_i \qquad (i = 0, \ldots, n).$$

The *remainder system* $(\bar{k}_0, \ldots, \bar{k}_n)$ *is different from the remainder system* $(\bar{\bar{k}}_0, \ldots, \bar{\bar{k}}_n)$. Otherwise $\bar{k}_i = \bar{\bar{k}}_i$ for all i and, therefore, $b_i/\bar{a} - \bar{\bar{a}}$ for all i. Since the b_i's are relatively prime this gives that $b_0 b_1 \ldots b_n/\bar{a} - \bar{\bar{a}}$ which cannot possibly happen.

Since the numbers a with $0 \leqq a \leqq b_0 b_1 \ldots b_n - 1$ provide us with different remainder systems, we obtain altogether $b_0 b_1 \ldots b_n$ different remainder systems. Now there are altogether $b_0 b_1 \ldots b_n$ different sequences $b_0^*, \ldots, b_n^*$ of numbers with $b_0^* < b_0, \ldots, b_n^* < b_n$. Thus, each one of these sequences of numbers must occur once (and only once) as a remainder system of the above type. On the other hand the sequence $k_0, \ldots, k_n$ is a sequence of numbers of the required kind because $k_i \leqq m < m! = b < 1 + (i+1)\,b = b_i$. This shows that there exists an $a < b_0 \ldots b_n$ such that the assertion Section 3 (II) is valid.

References

GÖDEL, K.: Über formal unentscheidbare Sätze der Principia Mathematica und verwandter Systeme I. Mh. Math. Phys. **38**, 173–198 (1931).

— On Undecidable Propositions of Formal Mathematical Systems. Mimeographed. Institute for Advanced Study, Princeton, N. J. 1934. pp. 30.

TARSKI, A.: Der Wahrheitsbegriff in den formalisierten Sprachen. Studia Philosophica **1**, 261–405 (1935). Cf. also: Logic, Semantics, Metamathematics. Papers from 1923 to 1938 by A. TARSKI, translated by J. H. WOODGER, Oxford: Clarendon Press 1956. pp. 152–278.

CHURCH, A.: An Unsolvable Problem of Elementary Number Theory. Amer. J. Math. **58**, 345–363 (1936).

ROSSER, B.: Extensions of Some Theorems of Gödel and Church. J. symbolic Logic **1**, 87–91 (1936).

SKOLEM, TH.: Einfacher Beweis der Unmöglichkeit eines allgemeinen Lösungsverfahrens für arithmetische Probleme. Norske Vidensk. Selsk. Forhandl., Trondheim **13**, 1–4 (1940).

KALMÁR, L.: Egyszerü példa eldönthetetlen aritmetikai problémára. (Ein einfaches Beispiel für ein unentscheidbares arithmetisches Problem.) [Hungarian, with German abstract.] Mat. fiz. Lapok **50**, 1–23 (1943).

MOSTOWSKI, A.: Sentences Undecidable in Formalized Arithmetic. Amsterdam: North-Holland Publishing Company 1952.

TARSKI, A., A. MOSTOWSKI and R. M. ROBINSON: Undecidable Theories. Amsterdam: North-Holland Publishing Company 1953. (This book discusses especially the essential undecidability of the theories it deals with.)

GOODSTEIN, R. L.: Recursive Number Theory. Amsterdam: North-Holland Publishing Company 1957.

GRZEGORCZYK, A.: Fonctions Récursives. Collection de Logique Mathématique, Série A. Paris – Louvain: Gauthiers-Villars – E. Nauwelaerts 1961.

CHAPTER 7

MISCELLANEOUS

We shall show that every recursive predicate is arithmetical. Thus, the arithmetical predicates introduced in § 27.1 are generalizations of recursive predicates. We can divide (§ 29) the arithmetical predicates into classes (which have elements in common) where the smallest class is that of the recursive and a further class is that of the recursively enumerable predicates which we shall discuss in § 28.

In § 23 (in the introduction and in Sections 3 and 8) we hinted at the fact that one can give an account of special groups, semi-Thue systems and Thue systems with unsolvable word problems by the help of a universal Turing machine **U**. We shall construct such a machine in § 30. There we shall refer to the primitive recursive predicates which were introduced in § 18.

In §§ 31, 32 and 33 we shall discuss another few suggestions for precise replacements of constructive concepts.

Finally, in § 34, we shall show by simple examples how the theory of constructive concepts can be applied to analysis.

§ 28. Enumerable Predicates

In § 2.2 we introduced from an intuitive point of view enumerable sets and more generally enumerable predicates. According to the definitions given there a set of natural numbers is enumerable if it is either empty or the domain of values of a computable function; an n-ary predicate is enumerable if it is either empty or if there exist n computable singulary functions $f_1, \ldots, f_n$ such that, for all $\mathfrak{x}$, $P\mathfrak{x}$ is equivalent to the existence of a natural number y such that $x_1 = f_1(y), \ldots, x_n = f_n(y)$. Having made precise the concept of computable function we can proceed without any further difficulty to the concept of the so-called recursively enumerable predicates (Section 1). In Section 2 we shall prove a few theorems which establish a connection between the recursively enumerable predicates and the recursive predicates. The recursively enumerable predicates are, just like the recursive predicates, special arithmetical predicates (cf. § 29).

In this paragraph we shall use the word "*recursive*" to mean that we are dealing with a concept which serves as a precise replacement of an initially intuitively given concept, without committing ourselves to using the word "recursive" for a certain one of such precise replacements (cf. the remark in the introduction to Chapter 5).

1. Definition. An n-ary predicate P is called recursively enumerable if P is empty or if there exist n singulary recursive functions $f_1, \ldots, f_n$ such that for all $x_1, \ldots, x_n$

$$(1) \qquad P x_1 \ldots x_n \leftrightarrow \bigvee_y (f_1(y) = x_1 \wedge \ldots \wedge f_n(y) = x_n).$$

2. Theorems about recursively enumerable predicates.

Theorem 1. An n-ary predicate P is recursively enumerable if and only if there exists an $(n+1)$-ary predicate Q such that for all $\mathfrak{x}$

$$(2) \qquad P\mathfrak{x} \leftrightarrow \bigvee_y Q\mathfrak{x}y.$$

Proof. (a) First let P be recursively enumerable. If P is empty, then P, the empty predicate, is recursive (§ 11.5). Put $Q\mathfrak{x}y \leftrightarrow P\mathfrak{x} \wedge y = y$. Then we have that $P\mathfrak{x} \leftrightarrow \bigvee_y (P\mathfrak{x} \wedge y = y)$ and, therefore, $P\mathfrak{x} \leftrightarrow \bigvee_y Q\mathfrak{x}y$. — If P is not empty, then P can be represented in the form (1), in which $f_1(y) = x_1 \wedge \ldots \wedge f_n(y) = x_n$ is a recursive predicate $Q\mathfrak{x}y$, which completes the proof of (2).

(b) Let P be representable in the form (2) with recursive Q. We can assume that P is not empty (since otherwise the assertion is trivial). Let $\bar{\mathfrak{x}} = (\bar{x}_1, \ldots, \bar{x}_n)$ be an n-tuple of numbers (fixed for the time being) such that $P\bar{\mathfrak{x}}$. We define for $j = 1, \ldots, n$

$$(3) \qquad f_j(y) = \begin{cases} \sigma_{n+1,j}(y), \text{ if } Q\sigma_{n+1,1}(y) \ldots \sigma_{n+1,n+1}(y) \\ \bar{x}_j \text{ otherwise.} \end{cases}$$

This definition shows that the functions f_j are recursive (§ 11.6, § 12.4 and § 14.2). We shall show that assertion (1) is valid for these functions.

3_1) If $Px_1 \ldots x_n$, then there exists according to (2) a z such that $Q\mathfrak{x}z$. We put $y = \sigma_{n+1}(x_1, \ldots, x_n, z)$. Then we have that $Q\sigma_{n+1,1}(y) \ldots \sigma_{n+1,n+1}(y)$ and, therefore, $f_j(y) = \sigma_{n+1,j}(y) = x_j$ for $j = 1, \ldots, n$.

3_2) Now we start by assuming that there exists a y such that $f_1(y) = x_1$, $\ldots, f_n(y) = x_n$. We have to show that $Px_1 \ldots x_n$. According to the definition of the functions f_j we have to distinguish between two cases. (I) We have $Q\sigma_{n+1,1}(y) \ldots \sigma_{n+1,n+1}(y)$. Then $f_j(y) = \sigma_{n+1,j}(y)$. Thus we have $Qf_1(y) \ldots f_n(y)\, \sigma_{n+1,n+1}(y)$, i.e. $Qx_1 \ldots x_n\, \sigma_{n+1,n+1}(y)$. Thus, there exists a z such that $Qx_1 \ldots x_n z$, by which the left hand side of (1), namely $P\mathfrak{x}$, is proved. (II) We have not $Q\sigma_{n+1,1}(y) \ldots \sigma_{n+1,n+1}(y)$. Then, according to (3), $f_j(y) = \bar{x}_j$. Thus, according to our assumption, $x_j = \bar{x}_j$ $(j = 1, \ldots, n)$. Then we have that $Px_1 \ldots x_n$ because $P\bar{x}_1 \ldots \bar{x}_n$.

Thus, every predicate $\bigvee_y Q\mathfrak{x}y$ with recursive kernel Q represents a recursively enumerable predicate and every recursively enumerable predicate can be represented in this way. This explains the name we gave to a theorem which we already proved in § 18.6 and which we shall once more state in a somewhat different form:

Theorem 2. Kleene's enumeration theorem. For every n-ary recursively enumerable predicate P there exists a number t such that for all $\mathfrak{x}$

$$P\mathfrak{x} \leftrightarrow \bigvee_y T_n t\mathfrak{x}y. \tag{4}$$

Conversely, if we define a predicate P by the above bi-implication (with an arbitrary fixed t), then P is recursively enumerable.

Let us now assume that P is a non-empty recursively enumerable predicate. Then we can define, following part (b) of the above proof for Theorem 1, n functions $f_j(y)$ according to (3) (with $T_n t\mathfrak{x}y$ (with fixed t) instead of $Q\mathfrak{x}y$) and obtain in this way a representation of P of the form (1). The functions $f_j(y)$ under consideration are not only recursive but even *primitive recursive*. This follows (for the case considered here) quite easily from the definition (3), because T_n is primitive recursive. This proves

Theorem 3. An n-ary predicate P is recursively enumerable if and only if P is empty or there exist n singulary primitive recursive functions $f_1, \ldots, f_n$ *such that for all* $\mathfrak{x}$ *we have the representation* (1).

This theorem originates from Rosser.

There is an especially simple connection between recursive and recursively enumerable predicates.

Theorem 4. A predicate P is recursive if and only if both P and the complementary predicate (the negation) $\overline{P}$ (cf. § 11.2) *are recursively enumerable.*

Proof. (a) Let P be recursive. We introduce a new recursive predicate Q by the stipulation that, for all $\mathfrak{x}$ and y, $Q\mathfrak{x}y \leftrightarrow P\mathfrak{x} \wedge y = y$. Then we have for all $\mathfrak{x}$ that $\bigvee_y Q\mathfrak{x}y \leftrightarrow \bigvee_y (P\mathfrak{x} \wedge y = y) \leftrightarrow P\mathfrak{x}$. According to

Theorem 1 this representation shows that P is recursively enumerable. According to the theorem in § 14.2 $\overline{P}$ is recursive if P is, and, therefore, it is recursively enumerable according to the above considerations.

(b) Now we assume that both P and $\overline{P}$ are recursively enumerable. According to Theorem 1 there exist recursive predicates Q and R such that for all $\mathfrak{x}$

$$(5) \qquad P\mathfrak{x} \leftrightarrow \bigvee_y Q\mathfrak{x}y, \qquad \neg P\mathfrak{x} \leftrightarrow \bigvee_y R\mathfrak{x}y.$$

According to the law of excluded middle we have $P\mathfrak{x} \vee \neg P\mathfrak{x}$. Thus, according to (5), $\bigvee_y Q\mathfrak{x}y \vee \bigvee_y R\mathfrak{x}y$, i.e. $\bigvee_y (Q\mathfrak{x}y \vee R\mathfrak{x}y)$: Thus, the forming of $\mu y(Q\mathfrak{x}y \vee R\mathfrak{x}y)$ is an application of the μ-operator to a *regular* predicate and therefore, according to the definition of μ-recursive functions, $\mu y(Q\mathfrak{x}y \vee R\mathfrak{x}y)$ is a recursive function. Now, according to § 14.2, $Q\mathfrak{x}\mu y(Q\mathfrak{x}y \vee R\mathfrak{x}y)$ is a recursive predicate. Therefore, once we have shown that for all $\mathfrak{x}$

$$(6) \qquad P\mathfrak{x} \leftrightarrow Q\mathfrak{x}\mu y(Q\mathfrak{x}y \vee R\mathfrak{x}y),$$

then the proof is completed.

(a) We assume that $Q\mathfrak{x}\mu y(Q\mathfrak{x}y \vee R\mathfrak{x}y)$. Thus, there exists a y such that $Q\mathfrak{x}y$ (namely $\mu y(Q\mathfrak{x}y \vee R\mathfrak{x}y)$). Then we have, according to (5), that $P\mathfrak{x}$.

(b) We assume that $P\mathfrak{x}$. From that follows, according to (5), that $\neg \bigvee_y R\mathfrak{x}y$. Thus, for this $\mathfrak{x}$ and all y

$$Q\mathfrak{x}y \vee R\mathfrak{x}y \leftrightarrow Q\mathfrak{x}y.$$

We have therefore for this $\mathfrak{x}$ that

$$(7) \qquad \mu y(Q\mathfrak{x}y \vee R\mathfrak{x}y) = \mu y Q\mathfrak{x}y,$$

where in both cases the μ-operator is applied to a regular predicate. Now, from the regularity of $Q\mathfrak{x}y$ follows that $Q\mathfrak{x}\mu y Q\mathfrak{x}y$. Thus, according to (7), we also have that $Q\mathfrak{x}\mu y(Q\mathfrak{x}y \vee R\mathfrak{x}y)$, by which the proof of (6) is completed.

Theorem 5. Let the predicates Q and R be recursive. Let the two representations

$$(8) \qquad P\mathfrak{x} \leftrightarrow \bigvee_y Q\mathfrak{x}y, \qquad P\mathfrak{x} \leftrightarrow \bigwedge_y R\mathfrak{x}y$$

both be valid for the predicate P. Then P is recursive.

This theorem follows directly from the previous one, for $\neg P\mathfrak{x} \leftrightarrow \neg \bigwedge_y R\mathfrak{x}y \leftrightarrow \bigvee_y \neg R\mathfrak{x}y \leftrightarrow \bigvee_y \overline{R}\mathfrak{x}y$, and $\overline{R}$ is recursive since R is.

3. *Examples.* Now we shall give three examples. The first is an example of a non-enumerable predicate, which we used in § 27.2. The second is an example of an enumerable but non-decidable predicate. Finally, the third is an example of an enumerable predicate, about which we do not yet know whether or not it is decidable.

Example 1. We refer back to Kleene's ternary primitive recursive predicate T_1 (cf. § 18.3). We introduce a singulary predicate P by the stipulation that for every number x

$$Px \leftrightarrow \bigwedge_y \neg T_1 xxy.$$

We assert that P is *not recursively enumerable.* Proof by *reductio ad absurdum.* If P is recursively enumerable, then there exists, according to Theorem 2, a number t such that for all x

$$\bigwedge_y \neg T_1 xxy \leftrightarrow \bigvee_y T_1 txy.$$

From this follows especially for $x = t$ (diagonal procedure!)

$$\bigwedge_y \neg T_1 tty \leftrightarrow \bigvee_y T_1 tty.$$

Using the fact that $\bigwedge_y \neg \alpha$ is always equivalent to $\neg \bigvee_y \alpha$ we obtain the contradiction

$$\neg \bigvee_y T_1 tty \leftrightarrow \bigvee_y T_1 tty.$$

Example 2. Let the singulary predicate Q be defined by the stipulation that for all x

$$Qx \leftrightarrow \bigvee_y T_1 xxy.$$

According to Theorem 1 Q is recursively enumerable. Further we have that $Qx \leftrightarrow \neg Px$ for every x (cf. the previous example). Thus, Q and P are complementary predicates. If Q were recursive, then, according to Theorem 4, the predicate P would be recursively enumerable, which is not the case. *Thus, Q is recursively enumerable but not recursive.*

Example 3. We introduce a singulary predicate F which is connected with the unsolved problem of Fermat. We stipulate that for all n

$$Fn \leftrightarrow \bigvee_x \bigvee_y \bigvee_z (xyz \neq 0 \wedge x^n + y^n = z^n).$$

We have that $F1$ and $F2$ and, among others, $\neg F3$ and $\neg F4$. We do not know whether $\neg Fn$ for all remaining n. If this were the case then F would be decidable. One does not know whether F is decidable. However, F is recursively enumerable. In order to prove this we must find a representation of F which has only *one* existential operator and where the kernel is recursive. This can easily be done if we introduce $t = \sigma_3\,(x, y, z)$ since we obviously have that

$$Fn \leftrightarrow \bigvee_t (\sigma_{31}(t)\,\sigma_{32}(t)\,\sigma_{33}(t) \neq 0 \wedge \sigma_{31}(t)^n + \sigma_{32}(t)^n = \sigma_{33}(t)^n).$$

References

Post, E.: Recursively Enumerable Sets of Positive Integers and their Decision Problems. Bull. Amer. math. Soc. **50**, 284–316 (1944).

Robinson, R. M.: Arithmetical Representation of Recursively Enumerable Sets. J. symbolic Logic **21**, 162–186 (1956).

Rosser, B.: Extensions of Some Theorems of Gödel and Church. J. symbolic Logic **1**, 87–91 (1936).

Smullyan, R. M.: Theory of Formal Systems. Princeton, N. J.: Princeton University Press 1961, revised edition 1963.

§ 29. Arithmetical Predicates

In § 27.1 we introduced the arithmetical formulae and the arithmetical predicates defined by them. Now we shall show that the arithmetical predicates are closely connected with the recursive predicates. An essential proposition used here is Lemma 3 of § 27.2. By the help of recursive predicates Kleene and Mostowski introduced a denumerable sequence ("hierarchy") of classes of predicates. The smallest of these classes is the class of the recursive predicates, another one is that of the recursively enumerable predicates. The classes which come later in the sequence contain the earlier ones. In each step new predicates are added, which are "of a higher degree of undecidability" than the previous ones. — We shall conclude the paragraph with a few remarks about the unsolved Hilbert's tenth problem.

1. The Kleene-Mostowski hierarchy. Let $\bigcirc$, $\bigvee$, $\bigwedge$, $\bigvee\bigwedge$, $\bigwedge\bigvee$, $\bigvee\bigwedge\bigvee$, ... be classes of predicates of arbitrary number of arguments, which are defined as follows.

$P \in \bigcirc$ $=_{Df}$ there exists a recursive predicate R such that for all $\mathfrak{x}$: $P\mathfrak{x} \leftrightarrow R\mathfrak{x}$

$P \in \bigvee$ $=_{Df}$ there exists a recursive predicate R such that for all $\mathfrak{x}$: $P\mathfrak{x} \leftrightarrow \bigvee_{y_1} R\mathfrak{x}y_1$

$P \in \bigwedge$ $=_{Df}$ there exists a recursive predicate R such that for all $\mathfrak{x}$: $P\mathfrak{x} \leftrightarrow \bigwedge_{y_1} R\mathfrak{x}y_1$

$P \in \bigvee\bigwedge$ $=_{Df}$ there exists a recursive predicate R such that for all $\mathfrak{x}$: $P\mathfrak{x} \leftrightarrow \bigvee_{y_1}\bigwedge_{y_2} R\mathfrak{x}y_1y_2$

$P \in \wedge\vee$ $=_{Df}$ there exists a recursive predicate R such that for all $\mathfrak{x}$: $P\mathfrak{x} \leftrightarrow \bigwedge_{y_1}\bigvee_{y_2} R\mathfrak{x}y_1y_2$

$P \in \vee\wedge\vee$ $=_{Df}$ there exists a recursive predicate R such that for all $\mathfrak{x}$: $P\mathfrak{x} \leftrightarrow \bigvee_{y_1}\bigwedge_{y_2}\bigvee_{y_3} R\mathfrak{x}y_1y_2y_3$

etc.

Thus, $\bigcirc$ is the class of recursive predicates, $\vee$ is the class of recursively enumerable predicates.

The classes can be ordered in a two line schema (where the class $\bigcirc$ appears in both lines) in the following way:

$$(*) \qquad \begin{cases} \bigcirc & \vee & \vee\wedge & \vee\wedge\vee & \vee\wedge\vee\wedge & \dots \\ \bigcirc & \wedge & \wedge\vee & \wedge\vee\wedge & \wedge\vee\wedge\vee & \dots \end{cases}$$

We shall call the classes which are in the same column of this schema *corresponding classes*. For example, $\bigcirc$ and $\bigcirc$, $\wedge$ and $\vee$ are corresponding classes. — We assert the

Theorem. (a) Every class occurring in schema (*) has only arithmetical predicates as members.

(b) Every arithmetical predicate belongs to at least one of the classes of the schema (*).

(c) Every class of the schema (*) is contained in every class of the schema which is on the right of it (above or below).

(d) Two corresponding classes have as members each the complementary predicates of the predicates of the other.

(e) In every class, with the exception of $\bigcirc$, there exists a predicate which occurs in none of the classes on the left of it (above or below) and also not in the corresponding class. Furthermore, such predicates exist with arbitrary number $\geqq 1$ of arguments.

(f) Let for all $\mathfrak{x}$

$$P_3\mathfrak{x} \leftrightarrow P_1\mathfrak{x} \wedge P_2\mathfrak{x}, \quad P_4\mathfrak{x} \leftrightarrow P_1\mathfrak{x} \vee P_2\mathfrak{x}.$$

If P_1 and P_2 are in a class of the schema (*), then P_3 and P_4 are also in the same class. Thus, every class is closed under conjunction and alternative.

(g) The intersection of the classes $\vee$ and $\wedge$ is $\bigcirc$.

(h) Apart from the case mentioned in (g) the intersection of any two corresponding classes has more elements than the union of the two preceding corresponding classes.

The assertions (a), (b), (c), (e), (g), (h) can be represented in the following *figure*.

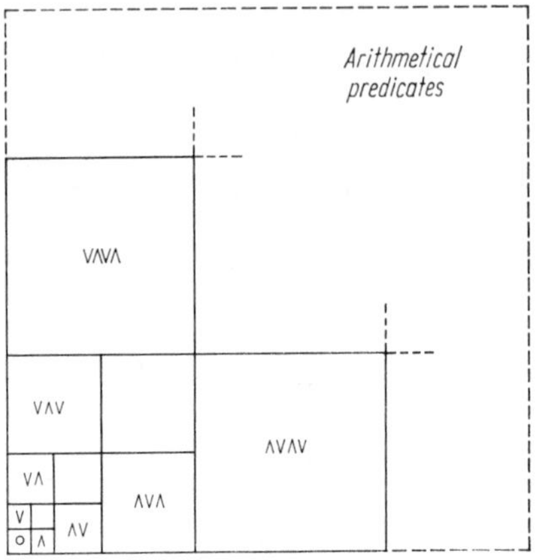

Fig. 29.1. The Kleene-Mostowski Hierarchy of the Arithmetical Predicates

In the diagram:

O is represented by the left bottom square.

The classes V ... are represented by the square which carries the symbol "V ...", *together* with the neighbouring square *below*.

The classes Λ ... are represented by the square which carries the symbol "Λ ...", *together* with the neighbouring square on the *left*.

2. *The proof* of the assertions (a), ..., (h) is carried out in a partially altered sequence.

About (a). We show

(1) *If the n-ary predicate P is arithmetical and the* $(n-1)$*-ary predicate Q is obtained by the i-th generalization or i-th particularization of P, then Q is also arithmetical.* Indeed, if $\alpha(\xi_1, \ldots, \xi_n)$ defines the predicate P, then $\bigwedge_{\xi_i} \alpha(\xi_1, \ldots, \xi_n)$ or $\bigvee_{\xi_1} \alpha(\xi_1, \ldots, \xi_n)$ respectively defines the predicate Q. We show this in detail only in the case of generalization. According to § 27 Lemma 1, we have to show that for an arbitrary interpretation $\mathfrak{I}$ the formula $\bigwedge_{\xi_i} \alpha$ is valid under $\mathfrak{I}$ if and only if

$$Q\,\mathfrak{I}(\xi_1) \ldots \mathfrak{I}(\xi_{i-1})\,\mathfrak{I}(\xi_{i+1}) \ldots \mathfrak{I}(\xi_n),$$

where a similar situation is assumed for α and P. We have that

$\bigwedge_{\xi_i} \alpha$ is valid under $\mathfrak{I}$ if and only if α is valid under every $\mathfrak{I}' \underset{\xi_1}{=} \mathfrak{I}$
if and only if $P\,\mathfrak{I}'(\xi_1) \ldots \mathfrak{I}'(\xi_n)$ for every $\mathfrak{I}' \underset{\xi_i}{=} \mathfrak{I}$
if and only if $P\,\mathfrak{I}(\xi_1) \ldots \mathfrak{I}'(\xi_i) \ldots \mathfrak{I}(\xi_n)$ for every $\mathfrak{I}' \underset{\xi_i}{=} \mathfrak{I}$
if and only if $P\,\mathfrak{I}(\xi_1) \ldots r_i \ldots \mathfrak{I}(\xi_n)$ for every r_i
if and only if $Q\,\mathfrak{I}(\xi_1) \ldots \mathfrak{I}(\xi_{i-1})\,\mathfrak{I}(\xi_{i+1}) \ldots \mathfrak{I}(\xi_n)$.

(2) *Every primitive recursive predicate P is arithmetical.* If P is primitive recursive, then there exists a primitive recursive function f such that $P\mathfrak{x}$ if and only if $f(\mathfrak{x}) = 0$. It follows from Lemma 3 and Definition 1 of § 27 that the predicate Q for which $Q\mathfrak{x}y$ if and only if $f(\mathfrak{x}) = y$ is defined

by an arithmetical formula $\alpha_f(\xi_1, \ldots, \xi_n, \xi)$. P is then defined by the formula $\bigvee_{\xi}(\alpha_f(\xi_1, \ldots, \xi_n, \xi) \wedge \xi = 0)$ and so it is arithmetical.

(3) *Every recursive predicate P is arithmetical.* Let P be a recursive predicate. According to § 28.2, Theorem 4, P is recursively enumerable. Therefore, there exists according to Kleene's enumeration theorem a number t such that $P\mathfrak{x} \leftrightarrow \bigvee_y T_n t\mathfrak{x}y$ for all $\mathfrak{x}$. The $(n+1)$-ary predicate Q which is defined by $Q\mathfrak{x}y \leftrightarrow T_n t\mathfrak{x}y$ (for this fixed t) is arithmetical by (2). (As a matter of fact, Q is primitive recursive, since it is obtained from T by substitution of a primitive recursive function for one argument of T.) We have that $P\mathfrak{x} \leftrightarrow \bigvee_y Q\mathfrak{x}y$. Thus, P is arithmetical by (1).

(4) *Every predicate of* $\bigcirc$ *is arithmetical according to* (3). That *the predicates of the other classes are also arithmetical* follows from this by (1).

About (c). It is sufficient to show that a predicate P which is in any one of the Kleene-Mostowski classes is also in both classes which are immediately on the right (above or below) of the class in question. We carry out the proof for the class $\wedge\vee\wedge$. (In all other cases the proofs are similar.) We have for all $\mathfrak{x}$ that

$$\begin{aligned} P\mathfrak{x} &\leftrightarrow \bigwedge_{y_1}\bigvee_{y_2}\bigwedge_{y_3} R\mathfrak{x}y_1y_2y_3 \\ &\leftrightarrow \bigwedge_{y_2}\bigvee_{y_3}\bigwedge_{y_4} R\mathfrak{x}y_2y_3y_4 \\ &\leftrightarrow \bigvee_{y_1}\bigwedge_{y_2}\bigvee_{y_3}\bigwedge_{y_4}(y_1 = y_1 \wedge R\mathfrak{x}y_2y_3y_4)\,. \end{aligned}$$

Since $y_1 = y_1 \wedge R\mathfrak{x}y_2y_3y_4$ is recursive, this representation shows that $P \in \vee\wedge\vee\wedge$. That $P \in \wedge\vee\wedge\vee$ as well follows from the fact that for all $\mathfrak{x}$

$$\begin{aligned} P\mathfrak{x} &\leftrightarrow \bigwedge_{y_1}\bigvee_{y_2}\bigwedge_{y_3} R\mathfrak{x}y_1y_2y_3 \\ &\leftrightarrow \bigwedge_{y_1}\bigvee_{y_2}\bigwedge_{y_3}\bigvee_{y_4}(R\mathfrak{x}y_1y_2y_3 \wedge y_4 = y_4)\,. \end{aligned}$$

About (d). For the class $\bigcirc$ this assertion means that a predicate is the complement of a recursive predicate if and only if it is recursive itself. This follows directly from the Theorem in § 14.2. Let us now take any other class. We continue with our considerations using the example $\vee\wedge\vee$. (In all other cases the proofs are similar.) We have to show that for an arbitrary $P \in \vee\wedge\vee$ the complement $\overline{P}$ is in $\wedge\vee\wedge$ and, conversely, that for an arbitrary $Q \in \wedge\vee\wedge$ the complement $\overline{Q}$ is in $\vee\wedge\vee$. We shall only prove the first half (the second follows similarly). If $P \in \vee\wedge\vee$, then there exists a recursive predicate R such that for all $\mathfrak{x}$

$$P\mathfrak{x} \leftrightarrow \bigvee_{y_1}\bigwedge_{y_2}\bigvee_{y_3} R\mathfrak{x}y_1y_2y_3$$

and so
$$\neg P\mathfrak{x} \leftrightarrow \neg \bigvee_{y_1} \bigwedge_{y_2} \bigvee_{y_3} R\mathfrak{x}y_1y_2y_3$$
$$\leftrightarrow \bigwedge_{y_1} \bigvee_{y_2} \bigwedge_{y_3} \neg R\mathfrak{x}y_1y_2y_3.$$

Since $\bar{R}$ is recursive, this representation shows that $\bar{P} \in \wedge\vee\wedge$.

About (b). The arithmetical predicates are defined by arithmetical formulae. We shall prove by induction on the structure of these formulae that every such formula defines a predicate which is in one of the classes of the schema (*).

(b_1) For every term t which contains at most the variables $\xi_1, \ldots, \xi_n$ in any sequence whatsoever, there obviously exists a primitive recursive function f such that for every interpretation $\mathfrak{I}(t) = f(\mathfrak{I}(\xi_1), \ldots, \mathfrak{I}(\xi_n))$. Let us consider a given arithmetical formula $t_1 = t_2$, in which the variables $\xi_1, \ldots, \xi_n$ (ordered according to their position in the natural sequence) may occur. This equation defines an arithmetical predicate P for which

$P r_1 \ldots r_n$ if and only if
there exists an $\mathfrak{I}$ such that $\mathfrak{I}(\xi_1) = r_1, \ldots, \mathfrak{I}(\xi_n) = r_n$ under which $t_1 = t_2$ is valid.

That $t_1 = t_2$ is valid under $\mathfrak{I}$ means that $\mathfrak{I}(t_1) = \mathfrak{I}(t_2)$, i.e. that $f_1(\mathfrak{I}(\xi_1), \ldots, \mathfrak{I}(\xi_n)) = f_2(\mathfrak{I}(\xi_1), \ldots, \mathfrak{I}(\xi_n))$, where f_1 and f_2 are the functions associated with the terms t_1 and t_2 respectively according to the previous remark. This gives that

$$P r_1 \ldots r_n \text{ if and only if } f_1(r_1, \ldots, r_n) = f_2(r_1, \ldots, r_n).$$

This representation shows that P is (even primitive) recursive and, therefore, belongs to the class O.

(b_2) Let the arithmetical predicate P defined by α belong to one of the classes of the schema (*). The predicate Q defined by $\neg\alpha$ is the complement of P and so it belongs to the complementary class, as we have already shown under *About* (d).

(b_3) Let the predicates P_1 and P_2 defined by the formulae α_1 and α_2 respectively belong to classes of the schema (*). We have to show that the predicate Q defined by the formula $\alpha_1 \wedge \alpha_2$ also belongs to such a class. Let $\xi_1, \ldots, \xi_n$ (given according to their order in the natural sequence) be the free variables occurring in α_1 or α_2. We shall use the notation $\alpha_i[\xi_1, \ldots, \xi_n]$ to denote that not necessarily all of the variables $\xi_1, \ldots, \xi_n$ occur free in α_i. Now we have, according to Lemma 1 of § 27.1, that for all $\mathfrak{I}$

$$\alpha_1[\xi_1, \ldots, \xi_n] \text{ is valid under } \mathfrak{I} \text{ if and only if } P_1[\mathfrak{I}(\xi_1) \ldots \mathfrak{I}(\xi_n)],$$
$$\alpha_2[\xi_1, \ldots, \xi_n] \text{ is valid under } \mathfrak{I} \text{ if and only if } P_2[\mathfrak{I}(\xi_1) \ldots \mathfrak{I}(\xi_n)],$$

where the notation $P_i[\mathfrak{J}(\xi_1) \ldots \mathfrak{J}(\xi_n)]$ means that $\mathfrak{J}(\xi_j)$ is to be omitted for those ξ_j which do not occur free in α_i. Finally we have that for all $\mathfrak{J}$

$$(\alpha_1 \wedge \alpha_2)\ (\xi_1, \ldots, \xi_n) \text{ is valid under } \mathfrak{J} \text{ if and only if } Q\mathfrak{J}(\xi_1) \ldots \mathfrak{J}(\xi_n).$$

From these three relations follows that for all $\mathfrak{J}$

$$Q\mathfrak{J}(\xi_1) \ldots \mathfrak{J}(\xi_n) \text{ if and only if } P_1[\mathfrak{J}(\xi_1) \ldots \mathfrak{J}(\xi_n)] \text{ and } P_2[\mathfrak{J}(\xi_1) \ldots \mathfrak{J}(\xi_n)].$$

This is equivalent to saying that for all $\mathfrak{x}$

$$Q\mathfrak{x} \leftrightarrow P_1[\mathfrak{x}] \wedge P_2[\mathfrak{x}].$$

Because of (c) we can assume that P_1 is in a class $\ldots \vee$ and that P_2 is in a class $\wedge \ldots$. In order to avoid complicated formulae we assume that $P_1 \in \wedge\vee$ and $P_2 \in \wedge\vee\wedge$. Then there exist recursive predicates R_1 and R_2 such that for all $\mathfrak{x}$

$$P_1[\mathfrak{x}] \leftrightarrow \bigwedge_{y_1} \bigvee_{y_2} R_1[\mathfrak{x}] y_1 y_2 \qquad P_2[\mathfrak{x}] \leftrightarrow \bigwedge_{y_1} \bigvee_{y_2} \bigwedge_{y_3} R_2[\mathfrak{x}] y_1 y_2 y_3.$$

It follows that

$$Q\mathfrak{x} \leftrightarrow P_1[\mathfrak{x}] \wedge P_2[\mathfrak{x}] \leftrightarrow \bigwedge_{y_1} \bigvee_{y_2} R_1[\mathfrak{x}] y_1 y_2 \wedge \bigwedge_{y_3} \bigvee_{y_4} \bigwedge_{y_5} R_2[\mathfrak{x}] y_3 y_4 y_5$$

$$\leftrightarrow \bigwedge_{y_1} \bigvee_{y_2} \bigwedge_{y_3} \bigvee_{y_4} \bigwedge_{y_5} (R_1[\mathfrak{x}] y_1 y_2 \wedge R_2[\mathfrak{x}] y_3 y_4 y_5).$$

This representation shows that $Q \in \wedge\vee\wedge\vee\wedge$, since the kernel describes a predicate which is recursive by § 14.2.

(b_4) Let the predicate P defined by α be in one of the classes of the schema (*). Because of (c) we can assume that $P \in \vee\ldots$. Obviously the predicate defined by $\bigwedge_{\xi} \alpha$ is in the class $\wedge\vee \ldots$.

About (e). According to (c) it is sufficient to show that in every class which is different from O there exists a predicate P which does not occur in the corresponding class. We show that there exists a predicate in the class $\wedge\vee\wedge$ which is not in the corresponding class $\vee\wedge\vee$. We can give a similar proof for every class $\ldots\wedge$. For the classes $\ldots\vee$ cf. the concluding remarks. We have already discussed a special case in § 28.3, Example 2.

We show that the predicate given by $\bigwedge_{y_1} \bigvee_{y_2} \bigwedge_{y_3} \neg T_3 x x y_1 y_2 y_3$, which is obviously in the class $\wedge\vee\wedge$, does not belong to $\vee\wedge\vee$. We do this by *reductio ad absurdum*. We assume that there exists a recursive predicate R such that for all x

$$\bigwedge_{y_1} \bigvee_{y_2} \bigwedge_{y_3} \neg T_3 x x y_1 y_2 y_3 \leftrightarrow \bigvee_{y_1} \bigwedge_{y_2} \bigvee_{y_3} R x y_1 y_2 y_3.$$

According to Kleene's enumeration theorem (in the form of § 18.6) there exists a number t such that for all x, y_1 and y_2

$$\bigvee_{y_3} R x y_1 y_2 y_3 \leftrightarrow \bigvee_{y_3} T_3 t x y_1 y_2 y_3.$$

It follows that for such a t and for all x

$$\bigwedge_{y_1}\bigvee_{y_2}\bigwedge_{y_3}\neg T_3xxy_1y_2y_3 \leftrightarrow \bigvee_{y_1}\bigwedge_{y_2}\bigvee_{y_3} T_3txy_1y_2y_3.$$

We have this relation for all x and, therefore, especially for $x=t$ (*diagonal procedure*!). This leads to the contradiction

$$\bigwedge_{y_1}\bigvee_{y_2}\bigwedge_{y_3}\neg T_3tty_1y_2y_3 \leftrightarrow \bigvee_{y_1}\bigwedge_{y_2}\bigvee_{y_3} T_3tty_1y_2y_3.$$

In this way we can show that in every class $\ldots\bigwedge$ there exists a predicate P which is not in the corresponding class $\ldots\bigvee$. Then, the complement $\overline{P}$ of such a predicate P is in $\ldots\bigvee$ by (d), but not in $\ldots\bigwedge$, since otherwise $\overline{\overline{P}}$, i.e. P, would be in $\ldots\bigvee$.

With this the existence of a *singulary* predicate with the required property is shown. The existence of such predicates with more arguments follows from this, as shown below, where we shall deal with the case of binary predicates as a typical example. Let the singulary predicate already obtained be P, where, for all x, Px if and only if

$$\prod_{y_1,\ldots,y_n} Rxy_1\ldots y_n.$$

In this representation the prefix $\prod$ characterizes the hierarchy class of P, and R is a recursive predicate. Now we consider the binary predicate Q where, for all x and z, Qxz if and only if

$$\prod_{y_1,\ldots,y_n} Rxy_1\ldots y_n\wedge z=z.$$

Trivially, this predicate can be written in the form

$$\prod_{y,\ldots,y_n} (Rxy_1\ldots y_n\wedge z=z).$$

This shows that the predicate Q is in the hierarchy class $\prod$. If Q where also in a hierarchy class $\prod'$ which occurs on the left of $\prod$ or which corresponds to $\prod$, then we would have that, for all x and z, Qxz if and only if

$$\prod'_{u_1,\ldots,u_r} Sxzu_1\ldots u_r,$$

where S is recursive. From that follows that, for all x, Px if and only if

$$\prod'_{u_1,\ldots,u_r} Sxxu_1\ldots u_r.$$

Thus P would also be in the class $\prod'$, which is not the case.

About (g). This is essentially Theorem 5 of § 28.2.

About (f). We carry out the proof for the class $\bigvee\bigwedge\bigvee$. (For other classes the assertion can be proved in a similar way.) First of all, we have for arbitrary predicates Q_1 and Q_2 (we only write down the essential arguments) that

$$\bigwedge_y Q_1 y \wedge \bigwedge_y Q_2 y \leftrightarrow \bigwedge_y (Q_1 y \wedge Q_2 y)$$

$$\bigvee_y Q_1 y \vee \bigvee_y Q_2 y \leftrightarrow \bigvee_y (Q_1 y \vee Q_2 y)$$

$$\bigvee_y Q_1 y \wedge \bigvee_y Q_2 y \leftrightarrow \bigvee_y (Q_1 \sigma_{21}(y) \wedge Q_2 \sigma_{22}(y))$$

$$\bigwedge_y Q_1 y \vee \bigwedge_y Q_2 y \leftrightarrow \bigwedge_y (Q_1 \sigma_{21}(y) \vee Q_2 \sigma_{22}(y)).$$

The first two bi-implications are purely logical. We show the last but one (the last can be shown in a similar way). We need only provide an argument to show that if $\bigvee_y Q_1 y$ and $\bigvee_y Q_2 y$, then $\bigvee_y (Q_1 \sigma_{21}(y) \wedge Q_2 \sigma_{22}(y))$. For this purpose we take a y_1 such that $Q_1 y_1$ and a y_2 such that $Q_2 y_2$ and put $y = \sigma_2(y_1, y_2)$.

Now we show the assertion of (f) for P_3 (for P_4 we can give a similar argument).

Because $P_1 \in \bigvee\bigwedge\bigvee$ and $P_2 \in \bigvee\bigwedge\bigvee$ we have the representations

$$P_1 \mathfrak{x} \leftrightarrow \bigvee_{y_1} \bigwedge_{y_2} \bigvee_{y_3} R_1 \mathfrak{x} y_1 y_2 y_3, \qquad P_2 \mathfrak{x} \leftrightarrow \bigvee_{y_1} \bigwedge_{y_2} \bigvee_{y_3} R_2 \mathfrak{x} y_1 y_2 y_3$$

where R_1 and R_2 are recursive predicates. It follows that for every $\mathfrak{x}$

$$\begin{aligned} P_3 \mathfrak{x} &\leftrightarrow \bigvee_{y_1} \bigwedge_{y_2} \bigvee_{y_3} R_1 \mathfrak{x} y_1 y_2 y_3 \wedge \bigvee_{y_1} \bigwedge_{y_2} \bigvee_{y_3} R_2 \mathfrak{x} y_1 y_2 y_3 \\ &\leftrightarrow \bigvee_{y_1} (\bigwedge_{y_2} \bigvee_{y_3} R_1 \mathfrak{x} \sigma_{21}(y_1) y_2 y_3 \wedge \bigwedge_{y_2} \bigvee_{y_3} R_2 \mathfrak{x} \sigma_{22}(y_1) y_2 y_3) \\ &\leftrightarrow \bigvee_{y_1} \bigwedge_{y_2} (\bigvee_{y_3} R_1 \mathfrak{x} \sigma_{21}(y_1) y_2 y_3 \wedge \bigvee_{y_3} R_2 \mathfrak{x} \sigma_{22}(y_1) y_2 y_3) \\ &\leftrightarrow \bigvee_{y_1} \bigwedge_{y_2} \bigvee_{y_3} (R_1 \mathfrak{x} \sigma_{21}(y_1) y_2 \sigma_{21}(y_3) \wedge R_2 \mathfrak{x} \sigma_{22}(y_1) y_2 \sigma_{22}(y_3)). \end{aligned}$$

The last formula contains a recursive kernel and, therefore, shows that $P_3 \in \bigvee\bigwedge\bigvee$.

About (h). We shall show that there exists a predicate Q which is both in the class $\bigvee\bigwedge\cdots$ and in the corresponding class $\bigwedge\bigvee$--- (where --- is obtained from $\cdots$ by interchanging the symbols $\bigwedge$ and $\bigvee$), but neither in the class $\bigvee$--- (which precedes $\bigvee\bigwedge\cdots$) nor in the corresponding class $\bigwedge\cdots$ (which precedes $\bigwedge\bigvee$---). For this purpose we start with a singulary predicate P which is in $\bigvee$--- but not in $\bigwedge\cdots$ (see (e)). Then the complement $\bar{P}$ is in $\bigwedge\cdots$ but not in $\bigvee$--- (see (d)). Thus, we have the representations

$$Px \leftrightarrow \bigvee_{y_1} \text{---} Rxy_1 \ldots, \qquad \neg Px \leftrightarrow \bigwedge_{y_1} \cdots \neg Rxy_1 \ldots,$$

where R is recursive. Now we define the binary predicate Q by the stipulation that, for all x and z,

$$Qxz \leftrightarrow (Px \wedge z = 0) \vee (\neg Px \wedge z = 1).$$

Because $Px \wedge z = 0 \leftrightarrow \bigvee_{y_1} \text{---} (Rxy_1 \cdots \wedge z = 0)$ the binary predicate P_1 defined by $Px \wedge z = 0$ is in $\bigvee \text{---}$. Because

$$\neg Px \wedge z = 1 \leftrightarrow \bigwedge_{y_1} \cdots (\neg Rxy_1 \cdots \wedge z = 1)$$

the binary predicate P_2 defined by $\neg Px \wedge z = 1$ is in $\bigwedge \cdots$. According to (c) P_1 and P_2 are in both $\bigvee\bigwedge \cdots$ and $\bigwedge\bigvee \text{---}$. Because

$$Qxz \leftrightarrow P_1 xz \vee P_2 xz$$

the predicate Q is also in both $\bigvee\bigwedge \cdots$ and $\bigwedge\bigvee \text{---}$ by (f). Thus, Q is in the intersection of these two classes.

It remains to be shown that Q is neither in $\bigvee \text{---}$ nor in $\bigwedge \cdots$. By symmetry we can confine ourselves to showing that $Q \notin \bigvee \text{---}$. Proof by *reductio ad absurdum*. If Q is in $\bigvee \text{---}$, then there exists a recursive predicate R_0 such that for all x and z

$$Qxz \leftrightarrow \bigvee_{y_1} \text{---} R_0 xzy_1 \ldots.$$

It follows that $Qx1 \leftrightarrow \bigvee_{y_1} \cdots R_0 x1y_1 \cdots$ and so the singulary predicate given by $Qx1$ is also in the class $\bigvee \text{---}$. On the other hand, it is obvious that $Qx1 \leftrightarrow \neg Px$. Thus, $\overline{P} \in \bigvee \text{---}$, in contradiction to our assumption.

3. *Hilbert's tenth problem* is that of finding an algorithm by the help of which we can decide for every diophantine equation whether or not it is solvable. We shall classify this until now unsolved problem in the complex of questions dealt with in this book.

A *diophantine equation* is an equation of the form $P = 0$, where P is a polynomial of n variables $x_1, \ldots, x_n$ with *integer* coefficients. We say that $P = 0$ is *solvable* if there exist *integers* $g_1, \ldots, g_n$ for which P has the value zero.

First we shall show that the problem can be reduced to a problem in which only polynomials with *natural number* coefficients are considered and in which we require a solution in *natural numbers*. Let us start with the latter. Let $(\varepsilon_1, \ldots, \varepsilon_n)$ be a sequence of numbers ε_i, with $\varepsilon_i = \pm 1$. There are 2^n such sequences and with every one of them we associate a polynomial $P_{\varepsilon_1, \ldots, \varepsilon_n}$ by the definition.

$$P_{\varepsilon_1, \ldots, \varepsilon_n}(z_1, \ldots, z_n) = P(\varepsilon_1 z_1, \ldots, \varepsilon_n z_n).$$

Now, if $P_{\varepsilon_1, \ldots, \varepsilon_n}$ has a solution in natural numbers $r_1, \ldots, r_n$, then P has the solution $\varepsilon_1 r_1, \ldots, \varepsilon_n r_n$ in integers. If on the other hand P has a solution $g_1, \ldots, g_n$ in integers and if ε_i is chosen so that $\varepsilon_i g_i \geqq 0$ $(i = 1, \ldots, n)$, then $P_{\varepsilon_1,\ldots,\varepsilon_n}$ has the solution $\varepsilon_1 g_1, \ldots, \varepsilon_n g_n$ in natural numbers. From these remarks follows that P has a solution in integers if and only if there exists at least one polynomial $P_{\varepsilon_1,\ldots,\varepsilon_n}$ which has a solution in natural numbers. Instead of this we can also say that the polynomial $Q = \prod_{\varepsilon_1,\ldots,\varepsilon_n} P_{\varepsilon_1,\ldots,\varepsilon_n}$ has a solution in *natural numbers*. Therefore we need only consider solvability in natural numbers.

We can eliminate the negative coefficients occurring in a polynomial equation $Q_1 = 0$ in a trivial way by taking the parts in question over to the right hand side. This way we obtain an equation of the form $P = Q$, where P and Q are terms which can be formed from variables and symbols for the constants 0, 1, 2, ... We shall call such equations *polynomial equations.* If $x_1, \ldots, x_n$ are the variables occurring in $P = Q$, then this polynomial equation has a solution if and only if $\bigvee_{x_1} \ldots \bigvee_{x_n} P = Q$ is true. *Thus, Hilbert's tenth problem is that of finding an algorithm by the help of which we can decide whether or not an arbitrary arithmetical sentence of the form*

$$(*) \qquad \bigvee_{x_1} \ldots \bigvee_{x_n} P = Q$$

is true.

We can formulate the connection with arithmetical sentences even more pointedly. Namely, we can *effectively associate with every arithmetical sentence α a polynomial equation $P = Q$ and a prefix of the form* $\mathbb{X}_{x_1} \cdots \mathbb{X}_{x_n}$ (where $\mathbb{X}$ stands for either $\bigwedge$ or $\bigvee$) *such that α is true if and only if*

$$(**) \qquad \mathbb{X}_{x_1} \ldots \mathbb{X}_{x_n} P = Q$$

is true. We show, more generally, that we can associate with every arithmetical *formula* an equivalent formula of the given form (where variables can occur in P and Q which are not bounded by the "prefix" $\mathbb{X}_{x_1} \ldots \mathbb{X}_{x_n}$). We confine ourselves to giving the essential steps.

In the proof we make use of the following relations.

$$(1) \qquad P_1 = Q_1 \wedge P_2 = Q_2 \leftrightarrow (P_1 - Q_1) = 0 \wedge (P_2 - Q_2) = 0$$
$$\leftrightarrow (P_1 - Q_1)^2 + (P_2 - Q_2)^2 = 0$$
$$\leftrightarrow P_1^2 + Q_1^2 + P_2^2 + Q_2^2 = 2P_1Q_1 + 2P_2Q_2$$

$$(2) \qquad P_1 = Q_1 \vee P_2 = Q_2 \leftrightarrow (P_1 - Q_1) = 0 \vee (P_2 - Q_2) = 0$$
$$\leftrightarrow (P_1 - Q_1)(P_2 - Q_2) = 0$$
$$\leftrightarrow P_1P_2 + Q_1Q_2 = P_1Q_2 + Q_1P_2$$

$$(3) \qquad z \neq 0 \leftrightarrow \bigvee_u z = u + 1$$

$$(4) \qquad P \neq Q \leftrightarrow \bigvee_z (z \neq 0 \wedge (P + z = Q \vee P = Q + z))$$

(5) $$\neg \bigwedge_x \delta \leftrightarrow \bigvee_x \neg \delta$$
$$\neg \bigvee_x \delta \leftrightarrow \bigwedge_x \neg \delta$$

(6) $\underset{x_1}{\text{XX}} \dots \underset{x_n}{\text{XX}} \delta_1 \wedge \underset{y_1}{\text{XX}} \dots \underset{y_m}{\text{XX}} \delta_2 \leftrightarrow \underset{x_1}{\text{XX}} \dots \underset{x_n}{\text{XX}} \underset{z_1}{\text{XX}} \dots \underset{z_m}{\text{XX}} (\delta_1 \wedge \delta_2')$, where δ_2' is obtained from δ_2 by renaming of the y_i by such z_i which are different from the x_k.

The proof is carried out by induction on the structure of α. If α has the form $t_1 = t_2$, then we have nothing to prove. If $\alpha \equiv \neg \beta$, where $\beta \leftrightarrow \underset{x_1}{\text{XX}} \dots \underset{x_n}{\text{XX}} P = Q$, then we can bring the negation sign immediately in front of $P = Q$ (by (5)) and then deal with the formula $P \neq Q$ using (1), (2), (3), (4), (6). If

$$\alpha \equiv \beta_1 \wedge \beta_2, \quad \beta_1 \leftrightarrow \underset{x_1}{\text{XX}} \dots \underset{x_n}{\text{XX}} P_1 = Q_1, \quad \beta_2 \leftrightarrow \underset{y_1}{\text{XX}} \dots \underset{y_m}{\text{XX}} P_2 = Q_2,$$

then (6) together with (1) provides us with the required formula. Finally, if $\alpha \equiv \bigwedge_x \beta$, $\beta \leftrightarrow \underset{x_1}{\text{XX}} \dots \underset{x_n}{\text{XX}} P = Q$, then $\alpha \leftrightarrow \bigwedge_x \underset{x_1}{\text{XX}} \dots \underset{x_n}{\text{XX}} P = Q$.

References

KLEENE, S. C.: Recursive Predicates and Quantifiers. Trans. Amer. math. Soc. **53**, 41–73 (1943).

MOSTOWSKI, A.: On Definable Sets of Positive Integers. Fundam. Math. **34**, 81–112 (1947).

— On a Set of Integers not Definable by Means of One-Quantifier Predicates. Ann. Soc. Polonaise Math. **21**, 114–119 (1948).

KLEENE, S. C.: Introduction to Metamathematics. Amsterdam: North-Holland Publishing Company [3]1959.

MOSTOWSKI, A.: Development and Applications of the "Projective" Classification of Sets of Integers. Proc. internat. Congr. Math. Amsterdam **1** (1954).

KLEENE, S. C.: Hierarchies of Number-Theoretic Predicates. Bull. Amer. math. Soc. **61**, 193–213 (1955).

DAVIS, M.: Computability & Unsolvability. New York-Toronto-London: McGraw-Hill Book Company 1958.

On Hilbert's tenth problem cf. besides the book by Davis cited above

HILBERT, D.: Mathematische Probleme. Vortrag, gehalten auf dem internationalen Mathematiker-Kongreß zu Paris 1900. Nachr. Ges. Wiss. Göttingen, math.-phys. Kl., 253–297 (1900).

DAVIS, M.: Arithmetical Problems and Recursively Enumerable Predicates. J. symbolic Logic **18**, 33–41 (1953).

DAVIS, M., and H. PUTNAM: Reductions of Hilbert's Tenth Problem. J. symbolic Logic **23**, 183–187 (1958).

DAVIS, M., and H. PUTNAM: Research on Hilbert's Tenth Problem. Rensselaer Polytechnic Institute, Troy, N. Y., 3–1 to 3–31 (1959).

DAVIS, M.: Extensions and Corollaries of Recent Work on Hilbert's Tenth Problem. Illinois J. Math. **7**, 246–250 (1963).

§ 30. Universal Turing Machines

We shall show in this paragraph that there exists a "universal" Turing machine **U**, which can in a certain sense be trusted to do the work of an arbitrary Turing machine **M**. We could say in the terminology used in the domain of electronic computers that **U** is capable of *simulating* any arbitrary Turing machine **M**. We shall presume that the alphabet of a machine **M** is an initial part of the infinite alphabet $\{a_1, a_2, a_3, \ldots\}$ (cf. § 5.7). We shall construct a machine **U** which is over the alphabet $\{a_1\}$. We use the notation introduced at the end of § 6.5.

1. Preliminary remarks. In this paragraph we shall use the fundamental concepts of the theory of Turing machines. We introduced these concepts in § 5. In § 17.3 we characterized the Turing machine **M** in a certain way by its Gödel number t. We wrote $t = G(\mathsf{M})$. On the basis of a special numbering of the squares of the computing tape we represented in § 17.2 the tape expression B by a number b. In § 18.1 we characterized the configuration $K = (A, B, C)$ of **M** by a Gödel number $k = \sigma_3(a, b, c)$, where a was the Gödel number of the square A and $c = C$. We called k the Gödel number of the configuration (A, B, C). We wrote $k = g(A, B, C)$. We also spoke (in a figurative sense) of the configuration k.

In § 18.1 we introduced the predicate E. If t is the Gödel number of a Turing machine, then Etk if and only if k is a terminal configuration of the Turing machine which has the Gödel number t. Let T^*t mean that t is the Gödel number of a Turing machine. That T^* is primitive recursive can easily be shown using the results of § 17.3. Let $E'tk$ if and only if $T^*t \wedge Etk$. Then $E'tk$ if and only if t is the Gödel number of a Turing machine and k is a terminal configuration of this machine. E' is primitive recursive. Thus, there exists a primitive recursive function e which assumes only 0 and 1 as values and which is such that

$$E'tk \leftrightarrow e(t, k) = 0.$$

According to § 16 there exists a Turing machine **E** which standard computes e. Further, in § 18.1 we considered a function F which gives for a machine with the Gödel number t the consecutive configuration of the configuration k. F is primitive recursive and is standard computed by the Turing machine **F**.

2. Definition. We call a Turing machine **U** a *universal Turing machine* if **U** operates as follows:

Let B_0 be an arbitrary (naturally finite) tape expression. Let A_0 be an arbitrary square of the computing tape. Let **M** be an arbitrary Turing machine. We place **M** on B_0 over A_0. This way we obtain the initial configuration (A_0, B_0, C_0), where $C_0 = c_{\mathsf{M}}$. Now **M** will run through a sequence (A_1, B_1, C_1), $(A_2, B_2, C_2), \ldots, (A_n, B_n, C_n), \ldots$

of configurations which, provided that M stops operating in finitely many steps, eventually terminates with a terminal configuration $(A_{n_0}, B_{n_0}, C_{n_0})$ (cf. the left column of the schema given at the end of this section).

We place the machine U, which is to simulate M, on the tape expression $t * k_0$, where $t = G(\mathsf{M})$ and $k_0 = g(A_0, B_0, C_0)$. This means, more precisely, that first we print onto the otherwise empty computing tape $t+1$ strokes to represent the number t (cf. 1.3), after which we print, leaving a one-square gap (represented by $*$), $k_0 + 1$ strokes to represent the number k_0, and then we choose the square behind the last marked square to be the original scanned square of U. Let C_0^{U} be the initial state of U. The initial configuration $(A_0^{\mathsf{U}}, B_0^{\mathsf{U}}, C_0^{\mathsf{U}})$ of U is now determined.

Now we assume that M accomplishes a step which leads from the 0-th configuration (A_0, B_0, C_0) to the first configuration (A_1, B_1, C_1). Then we require that U accomplishes at least one step and that U after a finite number r_1 of steps reaches a configuration $(A_{r_1}^{\mathsf{U}}, B_{r_1}^{\mathsf{U}}, C_{r_1}^{\mathsf{U}})$, in which the scanned square is the square behind the last marked square and the tape expression is given by $t * k_1$, where $k_1 = g(A_1, B_1, C_1)$. We shall denote this fact in short by saying that the r_1-th configuration of U *corresponds* to the first configuration of M. (We see that in this terminology the 0-th configuration of U corresponds to the 0-th configuration of M).

If M now accomplishes a further step which leads to the configuration (A_2, B_2, C_2), then there should exist a corresponding configuration $(A_{r_2}^{\mathsf{U}}, B_{r_2}^{\mathsf{U}}, C_{r_2}^{\mathsf{U}})$ of U with $r_2 > r_1$, etc. Finally, if M eventually reaches a terminal configuration $(A_{n_0}, B_{n_0}, C_{n_0})$ after finitely many steps, then we require that U stops operating in a corresponding terminal configuration $(A_{r_{n_0}}^{\mathsf{U}}, B_{r_{n_0}}^{\mathsf{U}}, C_{r_{n_0}}^{\mathsf{U}})$.

If (A_0, B_0, C_0) is already a terminal configuration, i.e. $n_0 = 0$, then we shall allow that $r_{n_0} = r_0 > 0$.

Machine M	*Machine* U (simulating M)
(A_0, B_0, C_0)	$(A_0^{\mathsf{U}}, B_0^{\mathsf{U}}, C_0^{\mathsf{U}})$, with $B_0^{\mathsf{U}} = t * k_0$ and $A_0^{\mathsf{U}} =$ square behind the last marked square
(A_1, B_1, C_1)	$(A_{r_1}^{\mathsf{U}}, B_{r_1}^{\mathsf{U}}, C_{r_1}^{\mathsf{U}})$, with $B_{r_1}^{\mathsf{U}} = t * k_1$ and $A_{r_1}^{\mathsf{U}} =$ square behind the last marked square
(A_2, B_2, C_2)	$(A_{r_2}^{\mathsf{U}}, B_{r_2}^{\mathsf{U}}, C_{r_2}^{\mathsf{U}})$, with $B_{r_2}^{\mathsf{U}} = t * k_2$ and $A_{r_2}^{\mathsf{U}} =$ square behind the last marked square
. . . . . .	. . . .

The method of operation of a universal machine U

3. The construction of a universal Turing machine U_0. It is easy to give an account of a machine U_0 which operates as required in Section 2.

If we place U_0 behind $t * k_0$, then first we have to check whether M carries out a step at all. M does not carry out any step at all if and only if $E' t k_0$, i.e. if and only if $e(t, k_0) = 0$. We construct U_0 from simple machines, beginning with the machine E, which according to Section 1 standard computes e.

$$\underset{\uparrow\!\_\_\_\_\_\_\_\_\_\_\_\_\_\_\_\_\_\_\_\_\_\_\_\_\_|}{\mathsf{E}\,\mathfrak{l} * \mathfrak{l} \overset{1}{\rightarrow} * \mathfrak{l}\,\mathsf{F}\,\boldsymbol{\sigma}}$$

The universal Turing machine U_0

After the computation carried out by E we have on the computing tape the inscription $t * k_0 * e(t, k_0)$. By $\mathfrak{l} * \mathfrak{l}$ we erase the next square on the left and move one more square to the left. Now, $e(t, k_0) = 0$ or 1 depending on whether the square scanned after this step is empty or marked respectively.

First we assume that M does not accomplish any step at all, i.e. $e(t, k_0) = 0$. In this case the computation is at an end.

If, on the other hand, M accomplishes a step, then $e(t, k_0) = 1$. Now we erase the value $e(t, k_0)$ and return to the square behind the right hand end of the original inscription by $\overset{1}{\rightarrow} * \mathfrak{l}$. Now we have the task of computing k_1 and printing it in place of k_0. We have that $k_1 = F(t, k_0)$. According to Section 1 F is standard computed by F. After the computation carried out by F we have on the computing tape the inscription $t * k_0 * k_1$.

Now we must erase k_0 and push k_1 back to t. This is done by the shifting machine $\boldsymbol{\sigma}$, which we introduced in § 8.6.

This way we have after a finite number r_1 of steps reached a configuration of U_0 which corresponds to (A_1, B_1, C_1).

Now we have to determine whether k_1 is the Gödel number of a terminal configuration, etc. This can be done without any further difficulty by a feedback to E. Thus, the construction of a universal Turing machine is complete.

4. Consequences. First, to sum up the results of the previous section, we can characterize an essential property of a universal Turing machine U by

Theorem 1. If U *is a universal Turing machine, then we have: an arbitrary Turing machine* M *with Gödel number* t, *placed on a tape expression* B_0 *over the square* A_0 (which together with $C_0 = c_{\mathsf{M}}$ determine a configuration with the Gödel number k_0), *stops operating after finitely many steps if and only if* U, *placed behind* $t * k_0$, *stops operating after finitely many steps.*

Now we turn our attention to § 22.2(4). There we considered the property E_2, which is satisfied by a Turing machine M if and only if M, placed behind its Gödel number $G(\mathsf{M})$, stops operating after finitely many steps. We showed that E_2 is not decidable.

We shall now show that for an arbitrary universal Turing machine U there exists no algorithm by the help of which we can decide whether or not U, placed on an arbitrary tape expression over an arbitrary square, stops operating after finitely many steps. Proof by *reductio ad absurdum*. We assume that there exists such an algorithm. Then we can decide, by the help of this algorithm, the undecidable property E_2 as follows. Let M be an arbitrary Turing machine. Let $t = G(\mathsf{M})$. We print t onto the otherwise empty computing tape and obtain in this way a tape expression B_0. Let A_0 be the square behind t. Then $k_0 = G(A_0, B_0, C_0)$ is the Gödel number of the initial configuration of M. Now we place U behind $t * k_0$. Then, according to Theorem 1, the machine U stops operating after finitely many steps if and only if M, placed behind t, stops operating after finitely many steps, i.e. if and only if M has the property E_2. According to our assumption we can decide whether U stops operating after finitely many steps and so we can decide whether M has the property E_2.

We sum up the result in

Theorem 2. It is not decidable whether or not a universal Turing machine U, *placed on an arbitrary tape expression over an arbitrary square, stops operating after finitely many steps.*

For the sake of completeness we finally state the main result of Section 3.

Theorem 3. We can give an account of a universal Turing machine U_0 *explicitly.*

Reference

TURING, A. M.: On Computable Numbers, with an Application to the Entscheidungsproblem. Proc. London math. Soc. (2), **42**, 230—265 (1937).

§ 31. λ-K-Definability

To operate with functions CHURCH developed the so-called λ-calculus and the so-called λ-K-calculus. We shall, after a few preliminary remarks in Section 1, build up the λ-K-calculus in Sections 2 and 3[1]. There exists an obvious representation of the natural numbers in the λ-K-calculus (Section 4). On this basis we can easily introduce the concept of λ-K-definable function (Section 5). We can show that the class of the λ-K-definable functions coincides with that of the computable functions. Since the values of the λ-K-definable functions are provided

---

[1] For the relation between the λ-calculus and the λ-K-calculus cf. the second note of Section 2.

by a calculus, these functions must be computable (Section 6). Conversely, in Section 7 we show that every μ-recursive function (and so every computable function) is λ-K-definable. This last fact is especially remarkable, for we could not expect before that every computable function belongs to the class of the λ-K-definable functions, because this class is defined from a point of view which on no account aims at the comprehension of *all* computable functions.

1. We begin with a few *preliminary remarks* on functions. These remarks motivate the formulation of the λ-K-calculus in Sections 2 and 3.

According to DIRICHLET's method we shall, quite generally, identify functions with unambiguous assignments. If f is a singulary function and x is an argument of f, then we shall denote (differing from usual mathematical usage) the value of the function f by (fx). It can happen that x is also a function and that (xf) is also defined. Then, in general, $(fx) \neq (xf)$. Further, the associative law is not valid; mostly $((fg)h) \neq (f(gh))$. We agree upon the abbreviation (fgh) for $((fg)h)$, and in general $(f_1 f_2 f_3 \ldots f_n)$ for $(\ldots((f_1 f_2) f_3) \ldots f_n)$ (*left parentheses*).

In principle it is sufficient to consider *singulary* functions (following a method of SCHÖNFINKEL). We shall explain this using addition as an example. We consider $+$ to be a singulary function which applied to the argument a provides as value $(+a)$ that function which on its part has as value, when applied to the argument b, the sum of a and b. In this way we can quite generally comprehend the formation of $f(x_1, \ldots, x_n)$ as the formation of $(\ldots((fx_1)x_2) \ldots x_n)$ — which is the same as $(fx_1 \ldots x_n)$ —, where only the values of singulary functions are considered.

Today it is not yet quite general in mathematics to distinguish in the notation between a function f and the value $f(x)$ of this function for the argument x. This sometimes leads to confusion and to fruitless discussions, e.g. on whether or not $f(x)$ and $f(y)$ are the same function. If we want to put the notation "$f(x)$" in order, then we should consider x, the variable in "$f(x)$", as a *bounded* variable and we should express this by writing "x" together with an *operator* before "(fx)". As such an operator CHURCH used the symbol "λ". Thus, we have to write "$\lambda x(fx)$". Then $\lambda x(fx) = \lambda y(fy)$, whereas in general $(fx) \neq (fy)$. The reader should remember here the similar cases of bounded variables in the established mathematical notations

$$\int_0^1 f(x)\,dx \left(= \int_0^1 f(y)\,dy\right) \quad \text{or} \quad \sum_i a_i \left(= \sum_k a_k\right).$$

We often use the λ-operator with advantage when we want to define functions by means of substitutions into other functions. For instance, if we start with the functions sin and cos, then we can define $\tan = \lambda x \frac{\sin x}{\cos x}$, or more systematically, if we introduce a function symbol Q for the quotient, $\tan = \lambda x(Q(\sin x)(\cos x)) = \lambda y(Q(\sin y)(\cos y))$.

From the meaning of the λ-operator follows the computational rule $\lambda x(fx)a = (fa)$ and, in general, for a function term $\lambda x F$, where F is a compound expression, the rule $\lambda x F a = G$, where G is obtained from F by replacing x by a everywhere where x "occurs free" in F. Thus, especially $(\tan 3) = (\lambda x(Q(\sin x)(\cos x))3) = (Q(\sin 3)(\cos 3))$.

2. *Terms and equations of the λ-K-calculus.* In the previous remarks the operation of application of a function to an argument and the λ-operation occur. If we consider these operations and the rules trivially valid for them, then we can develop a calculus in a natural way. We shall build up this calculus in this and the following section.

We start from a denumerable set of variables $x_1, x_2, x_3, \ldots$ and from the symbols (,), λ, $=$. As symbols for the variables we use $x, y, z, \ldots$ and sometimes, when it is suggestive, also the symbols $f, g, h, \ldots$. We can produce words from the variables and the above mentioned symbols. $W_1 \equiv W_2$ shall mean that W_1 and W_2 are the same word[1]. We define inductively what the word *term* means and, at the same time, we define which variables occur *free* in the terms.

Every variable x is a term. x occurs free in x. No other variable occurs free in x.

If T_1 and T_2 are terms, then $(T_1 T_2)$ is also a term. A variable x occurs free in $(T_1 T_2)$ if and only if x occurs free in at least one of T_1 and T_2.

If T is a term and x is an arbitrary variable, then $\lambda x T$ is a term[2]. A variable y occurs free in $\lambda x T$ if and only if y occurs free in T and is different from x.

If $T_1, \ldots, T_n$ are terms, then we use the abbreviation

$$(T_1 T_2 T_3 \ldots T_n) \text{ for } (\ldots((T_1 T_2) T_3) \ldots T_n) \textit{ (left parentheses)}.$$

If T_1 and T_2 are terms, then $T_1 = T_2$ is an *equation*.

Further, we need the *substitution operation* ${}^x/U$. This operation can be applied to any term T and provides as result the term T^x/U. We can define T^x/U by induction on the structure of T:

---

[1] The metasymbol $\equiv$ must not be confused with the symbol $=$ of the calculus.

[2] The λ-calculus is different from the λ-K-calculus insofar that in the λ-calculus we have the additional requirement that x should occur free in T. The notation "λ-K-calculus" can be explained as follows. In the λ-K-calculus we can define (in contrast to the λ-calculus) the so-called *constancy function* $K \equiv \lambda f \lambda x f$ for which $K x y = x$.

If T is a variable, then $T^x/U \equiv U$ or T according to whether $T \equiv x$ or $T \not\equiv x$ respectively.

$(T_1 T_2)^x/U \equiv (T_1^x/U \; T_2^x/U)$.

$[\lambda y T]^x/U \equiv \lambda y T$ or $\lambda y\, [T]^x/U$ according to whether $y \equiv x$ or $y \not\equiv x$.[1]

3. Derivability in the λ-K-calculus. In this section we shall give a few rules by the help of which equations of the λ-K-calculus can be derived. If $T_1 = T_2$ is derivable, then T_1 and T_2 represent, in the sense of the preliminary remarks of Section 1, the same function[2]. This fact is important for the intuitive meaning of the λ-K-calculus; however, it is unimportant for the formal considerations of this paragraph[3].

The *derivability* is given by the following *rules*.

(K 1) *Renaming of bounded variables:*

$$\lambda x T = \lambda y [T]^x/y,$$

provided that y does not occur in T.[4]

(K 2) *Elimination of the λ-operator.*

$$(\lambda x T\, U) = T^x/U,$$

provided that for no variable y which occurs free in U is the word λy a part of T.[5]

(K 1) expresses the fact that it does not matter which bounded variable is used. (K 2) says that λ is an operator which denotes functions.

The following two rules serve the purpose of accomplishing the alterations described by the operations (K 1) and (K 2) "in the inside" of the terms as well.

---

[1] The square brackets [ ] are not symbols of the calculus but metasymbols, which only serve to determine, in case of doubt, the term to which the operator $^x/U$ is applied.

[2] On the other hand, we do not assert that if T_1 and T_2 represent the same function, then the equation $T_1 = T_2$ is always derivable. Thus, the situation here is similar to that which we had in the case of the equation calculus by which we defined the recursive functions. Cf. § 21.7.

[3] CHURCH uses instead of the derivability of equations introduced here a relation between terms, which he calls λ-K-convertibility. T_1 is λ-K-convertible into T_2 if and only if the equation $T_1 = T_2$ is derivable.

[4] Without this provision we would have for instance the equation $\lambda x (xy) = \lambda y\, (yy)$, which is materially undesirable.

[5] Without this restriction we could for instance derive the equation $(\lambda x\, \lambda y\, (xy)\, y) = \lambda y\, (yy)$, which is materially undesirable, since on the left hand side the function $\lambda x\, \lambda y\, (xy)$ is applied to the term y with the free variable y, whereas y does not at all occur free on the right hand side.

(K 3) It is allowed to proceed from the equations

$$T_1 = T_2 \qquad \text{and} \qquad U_1 = U_2$$

to the equation

$$(T_1\, U_1) = (T_2\, U_2)\,.$$

(K 4) It is allowed to proceed from the equation

$$T_1 = T_2$$

to the equation

$$\lambda x\, T_1 = \lambda x\, T_2\,.$$

Finally we give a few more rules which represent purely formal equality laws.

(K 5) *Reflexiveness of the equality relation:*

$$T = T\,.$$

(K 6) *Symmetry of the equality relation:*
It is allowed to proceed from the equation

$$T_1 = T_2$$

to the equation

$$T_2 = T_1\,.$$

(K 7) *Transitiveness of the equality relation:*
It is allowed to proceed from the equations

$$T_1 = T_2 \qquad \text{and} \qquad T_2 = T_3$$

to the equation

$$T_1 = T_3\,.$$

4. Representation of the natural numbers in the λ-K-calculus. Most often the natural numbers are comprehended in mathematics as cardinal numbers or ordinal numbers. This way is not practicable if we want to represent the natural numbers in the frame of the λ-K-calculus. In order to obtain such a representation we first introduce the concept of *n-th iteration* of a function f.

2nd iteration of $f \equiv \lambda x\,(f\,(f\,x))$

3rd iteration of $f \equiv \lambda x\,(f\,(f\,(f\,x)))$

4th iteration of $f \equiv \lambda x\,(f\,(f\,(f\,(f\,x))))$

... etc.

We complete this list in an obvious way by the stipulations:

$$\text{1st iteration of } f \equiv \lambda x(fx)$$
$$\text{0-th iteration of } f \equiv \lambda xx.$$

Now we consider the following terms $\underline{0}, \underline{1}, \underline{2}, \ldots$:

$$\underline{0} \equiv \lambda f \lambda xx$$
$$\underline{1} \equiv \lambda f \lambda x(fx)$$
$$\underline{2} \equiv \lambda f \lambda x(f(fx))$$
$$\underline{3} \equiv \lambda f \lambda x(f(f(fx)))$$
$$\cdots\cdots\cdots\cdots\cdots\cdots$$

We shall say that the term $\underline{n}$ *of the above list represents the natural number* n[1].

For application later on we note that the following equations are derivable in the λ-K-calculus.

(1) $$(\underline{n}f) = \lambda x\underbrace{(f \ldots (f}_{n \text{ times}}x) \ldots) \qquad \text{(K 2)}$$

(2) $$(\underline{n}fx) = \underbrace{(f \ldots (f}_{n \text{ times}}x(\ldots) \qquad \text{(1), (K 5), (K 3), (K 2), (K7)}$$

(3) $$(f(\underline{n}fx)) = \underbrace{(f \ldots (f}_{n+1 \text{ times}}x) \ldots) \qquad \text{(2), (K 5), (K 3)}$$

(4) $$\lambda f \lambda x(f(\underline{n}fx)) = \underline{n'} \qquad \text{(3), (K 4)}$$

(1) shows that $(\underline{n}f) = n$-th iteration of f is derivable in the λ-K-calculus and so justifies saying that $\underline{n}$ represents the number n.

5. *λ-K-definable functions.* Let φ be a function which is defined for all n-tuples of natural numbers and whose values are natural numbers ($n \geqq 0$). We call φ *λ-K-definable if there exists a term F in the λ-K-calculus such that for all n-tuples $r_1, \ldots, r_n$ of natural numbers the equation*

$$(F\underline{r_1} \ldots \underline{r_n}) = \underline{\varphi(r_1, \ldots, r_n)}$$

is derivable in the λ-K-calculus[2]. We call F *a term defining* φ.

At first glance we might think that we must also require that the equation $(F\underline{r_1} \ldots \underline{r_n}) = \underline{s}$ is not derivable for any natural number s

---

[1] We see that $\underline{0}$ is indeed a term of the λ-K-calculus but not a term of the λ-calculus, since f does not occur free in λxx (cf. Section 2, second Note).

[2] For $n = 0$ this is defined to mean that the equation $F = \underline{r}$ is derivable in the λ-K-calculus.

different from $\underline{\varphi(r_1, \ldots, r_n)}$. But then the equation $\underline{\varphi(r_1, \ldots, r_n)} = \underline{s}$ would also be derivable. However, we can show that this is not the case. For a proof of this assertion we must refer the reader to the literature on the subject[1].

6. Theorem. A function f is computable if and only if it is λ-K-definable.

We deal briefly with the first half of the proof. Let φ be λ-K-definable. Let F be a term defining φ. Then the set of all derivable equations of the form $(F\underline{r_1} \ldots \underline{r_n}) = \underline{r}$ is enumerable (cf. § 2.4 (d)). But according to the definition of λ-K-definability there exists for every $r_1, \ldots, r_n$ at least *one* derivable equation of this form. According to the concluding remark of the last section there exists *only one* such equation. By systematical tabulation of all derivations (cf. again § 2.4 (d)) we can find this equation and read off from $\underline{r}$ the value of the function. Thus, φ is computable.

7. λ-K-definability of computable functions. In order to show that every computable function is λ-K-definable we make use of the fact that every computable function is μ-recursive (see § 18). Thus, we need only show that every μ-recursive function is λ-K-definable. This follows from the assertions (a), (b), (c), (d), which we shall instantly prove.

The proof is *constructive* in the sense that for a μ-recursive function φ, given by substitutions, inductive definitions and applications of the μ-operator to regular functions, we can give effectively a term which defines φ. This term can be chosen, as we shall show, even without free variables.

(a) *The functions* S, U_n^i *and* C_0^0 (cf. § 10) *are λ-K-definable and, indeed, by terms without free variables,* namely

(a$_1$) S by the term $\lambda r \lambda f \lambda x (f(r f x))$

(a$_2$) U_n^i by the term $\lambda x_1 \ldots \lambda x_n x_i$

(a$_3$) C_0^0 by the term $\underline{0}$.

[1] That for $n \neq m$ the equation $\underline{n} = \underline{m}$ cannot be derived is *materially* plausible, since according to the definitions $\underline{n}$ and $\underline{m}$ are different functions and since the rules (K 1), ..., (K 7) — speaking materially — can only provide us with a proof of equality of terms if these terms represent the same function. Formal proof of the non-derivability of $\underline{n} = \underline{m}$ (provided $n \neq m$) goes as follows. A term is called a *normal form* if it contains no subterm of the form $(\lambda x T_1 T_2)$. All terms $\underline{n}$ are obviously normal forms. We can show that for normal forms N_1 and N_2 the equation $N_1 = N_2$ is derivable only if N_2 can be obtained from N_1 by renaming of bounded variables. (This follows from a theorem of CHURCH and ROSSER.) But there is no possibility for such renaming for the terms $\underline{n}$ and $\underline{m}$ if $n \neq m$.

Proof.[1] The following equations are derivable:

(a$_1$) $$(\lambda r \lambda f \lambda x (f (r f x)) \underline{n}) = \lambda f \lambda x (f (\underline{n} f x)) = \underline{n'}$$ by (4).

(a$_2$) $$\begin{aligned}(\lambda x_1 \ldots \lambda x_n x_i \underline{r_1} \ldots \underline{r_n}) &= (\lambda x_2 \ldots \lambda x_n x_i \underline{r_2} \ldots \underline{r_n})\\ &= \ldots\\ &= (\lambda x_i \ldots \lambda x_n x_i \underline{r_i} \ldots \underline{r_n})\\ &= (\lambda x_{i+1} \ldots \lambda x_n \underline{r_i} \underline{r_{i+1}} \ldots \underline{r_n})\\ &= \ldots\\ &= \underline{r_i}\end{aligned}$$

(a$_3$) $$\underline{0} = \underline{0}.$$

(b) *Let* $\psi(r_1, \ldots, r_n) = \varphi(\varphi_1(r_1, \ldots, r_n), \ldots, \varphi_m(r_1, \ldots, r_n))$. *Let* $\varphi_1, \ldots, \varphi_m, \varphi$ *be λ-K-definable by the terms* $F_1, \ldots, F_m, F$ *without free variables. Then,* ψ *is λ-K-definable by the term*

$$G \equiv \lambda x_1 \ldots \lambda x_n (F (F_1 x_1 \ldots x_n) \ldots (F_m x_1 \ldots x_n))$$

which contains no free variables.

Proof. We can derive that[2]

$$\begin{aligned}(G \underline{r_1} \ldots \underline{r_n}) &= (F (F_1 \underline{r_1} \ldots \underline{r_n}) \ldots (F_m \underline{r_1} \ldots \underline{r_n}))\\ &= (F \underline{\varphi_1(r_1, \ldots, r_n)} \ldots \underline{\varphi_m(r_1, \ldots, r_n)})\\ &= \underline{\varphi(\varphi_1(r_1, \ldots, r_n), \ldots, \varphi_m(r_1, \ldots, r_n))}\\ &= \underline{\psi(r_1, \ldots, r_n)}.\end{aligned}$$

(c) *Let*

$$\begin{aligned}\varphi(0, r_1, \ldots, r_n) &= \psi(r_1, \ldots, r_n)\\ \varphi(r', r_1, \ldots, r_n) &= \chi(r, \varphi(r, r_1, \ldots, r_n), r_1, \ldots, r_n).\end{aligned}$$ [3]

Let ψ *and* χ *be λ-K-definable by the terms G and H without free variables. Then* φ *is also λ-K-definable by a term without free variables.*

Proof. We use the following propositions, which we shall prove in Section 8.

---

[1] Only the most important steps of the proof are given.

[2] The reader should note that here and in what follows the rule (K 2) is always applicable since the provision required for it is satisfied, because the substituted terms contain no free variables.

[3] It is convenient for the following proof to carry out the induction on the *first* argument and to choose the sequence of variables in the way given here. We obtain all primitive recursive functions in this way as well. (Cf. § 10.3(b).)

Proposition 1. We can give an account of terms V and M without free variables for which we have that

(') the predecessor function is defined by V,

('') the function $\min(x, y)$ is defined by M.

Proposition 2. For arbitrary terms A and B without free variables we can give an account of a term C without free variables such that

$$(C\,\underline{0}) = A$$
$$(C\,\underline{1}) = B$$

are derivable in the λ-K-calculus.

Proposition 3. For every term D without free variables we can give an account of a term E without free variables such that for every term T without free variables

$$(E\,T) = (D\,T\,E)$$

is derivable[1].

To prove (c) we introduce the following terms.

$A \equiv \lambda y \lambda f (y f G)$.

$B \equiv \lambda y \lambda f \lambda x_1 \ldots \lambda x_n (H (V y) (f (V y) x_1 \ldots x_n) x_1 \ldots x_n)$.

C is formed from A and B according to Proposition 2.

$D \equiv \lambda y (C (M y \underline{1}) y)$.

E is formed from D according to Proposition 3.

Now we shall show that the function φ is defined by E. For this purpose we consider the following equations. They are all derivable.

$$\begin{aligned}
(E\,\underline{0}\,\underline{r_1} \ldots \underline{r_n}) &= (D\,\underline{0}\,E\,\underline{r_1} \ldots \underline{r_n}) && \text{(Proposition 3)}\\
&= ((C (M\,\underline{0}\,\underline{1})\,\underline{0})\,E\,\underline{r_1} \ldots \underline{r_n}) && \text{(Definition of } D)\\
&= (C\,\underline{0}\,\underline{0}\,E\,\underline{r_1} \ldots \underline{r_n}) && \text{(Proposition 1 (''))}\\
&= (A\,\underline{0}\,E\,\underline{r_1} \ldots \underline{r_n}) && \text{(Proposition 2)}\\
&= (\underline{0}\,E\,G\,\underline{r_1} \ldots \underline{r_n}) && \text{(Definition of } A)\\
&= (G\,\underline{r_1} \ldots \underline{r_n}) && \text{(Definition of } \underline{0})\\
&= \underline{\psi(r_1, \ldots, r_n)} && \text{(Assumption about } G)\\
&= \underline{\varphi(0, r_1, \ldots, r_n)} && \text{(Definition of } \varphi)
\end{aligned}$$

[1] Proposition 3 is even valid for *arbitrary* terms T. The proof then is certainly not so easy to carry out as we shall do in Section 8, since the direct application of the rule (K 2) is hindered by the presence of free variables in T. In such a case we can help ourselves by untertaking a suitable renaming of bounded variables according to rule (K 1) before the substitution of T into a term. Later on we shall have to carry out the converse operation if necessary.

$$
\begin{aligned}
(E\underline{r'}\,\underline{r_1}\dots\underline{r_n}) &= (D\underline{r'}\,E\,\underline{r_1}\dots\underline{r_n}) && \text{(Proposition 3)}\\
&= (C\,(M\underline{r'}\,\underline{1})\,\underline{r'}\,E\,\underline{r_1}\dots\underline{r_n}) && \text{(Definition of } D)\\
&= (C\,\underline{1}\,\underline{r'}\,E\,\underline{r_1}\dots\underline{r_n}) && \text{(Proposition 1 ('')) }\\
&= (B\underline{r'}\,E\,\underline{r_1}\dots\underline{r_n}) && \text{(Proposition 2)}\\
&= (H\,(V\underline{r'})\,(E\,(V\underline{r'})\,\underline{r_1}\dots\underline{r_n})\,\underline{r_1}\dots\underline{r_n}) && \text{(Definition of } B)\\
&= (H\underline{r}\,(E\underline{r}\,\underline{r_1}\dots\underline{r_n})\,\underline{r_1}\dots\underline{r_n}) && \text{(Proposition 1 ('))}\\
&= (H\underline{r}\,\underline{\varphi(r, r_1, \dots, r_n)}\,\underline{r_1}\dots\underline{r_n}) && \text{(Induction hypothesis)}\\
&= \underline{\chi(r, \varphi(r, r_1, \dots, r_n), r_1, \dots, r_n)} && \text{(Assumption about } H)\\
&= \underline{\varphi(r', r_1, \dots, r_n)}\,. && \text{(Definition of } \varphi)
\end{aligned}
$$

(d) *Let there be for every* $r_1, \dots, r_n$ *an* r *such that* $\psi(r_1, \dots, r_n, r) = 0$. *Let* $\varphi(r_1, \dots, r_n) = \mu r(\psi(r_1, \dots, r_n, r) = 0)$. *Let* ψ *be λ-K-definable by the term G without free variables. Then* φ *is also λ-K-definable by a term F without free variables.*

Proof. We put

$A \equiv \lambda f \lambda y \lambda a y$.

$B \equiv \lambda f \lambda y \lambda a\,(f\,(a\,(Sy))(Sy)\;a)$, where S is the term which defines the successor function (see (a)).

C is formed from A and B according to Proposition 2.

$D \equiv \lambda x\,(C\,(Mx\,\underline{1}))$.

E is formed from D according to Proposition 3.

$F \equiv \lambda x_1 \dots \lambda x_n\,(E\,(Gx_1 \dots x_n\,\underline{0})\;\underline{0}\,(Gx_1 \dots x_n))$.

We shall show that the function φ *is defined by* F. This follows from the

Lemma.

$$
(E\,(G\underline{r_1}\dots\underline{r_n}\,\underline{r})\;\underline{r}\,(G\underline{r_1}\dots\underline{r_n})) = \begin{cases} \underline{r}, & \text{if } \psi(r_1, \dots, r_n, r) = 0\\ (E\,(G\underline{r_1}\dots\underline{r_n}\,\underline{r'})\;\underline{r'}\,(G\underline{r_1}\dots\underline{r_n})), & \text{if } \psi(r_1, \dots, r_n, r) \neq 0 \end{cases}
$$

is derivable.

Before giving a proof of the Lemma we shall make sure that the Lemma shows that φ is defined by F.

$$(F\underline{r_1}\dots\underline{r_n}) = (E\,(G\underline{r_1}\dots\underline{r_n}\,\underline{0})\;\underline{0}\,(G\underline{r_1}\dots\underline{r_n})) \qquad \text{(Definition of } F)$$

is derivable.

Now, if $\psi(r_1, \dots, r_n, 0) = 0$, then $\varphi(r_1, \dots, r_n) = 0$. From the Lemma follows, for $r = 0$, the derivability of the corresponding equation $(F\underline{r_1}\dots\underline{r_n}) = \underline{0}$.

However, if $\psi(r_1, \ldots, r_n, 0) \neq 0$, then, for $r = 0$, the derivability of

$$(F\underline{r_1} \ldots \underline{r_n}) = (E\,(G\underline{r_1} \ldots \underline{r_n}\,\underline{1})\,\underline{1}\,(G\underline{r_1} \ldots \underline{r_n}))$$

follows from the Lemma. Now, perhaps $\psi(r_1, \ldots, r_n, 1) = 0$. Then, for $r = 1$, the derivability of $(F\underline{r_1} \ldots \underline{r_n}) = \underline{1}$ follows from the Lemma. If, however, $\psi(r_1, \ldots, r_n, 1) \neq 0$, then the Lemma provides us, for $r = 1$, with the derivability of

$$(F\underline{r_1} \ldots \underline{r_n}) = (E\,(G\underline{r_1} \ldots \underline{r_n}\,\underline{2})\,\underline{2}\,(G\underline{r_1} \ldots \underline{r_n}))\,.$$

This way we see that if r is the smallest number for which $\psi(r_1, \ldots, r_n, r) = 0$, then the equation $(F\underline{r_1} \ldots \underline{r_n}) = \underline{r}$ is derivable corresponding to the fact that then $\varphi(r_1, \ldots, r_n) = r$.

Now *we prove the Lemma stated above.* The following equations are derivable.

$$\begin{aligned}(E\,(G\underline{r_1} \ldots \underline{r_n}\,\underline{r})\,\underline{r}\,(G\underline{r_1} \ldots \underline{r_n})) &= (E\,\underline{\psi(r_1, \ldots, r_n, r)}\,\underline{r}\,(G\underline{r_1} \ldots \underline{r_n})) \\ &\qquad \text{(Assumption about } G\text{)} \\ &= (D\,\underline{\psi(r_1, \ldots, r_n, r)}\;E\underline{r}\,(G\underline{r_1} \ldots \underline{r_n})) \\ &\qquad \text{(Proposition 3)} \\ &= (C\,(M\,\underline{\psi(r_1, \ldots, r_n, r)}\;\underline{1})\;E\underline{r}\,(G\underline{r_1} \ldots \underline{r_n})) \\ &\qquad \text{(Definition of } D\text{)}\end{aligned}$$

Now, let $\psi(r_1, \ldots, r_n, r) = 0$.

According to Proposition 1 ('') $(M\,\underline{\psi(r_1, \ldots, r_n, r)}\;\underline{1}) = \underline{0}$ is derivable and, therefore, so are

$$\begin{aligned}(E\,(G\underline{r_1} \ldots \underline{r_n}\,\underline{r})\,\underline{r}\,(G\underline{r_1} \ldots \underline{r_n})) &= (C\,\underline{0}\,E\underline{r}\,(G\underline{r_1} \ldots \underline{r_n})) \\ &= (A\,E\underline{r}\,(G\underline{r_1} \ldots \underline{r_n})) \quad \text{(Proposition 2)} \\ &= \underline{r}\,. \qquad\qquad \text{(Definition of } A\text{)}\end{aligned}$$

If, however, $\psi(r_1, \ldots, r_n, r) \neq 0$, then the minimum of $\psi(r_1, \ldots, r_n, r)$ and 1 is 1 and so $(M\,\underline{\psi(r_1, \ldots, r_n, r)}\;\underline{1}) = \underline{1}$ is derivable. Then we have the derivability of

$$\begin{aligned}(E\,(G\underline{r_1} \ldots \underline{r_n}\,\underline{r})\,\underline{r}\,(G\underline{r_1} \ldots \underline{r_n})) &= (C\,\underline{1}\,E\underline{r}\,(G\underline{r_1} \ldots \underline{r_n})) \\ &= (B\,E\underline{r}\,(G\underline{r_1} \ldots \underline{r_n})) \quad \text{(Proposition 2)} \\ &= (E\,(G\underline{r_1} \ldots \underline{r_n}\,(S\underline{r}))\,(S\underline{r})\,(G\underline{r_1} \ldots \underline{r_n})) \\ &\qquad \text{(Definition of } B\text{)} \\ &= (E\,(G\underline{r_1} \ldots \underline{r_n}\,\underline{r'})\,\underline{r'}\,(G\underline{r_1} \ldots \underline{r_n})) \\ &\qquad \text{(since } (S\underline{r}) = \underline{r'} \text{ is derivable).}\end{aligned}$$

8. Proof of the Propositions of Section 7.

About 1 ('). We define

$$\begin{aligned}
I &\equiv \lambda x x \\
P &\equiv \lambda x \lambda y \lambda z (z x y) \\
P_1 &\equiv \lambda a (a (\lambda b \lambda c (c I b))) \\
P_2 &\equiv \lambda a (a (\lambda b \lambda c (b I c))) \\
A &\equiv \lambda a (P (P_2 a) (S (P_2 a))) \quad \text{(for } S \text{ cf. Section 7(a))} \\
V &\equiv \lambda x (P_1 (x A (P \underline{0}\, \underline{0}))).
\end{aligned}$$

Let r and s be natural numbers. Then the following equations are derivable.

$$\begin{aligned}
(P_1 (P \underline{r}\, \underline{s})) &= (P_1 \lambda z (z \underline{r}\, \underline{s})) && \text{(Definition of P)} \\
&= (\lambda z (z \underline{r}\, \underline{s}) (\lambda b \lambda c (c I b))) && \text{(Definition of } P_1) \\
&= (\lambda b \lambda c (c I b)\, \underline{r}\, \underline{s}) && \text{(K 2)} \\
&= (\underline{s} I \underline{r}) && \text{(K 2)} \\
&= (\underbrace{I (I \ldots (I}_{s \text{ times}} \underline{r}) \ldots)) && \text{(Section 4 (2))} \\
(1) \qquad &= \underline{r} && \text{(Definition of } I).
\end{aligned}$$

Similarly we obtain that

(2) $$(P_2 (P \underline{r}\, \underline{s})) = \underline{s}.$$

is derivable.

Finally we have

$$\begin{aligned}
(A (P \underline{r}\, \underline{s})) &= (P (P_2 (P \underline{r}\, \underline{s})) (S (P_2 (P \underline{r}\, \underline{s})))) && \text{(Definition of } A) \\
&= (P \underline{s} (S \underline{s})) && \text{(by (2))} \\
(3) \qquad &= (P \underline{s}\, \underline{s'}) && \text{(Definition of } S).
\end{aligned}$$

Therefore the following equations are derivable.

$$\begin{aligned}
(V \underline{0}) &= (P_1 (\underline{0} A (P \underline{0}\, \underline{0}))) && \text{(K 2)} \\
&= (P_1 (P \underline{0}\, \underline{0})) && \text{(Definition of 0)} \\
&= \underline{0} && \text{(by (1))} \\
(V \underline{r'}) &= (P_1 (\underline{r'} A (P \underline{0}\, \underline{0}))) && \text{(K 2)} \\
&= (P_1 (\underbrace{A (A \ldots (A}_{r' \text{ times}} (P \underline{0}\, \underline{0})) \ldots))) && \text{(Section 4 (2))} \\
&= (P_1 (\underbrace{A (A \ldots (A}_{r \text{ times}} (P \underline{0}\, \underline{1})) \ldots))) && \text{(by (3))} \\
&= (P_1 (\underbrace{A (A \ldots (A}_{(r-1) \text{ times}} (P \underline{1}\, \underline{2})) \ldots))) && \text{(by (3))} \\
&\ldots\ldots\ldots\ldots\ldots\ldots\ldots\ldots \\
&= (P_1 (P \underline{r}\, \underline{r'})) && \text{(by (3))} \\
&= \underline{r} && \text{(by (1))}.
\end{aligned}$$

About 1 (″). First we define by the help of the term V introduced above

$$L \equiv \lambda x \lambda y (y V x).$$

Then we can derive for arbitrary natural numbers r and s

$$\begin{aligned} (L\underline{r}\,\underline{s}) &= (\underline{s}V\underline{r}) && \text{(Definition of } L) \\ &= (\underbrace{V(V \ldots (V}_{s \text{ times}}\underline{r}) \ldots)) && \text{(Section 4 (2))} \\ &= \underline{r \dot{-} s} && \text{(Property of } V). \end{aligned}$$

By the help of the difference term introduced in this way we can define

$$M \equiv \lambda x \lambda y (L x (L x y)).$$

Then

$$\begin{aligned} (M\underline{r}\,\underline{s}) &= (L\underline{r}(L\underline{r}\,\underline{s})) && \text{(Definition of } M) \\ &= (L\underline{r}\,\underline{r \dot{-} s}) && \text{(Property of } L) \\ &= \underline{r \dot{-} (r \dot{-} s)} && \text{(Property of } L) \\ &= \underline{\min(r, s)} \end{aligned}$$

are derivable.

About 2. Let A and B be given. We put

$$\begin{aligned} C^* &\equiv \lambda x \lambda y \lambda z (x z y) \\ C &\equiv \lambda a (a C^* \underline{0} B A). \end{aligned}$$

Then we can derive

$$\begin{aligned} (C\,\underline{0}) &= (\underline{0}\,C^*\underline{0}BA) && \text{(Definition of } C) \\ &= (\underline{0}BA) && \text{(Definition of } \underline{0}) \\ &= A && \text{(Definition of } \underline{0}) \\ (C\,\underline{1}) &= (\underline{1}\,C^*\underline{0}BA) && \text{(Definition of } C) \\ &= (C^*\underline{0}BA) && \text{(Definition of } \underline{1}) \\ &= (\underline{0}AB) && \text{(Definition of } C^*) \\ &= B && \text{(Definition of } \underline{0}). \end{aligned}$$

About 3. Let D be given. We put

$$\begin{aligned} E^* &\equiv \lambda b \lambda x (D x \lambda a (b b a)) \\ E &\equiv \lambda a (E^* E^* a). \end{aligned}$$

Let T be an arbitrary term. Then we can derive

$$\begin{aligned} (E T) &= (E^* E^* T) && \text{(Definition of } E) \\ &= (D T \lambda a (E^* E^* a)) && \text{(Definition of } E^*) \\ &= (D T E) && \text{(Definition of } E). \end{aligned}$$

References

Comprehensive presentations

CHURCH, A.: The Calculi of Lambda-Conversion. Princeton: Princeton University Press 1941.

CURRY, H. B., and R. FEYS: Combinatory Logic. Amsterdam: North-Holland Publishing Company 1958.

For the proofs that λ-K-definability and λ-definability are equivalent to other precise replacements of the concept of computability, compare also

KLEENE, S. C.: λ-Definability and Recursiveness. Duke math. J. 2, 340—353 (1936).

TURING, A. M.: Computability and λ-Definability. J. symbolic Logic 2, 153—163 (1937).

§ 32. The Minimal Logic of Fitch

In this paragraph we shall develop, following FITCH, a calculus which can be considered in an especially simple way as a "universal calculus". We shall build up special words from the alphabet {*, (,)} of three letters. These special words will be called "expressions". We shall give an account of rules of inference. $\vdash\alpha$ shall mean that the expression α is derivable by these rules of inference. Now we can express the above mentioned universality by the following

Theorem. For every enumerable n-ary relation R between expressions $(n \geqq 1)$ *there exists at least one expression* ϱ *such that for all expressions* $\alpha_1, \ldots, \alpha_n$

(0) $\qquad \vdash \varrho\,\alpha_1 \ldots \alpha_n$ *if and only if* $R\alpha_1 \ldots \alpha_n$.

Definition 1. We shall say that *an expression* ϱ *represents the relation*[1] *R between expressions if* (0) is valid for all expressions $\alpha_1, \ldots, \alpha_n$.

The calculus given by Fitch is especially remarkable because special expressions like $=$, $\wedge$, and $\vee$ occur in it. The rules of inference associated with these expressions correspond to the rules which are connected with these symbols in constructive logics. These rules are not given *ad hoc* but are based on the postulated validity of (0). Because of the above properties Fitch calls his calculus a "*minimal basic logic*".

1. The expressions. We start from the alphabet {*, (,)}. The *expressions* are generated by a calculus which has * as its only axiom and the passage from α and β to $(\alpha\beta)$ as its only rule of inference. We shall represent expressions by small Greek letters. *Examples* for expressions are *, (**), (*(**)), ((**)*), ((**) (**)).

We shall now note down a few simple facts about expressions, some of which we shall formulate in propositions. We leave the proofs to the reader.

[1] In this paragraph we distinguish between *relations* and *predicates*; cf. the first note of Section 7.

If $\alpha_1, \ldots, \alpha_n$ are expressions ($n \geqq 2$), then $\alpha_1 \ldots \alpha_n$ is not an expression. However, we shall use $\alpha_1 \ldots \alpha_n$ to denote the expression

$$((\ldots((\alpha_1\alpha_2)\alpha_3)\ldots)\alpha_n).$$

This way we have a rule for omitting parentheses (left parentheses) and we shall use this rule in the inside of expressions, too. Then, e.g. $(***)**(**)$ is an abbreviation for the expression $(((((**)*)*)*)(**))$. Instead of $\alpha_1 \ldots \alpha_n$ we shall sometimes also write $(\alpha_1 \ldots \alpha_n)$ for the sake of greater clarity.

Proposition 1. If $(\alpha\beta) \equiv (\gamma\delta)$, then $\alpha \equiv \gamma$ and $\beta \equiv \delta$.[1]

Because of this proposition we can define the properties (relations) of words by defining them first for $\alpha \equiv *$, and then for $\alpha \equiv (\beta\gamma)$ under the hypothesis that they are already defined for β and γ ("induction on the structure of α").

First we define what it means that an expression τ is *a part of an expression* α. τ is a part of $*$ if and only if $\tau \equiv *$. τ is a part of $(\beta\gamma)$ if and only if $\tau \equiv (\beta\gamma)$ or if τ is a part of at least one of β and γ.

Proposition 2. If $\alpha_1 \ldots \alpha_r \equiv \beta_1 \ldots \beta_s$ (left parentheses), then α_1 is a part of β_1 or vice versa.

We shall make use of *an infinite sequence of expressions. This sequence is such that for any two expressions occurring in it neither is a part of the other.* If we denote (temporarily) the expression $*\cdots*$ (n stars) by σ_n, then the sequence $\sigma_2\sigma_2, \sigma_3\sigma_3, \sigma_4\sigma_4, \ldots$ of expressions obviously has the required property. We abbreviate $\sigma_2\sigma_2, \ldots, \sigma_5\sigma_5$ by the symbols $=$, $\wedge$, $\vee$, $\bigvee$ respectively. We abbreviate

$\sigma_6\sigma_6, \sigma_{10}\sigma_{10}, \sigma_{14}\sigma_{14}, \ldots$	by	$\lambda_1, \lambda_2, \lambda_3, \ldots,$
$\sigma_7\sigma_7, \sigma_{11}\sigma_{11}, \sigma_{15}\sigma_{15}, \ldots$	by	$x_1, x_2, x_3, \ldots,$
$\sigma_8\sigma_8, \sigma_{12}\sigma_{12}, \sigma_{16}\sigma_{16}, \ldots$	by	$y_1, y_2, y_3, \ldots,$
$\sigma_9\sigma_9, \sigma_{13}\sigma_{13}, \sigma_{17}\sigma_{17}, \ldots$	by	$z_1, z_2, z_3, \ldots.$

The x_n, y_n and z_n are called *variables*.

Let $2n$ expressions $\alpha_1, \ldots, \alpha_n; \gamma_1, \ldots, \gamma_n$ be given. We shall unambiguously associate with every expression β an expression $\beta\frac{\alpha_1, \ldots, \alpha_n}{\gamma_1, \ldots, \gamma_n}$. We shall say that this expression is obtained from β by *simultaneous substitution* of $\gamma_1, \ldots, \gamma_n$ for $\alpha_1, \ldots, \alpha_n$. We define the operation

[1] We use the symbol $\equiv$ to denote the equality of words. Cf. § 19.2.

by induction on the structure of β. First, let $\beta \equiv *$. Then we put

$$* \frac{\alpha_1, \ldots, \alpha_n}{\gamma_1, \ldots, \gamma_n} \equiv \begin{cases} *, \text{ if } * \text{ does not coincide with any of the } \alpha_1, \ldots, \alpha_n, \\ \gamma_i, \text{ if } \alpha_i \text{ is the first of } \alpha_1, \ldots, \alpha_n \text{ which coincides with } *. \end{cases}$$

Now, let $\beta \equiv (\beta_1 \beta_2)$. Then we define

$$(\beta_1 \beta_2) \frac{\alpha_1, \ldots, \alpha_n}{\gamma_1, \ldots, \gamma_n} \equiv \begin{cases} \left(\beta_1 \frac{\alpha_1, \ldots, \alpha_n}{\gamma_1, \ldots, \gamma_n} \beta_2 \frac{\alpha_1, \ldots, \alpha_n}{\gamma_1, \ldots, \gamma_n}\right), \text{ if } (\beta_1\beta_2) \text{ does not coincide with any of the } \alpha_1, \ldots, \alpha_n. \\ \gamma_i, \text{ if } \alpha_i \text{ is the first of the } \alpha_1, \ldots, \alpha_n \text{ which coincides with } (\beta_1\beta_2). \end{cases}$$

If $\alpha_1, \ldots, \alpha_n, \beta, \gamma_1, \ldots, \gamma_n$ are known, then we can produce $\beta \frac{\alpha_1, \ldots, \alpha_n}{\gamma_1, \ldots, \gamma_n}$ effectively according to the above instruction.

2. *Heuristic remarks. We presume in this section that we already have a derivability concept* $\vdash$ *which has the property* (0) *mentioned in the introduction to this paragraph.* By the help of this derivability concept we shall give an account of a sequence of enumerable relations R. These relations must be representable by an expression ϱ. We shall identify the expressions representing these relations with $=, \wedge, \vee, \bigvee, \lambda_1, \lambda_2, \lambda_3, \ldots$[1]. Because of (0) it follows that certain laws containing the expressions must be valid. We shall give these laws one by one. In Section 3 we shall state these laws as defining rules of Fitch's calculus.

(a) Let $R\alpha\beta$ mean that $\alpha \equiv \beta$. R is enumerable (even decidable). Thus, there must be an expression which represents R. We identify this expression with the expression $=$. Thus we have, because of (0), the relation

(1) $\vdash = \alpha\beta$ *if and only if* $\alpha \equiv \beta$.

(b) Let $R\alpha\beta$ mean that both α and β are derivable. R is enumerable (cf. § 2.4 (d)). Let R be represented by $\wedge$. Thus we have

(2) $\vdash \wedge \alpha\beta$ *if and only if* $\vdash \alpha$ and $\vdash \beta$.

(c) Let $R\alpha\beta$ mean that at least one of the α and β is derivable. R is enumerable. Let $\vee$ represent R. Thus we have

(3) $\vdash \vee \alpha\beta$ *if and only if* $\vdash \alpha$ or $\vdash \beta$.

[1] Naturally, there is a certain amount of arbitrariness in this identification. This arbitrariness could also be disposed of.

(d) Let $R\alpha$ mean that there exists a β such that $\vdash \alpha\beta$. R is enumerable. Let $\bigvee$ represent R. Thus

(4) $\vdash \bigvee\alpha$ *if and only if there exists a* β *such that* $\vdash \alpha\beta$.

(e) Let $n \geqq 1$. Let $R_n \alpha_1 \ldots \alpha_n \beta \gamma_1 \ldots \gamma_n$ mean that the expression $\beta\frac{\alpha_1, \ldots, \alpha_n}{\gamma_1, \ldots, \gamma_n}$ is derivable. R_n is enumerable (cf. § 2.4 (d)). Let R_n be represented by the expression λ_n. Thus we have[1]

(5) $\vdash \lambda_n \alpha_1 \ldots \alpha_n \beta \gamma_1 \ldots \gamma_n$ *if and only if* $\vdash \beta\frac{\alpha_1, \ldots, \alpha_n}{\gamma_1, \ldots, \gamma_n}$.

3. The axiom and rules of Fitch's calculus. Now we develop a calculus based upon the previous considerations. We begin with the axiom[2].

Axiom for $=$: $\quad = \alpha\alpha$.

We have the following *rules of inference*[3].

Rule for $\wedge$: $\quad \dfrac{\alpha, \beta}{\wedge\alpha\beta}$

Rules for $\vee$: $\quad \dfrac{\alpha}{\vee\alpha\beta} \qquad \dfrac{\beta}{\vee\alpha\beta}$

Rule for $\bigvee$: $\quad \dfrac{\alpha\beta}{\bigvee\alpha}$

Rule for λ_n: $\quad \dfrac{\beta\frac{\alpha_1, \ldots, \alpha_n}{\gamma_1, \ldots, \gamma_n}}{\lambda_n \alpha_1 \cdots \alpha_n \beta \gamma_1 \cdots \gamma_n} \quad (n = 1, 2, 3, \ldots)$.

We shall use the notation $\alpha \Leftrightarrow \beta$ to express that the derivability of α implies that of β and vice versa.

4. Proofs of Section 2, (1), ..., (5). Now we shall show that the relations (1), ..., (5) of Section 2, which were justified there materially, are valid in the calculus defined in Section 3. Each one of these relations is a bi-implication. That the *left hand sides* are consequences of the *right hand sides* follows directly from the axioms and the rules of inference.

[1] The reader shold compare (5) for $n = 1$ with the rule (K 2) of the λ-K-calculus (§ 31).

[2] Strictly speaking this is not a simple axiom but a so-called *axiom schema*; i.e. for *every* expression α the expression $= \alpha\alpha$ is an axiom.

[3] These rules are to be comprehended to mean that we may proceed from the expressions above to the expressions below.

It only remains to be shown that the *right hand sides* of the bi-implications in Section 2 are consequences of the *left hand sides*. Here we reason using a characteristic procedure which is called "inversion principle" (so called by LORENZEN). First we consider (1). We assume that $=\alpha\beta$ is derivable. We have to show that $\alpha \equiv \beta$. In order to do this we ask ourselves on the basis of which axioms or with which rules $=\alpha\beta$ can be obtained. In any case the axiom for $=$ is a case in question. But this only allows us to obtain expressions of the form $=\alpha\alpha$ and so, in this case, $\alpha \equiv \beta$. We can convince ourselves that no other axiom and no rule allows us to derive $=\alpha\beta$. The necessary considerations are in all cases based on the same idea. We shall demonstrate it using the rule for $\bigvee$. If $=\alpha\beta$ were obtainable using this rule, then there would have to be an expression γ such that $=\alpha\beta \equiv \bigvee\gamma$. But this is not possible, since if it were, then according to Proposition 2 of Section 1 either $=$ would have to be a part of $\bigvee$, or $\bigvee$ would have to be a part of $=$, which according to Section 1 is not the case. — In conclusion we can say that $=\alpha\beta$ can only be obtained on the basis of the axiom for $=$ and, indeed, only if $\alpha \equiv \beta$.

We can show in the same way that the right hand sides of (2), (4) and (5) are consequences of the left hand sides. We still have to look into (3).

About (3). If $\vee\alpha\beta$ is derivable, then it can only be obtained on the basis of one of the two rules for $\vee$. If $\vee\alpha\beta$ has been obtained by the first rule for $\vee$, then α must be derivable. If $\vee\alpha\beta$ has been obtained by the second rule for $\vee$, then β must be derivable.

5. The meta-rule of structure transformation. Let be $\nu_1, \ldots, \nu_r$ and $\mu_1, \ldots, \mu_s$ sets of names for specific expressions and $\alpha_1, \ldots, \alpha_n$ variables for expressions. $\nu_1, \ldots, \nu_r, \mu_1, \ldots, \mu_s$ are supposed to be different from each other. One or both of the sets $\nu_1, \ldots, \nu_r$ or $\mu_1, \ldots, \mu_s$ may be void. Let be $\mathsf{M} = \mathsf{M}(\nu_1, \ldots, \nu_r, \mu_1, \ldots, \mu_s, \alpha_1, \ldots, \alpha_n)$ a *metaexpression* built up out of the ν_i, μ_j, α_k and the signs $=, \wedge, \vee, \bigvee, \lambda_m$ using parantheses (and the rule for omitting parantheses). Any of the ν_i, μ_j and α_k may occur several times in M or may not occur at all. An *example* for a meta-expression of the above kind ($r = 1$, $s = 2$, $n = 3$) is

$$(\alpha_3\alpha_1) \wedge \bigvee \nu_1(\mu_2\alpha_2\alpha_1\nu_1) \,.$$

It is now possible to construct in an effective way, starting with M, an expression ϕ (where of course none of the $\alpha_1, \ldots, \alpha_n$ occurs) s.t. we have

(7) *For every* $\alpha_1, \ldots, \alpha_n$: $\vdash \mathsf{M}$ *if and only if* $\vdash \phi\mu_1 \ldots \mu_s\alpha_1 \ldots \alpha_n$.

Proof. Let be

$$\phi \equiv \lambda_{r+s+n}\, x_1 \ldots x_r y_1 \ldots y_s z_1 \ldots z_n\, \mathsf{M}(x_1, \ldots, x_n, y_1, \ldots, y_s, z_1, \ldots, z_n)\, \nu_1 \ldots \nu_r\,,$$

where we get $\mathsf{M}(x_1, \ldots, z_n)$ by replacing in $\mathsf{M}(\nu_1, \ldots, \alpha_n)$ all occurrences of ν_1 by x_1 and so on. For our example we get for ϕ the expression

$$\lambda_6 x_1 y_1 y_2 z_1 z_2 z_3 ((z_3 z_1) \wedge \vee\, x_1 (y_2 z_2 z_1 x_1))\, \nu_1\,.$$

We now have $\vdash \phi \mu_1 \ldots \mu_s \alpha_1 \ldots \alpha_n$ iff $\vdash \mathsf{M}(x_1, \ldots, z_n) \dfrac{x_1, \ldots, z_n}{\nu_1, \ldots, \alpha_n}$ by (5), and $\mathsf{M}(x_1, \ldots, z_n) \dfrac{x_1, \ldots, z_n}{\nu_1, \ldots, \alpha_n} \equiv \mathsf{M}(\nu_1, \ldots, \alpha_n)$.

We normally apply (7) for the case where no μ_j occurs. The advantage of (7) is the transition from M to a metaexpression where the variables $\alpha_1, \ldots, \alpha_n$ occur only once and ordered at the end.

Since the structure of the metaexpression M is relevant in order to get the expressions ϕ we refer to an application of (7) by (ST) ("*structure transformation*").

We apply (7) only finitely many times. We hence may omit the application of (7) and may give a separate proof for every instance of it.

6. The expressions $\overset{n}{\vee}$ *and* $\overset{n}{\wedge}$. We want to construct in an effective way expressions $\overset{n}{\vee}$ and $\overset{n}{\wedge}$ $(n \geqq 1)$ s. t. we have the following generalizations of (2) and (4):

(8) *For every* α: $\vdash \overset{n}{\vee} \alpha$ *if and only if there exist* $\beta_1, \ldots, \beta_n$, *s. t.* $\vdash \alpha \beta_1 \ldots \beta_n$.

(9) *For all* $\alpha_1, \ldots, \alpha_n$: $\vdash \overset{n}{\wedge} \alpha_1 \ldots \alpha_n$ *if and only if* ($\vdash \alpha_1$ *and* $\ldots$ *and* $\vdash \alpha_n$).

Proof of (8). We may take $\overset{1}{\vee} \equiv \vee$, since we have (4). For $n+1$ we have

there exist $\beta_1, \ldots, \beta_{n+1}$, s.t. $\vdash \alpha \beta_1 \ldots \beta_{n+1}$

iff there exists a β_1 s.t. there exist $\beta_2, \ldots, \beta_{n+1}$ s.t. $\vdash (\alpha\beta_1)\, \beta_2 \ldots \beta_{n+1}$

iff there exists a β_1 s.t. $\vdash \overset{n}{\vee} (\alpha\beta_1)$ (inductive hypothesis)

iff there exists a β_1 s.t. $\vdash \phi_1 \alpha \beta_1$ (where we get ϕ_1 by (ST))

iff $\vdash \vee (\phi_1 \alpha)$ (by (4))

iff $\vdash \phi_2 \alpha$ (where we get ϕ_2 by (ST)).

This expression ϕ_2 may be used for $\overset{n+1}{\vee}$.

Proof of (9). We may take $\lambda_1 x_1 x_1$ for $\overset{1}{\wedge}$ and $\wedge$ for $\overset{2}{\wedge}$, since we have (2). For $n+1$ $(n \geqq 2)$ we have

$$\vdash \alpha_1 \textit{ and } \ldots \textit{ and } \vdash \alpha_{n+1}$$

iff $\quad \vdash \overset{n}{\wedge} \alpha_1 \ldots \alpha_n$ and $\vdash \alpha_{n+1}$ (induction hypothesis)

iff $\quad \vdash \wedge (\overset{n}{\wedge} \alpha_1 \ldots \alpha_n) \alpha_{n+1}$ (by (2))

iff $\quad \vdash \phi \alpha_1 \ldots \alpha_{n+1}$ (where we get ϕ by (ST)).

This expression ϕ may be used for $\overset{n+1}{\wedge}$.

7. *Primitive recursive functions whose arguments and values are expressions.* These are the analogues of the ordinary primitive recursive functions. In place of the ordinary initial functions[1] s, u_n^i, c_0^0 (cf. § 10.1) we have here

(a) the *binary juxtaposition function* J which, for arguments α, β, has the value $J(\alpha, \beta) \equiv (\alpha\beta)$,

(b) the *identity functions* U_n^i $(n \geqq 1, 1 \leqq i \leqq n)$ which are defined by $U_n^i(\alpha_1, \ldots, \alpha_n) \equiv \alpha_i$,

(c) the 0-ary constant C_0^* with value $*$.

A function F is called *primitive recursive* if it is one of the initial functions, or if it is obtainable from one of these initial functions by finitely many applications of the processes of substitution and quasi-induction.

Substitution is defined as in § 10.1.

Quasi-induction leads from an n-ary function G and an $(n+4)$-ary function H to an $(n+1)$-ary function F, where

$$(10) \quad \begin{cases} F(\alpha_1, \ldots, \alpha_n, *) \equiv G(\alpha_1, \ldots, \alpha_n), \\ F(\alpha_1, \ldots, \alpha_n, (\beta_1\beta_2)) \equiv H(\alpha_1, \ldots, \alpha_n, \beta_1, \beta_2, H(\alpha_1, \ldots, \alpha_n, \beta_1), \\ \qquad\qquad H(\alpha_1, \ldots, \alpha_n, \beta_2)) . \end{cases}$$

[1] We want to use here *small* letters for the functions whose arguments and values are numbers and *capital* letters for functions whose arguments and values are expressions. Hence in place of S, U_n^i, C_0^0 (§ 10.1) we use here s, u_n^i, c_0^0.

It may be shown that under a certain Gödelization (e.g. under $^-$ as defined in no. 10, these primitive recursive functions F correspond to the ordinary primitive recursive functions f. Since we do not make use of this proposition we omit a proof of it.

8. Representation of primitive recursive functions F. We shall say that *the expression φ represents the n-ary function F* if for all $\alpha_1, \ldots, \alpha_n, \beta$

(11) $\quad \vdash \varphi\alpha_1 \ldots \alpha_n\beta \quad$ *if and only if* $\quad F(\alpha_1, \ldots, \alpha_n) \equiv \beta$.

Lemma 1. Every primitive recursive function F has at least one representation.

We show this first for the initial functions and then we deal with the processes of substitution and quasi-induction.

(a) *The initial functions.* We have $C_0^* \equiv \beta$ iff $* \equiv \beta$. This by (1) is equivalent to $\vdash = * \beta$. Hence the expression $= *$ represents C_0^*. — $U_n^i(\alpha_1, \ldots, \alpha_n) \equiv \beta$ iff $\alpha_i \equiv \beta$. This is equivalent to $\vdash = \alpha_i\beta$. By (ST) we have an expression ϕ s.t. $\vdash = \alpha_i\beta$ is equivalent to $\vdash \phi\alpha_1 \ldots \alpha_n\beta$. This ϕ represents U_n^i. — $J(\alpha_1, \alpha_2) \equiv \beta$ iff $(\alpha_1\alpha_2) \equiv \beta$. This is equivalent to $\vdash = (\alpha_1\alpha_2)\beta$. By (ST) there is a ϕ s.t. $\vdash \phi\alpha_1\alpha_2\beta$ iff $\vdash = (\alpha_1\alpha_2)\beta$. ϕ represents J.

(b) *The substitution process.* We assume that the functions $G, H_1, \ldots, H_m$ are represented by the expressions $\psi, \varphi_1, \ldots, \varphi_m$. Let for all $\alpha_1, \ldots, \alpha_n$

$$F(\alpha_1, \ldots, \alpha_n) \equiv G(H_1(\alpha_1, \ldots, \alpha_n), \ldots, H_m(\alpha_1, \ldots, \alpha_n)).$$

Then we have

$$F(\alpha_1, \ldots, \alpha_n) \equiv \beta$$

iff there are $\beta_1, \ldots, \beta_m$ s.t. $H_1(\alpha_1, \ldots, \alpha_n) \equiv \beta_1$ and ... and $H_m(\alpha_1, \ldots, \alpha_n) \equiv \beta_m$ and $G(\beta_1, \ldots, \beta_m) \equiv \beta$

iff there are $\beta_1, \ldots, \beta_m$ s.t. $\vdash \varphi_1\alpha_1 \ldots \alpha_n\beta_1$ and ... and $\vdash \varphi_m\alpha_1 \ldots \alpha_n\beta_m$ and $\vdash \psi\beta_1 \ldots \beta_m\beta$

iff there are $\beta_1, \ldots, \beta_m$ s.t. $\vdash \overset{m+1}{\wedge} (\varphi_1\alpha_1 \ldots \alpha_n\beta_1) \ldots (\varphi_m\alpha_1 \ldots \alpha_n\beta_m)(\psi\beta_1 \ldots \beta_m\beta)$ (by (9))

iff there are $\beta_1, \ldots, \beta_m$ s.t. $\vdash \phi_1\alpha_1 \ldots \alpha_n\beta\beta_1 \ldots \beta_m$ (for a suitable ϕ_1 by (ST)) *iff* $\vdash \overset{m}{\vee} (\alpha_1 \ldots \alpha_n \beta)$ (by (8))

iff $\vdash \phi_2 \alpha_1 \ldots \alpha_n \beta$ (for a suitable ϕ_2 by (ST)).

Hence ϕ_2 represents F.

(c) *The process of quasi-induction.* Let F be defined with the help of G and H by (10). Let G and H be represented by the expressions ψ and χ respectively. We look for an expression $\varphi\varphi$ which should represent F. In order to find such a φ we make the following observations *under the assumption* that $\varphi\varphi$ *represents* F.

$$F(\alpha_1, \ldots, \alpha_n, \alpha) \equiv \beta$$

iff $(\alpha \equiv *$ and $G(\alpha_1, \ldots, \alpha_n) \equiv \beta)$ or there exist $\beta_1, \beta_2, \delta_1, \delta_2$ s.t.
$(\alpha \equiv (\beta_1 \beta_2)$ and $F(\alpha_1, \ldots, \alpha_n, \beta_1) \equiv \delta_1$ and $F(\alpha_1, \ldots, \alpha_n, \beta_2) \equiv \delta_2$
and $H(\alpha_1, \ldots, \alpha_n, \beta_1, \beta_2, \delta_1, \delta_2) \equiv \beta)$.

First let us reformulate a part of the right side of this equivalence.

there exist $\beta_1, \beta_2, \delta_1, \delta_2$ s.t. $(\alpha \equiv (\beta_1 \beta_2)$
and $F(\alpha_1, \ldots, \alpha_n, \beta_1) \equiv \delta_1$ and $F(\alpha_1, \ldots, \alpha_n, \beta_2) \equiv \delta_2$
and $H(\alpha_1, \ldots, \alpha_n, \beta_1, \beta_2, \delta_1, \delta_2) \equiv \beta)$

iff there exist $\beta_1, \beta_2, \delta_1, \delta_2$ s.t. $(\vdash \; = \alpha(\beta_1 \beta_2)$
and $\vdash \varphi\varphi\alpha_1 \ldots \alpha_n \beta_1 \delta_1$ and $\vdash \varphi\varphi\alpha_1 \ldots \alpha_n \beta_2 \delta_2$
and $\vdash \chi\alpha_1 \ldots \alpha_n \beta_1 \beta_2 \delta_1 \delta_2 \beta)$ (by (1), (9))

iff there exist $\beta_1, \beta_2, \delta_1, \delta_2$ s.t.
$\vdash \overset{4}{\wedge} (= \alpha(\beta_1 \beta_2)) (\varphi\varphi\alpha_1 \ldots \alpha_n \beta_1 \delta_1) (\varphi\varphi\alpha_1 \ldots \alpha_n \beta_2 \delta_2)$
$(\chi\alpha_1 \ldots \alpha_n \beta_1 \beta_2 \delta_1 \delta_2 \beta)$ (by (9))

iff there exist $\beta_1, \beta_2, \delta_1, \delta_2$ s.t. $\vdash \phi_1 \varphi\alpha_1 \ldots \alpha_n \alpha\beta\beta_1\beta_2\delta_1\delta_2$
(for a suitable expression ϕ_1 by (ST))

iff $\vdash \overset{4}{\vee} (\phi_1 \varphi\alpha_1 \ldots \alpha_n \alpha\beta)$.

Now we go back to our original equivalence. We have

$$F(\alpha_1, \ldots, \alpha_n, \alpha) \equiv \beta$$

iff $\vdash \vee(\wedge(= \alpha *) (\psi\alpha_1 \ldots \alpha_n \beta)) (\overset{4}{\vee} (\phi_1 \varphi\alpha_1 \ldots \alpha_n \alpha\beta))$ (by (1), (2), (3))

iff $\vdash \phi_2 \varphi\alpha_1 \ldots \alpha_n \alpha\beta$ (for a suitable expression ϕ_2 by (ST)).

Note that in the definition of ϕ_2 no use is made of the expression φ which is not yet known. *The statement* $\vdash \phi_2 \varphi \alpha_1 \ldots \alpha_n \alpha \beta$ *would be equivalent to* $F(\alpha_1, \ldots, \alpha_n, \alpha) \equiv \beta$ *if* $\varphi \equiv \phi_2$. *This now will be our definition of* φ. It is easy to prove that for this φ the expression $\varphi\varphi$ represents F, i.e. that $F(\alpha_1, \ldots, \alpha_n, \alpha) \equiv \beta$ iff $\vdash \varphi\varphi\alpha_1 \ldots \alpha_n \alpha \beta$, by induction on the length of α, using exactly the transitions which we have carried through a moment ago.

9. Representation of natural numbers by expressions. Correspondence of functions. We associate with each natural number r an expression $\underline{r}$ recursively defined by

$$\text{(12)} \qquad \begin{cases} \underline{0} \equiv * \\ \underline{r'} \equiv (\underline{r} *) . \end{cases}$$

Obviously $\underline{\ }$ is one-one. We shall say that a *function F corresponds to the ordinary function f* $(F \sim f)$ if F and f are both n-ary functions and if for all numbers $r_1, \ldots, r_n$ we have

$$\text{(13)} \qquad F(\underline{r_1}, \ldots, \underline{r_n}) \equiv \underline{f(r_1, \ldots, r_n)} .$$

Lemma 2. For every primitive recursive function f there is a primitive recursive function F s.t. $F \sim f$.

We first consider the ordinary initial functions and proceed by treating the processes of substitution and induction.

(a) *The initial functions.* $C_0^* \sim c_0^0$, since $C_0^* \equiv * \equiv \underline{0} \equiv \underline{c_0^0}$. — $U_n^i \sim u_n^i$, since $U_n^i(\underline{r_1}, \ldots, \underline{r_n}) \equiv \underline{r_i} \equiv \underline{u_n^i(r_1, \ldots, r_n)}$. — We define $S(\alpha) \equiv (\alpha *)$. Then $S \sim s$, since $S(\underline{r}) \equiv \underline{r} * \equiv \underline{r'} \equiv \underline{s(r)}$. Obviously S is primitive recursive[1].

(b) *Substitution.* Let $f(r_1, \ldots, r_n) = g(h_1(r_1, \ldots, r_n), \ldots, h_m(r_1, \ldots, r_n))$, $H_1 \sim h_1, \ldots, H_m \sim h_m$ and $G \sim g$. Let $F(\alpha_1, \ldots, \alpha_n) \equiv G(H_1(\alpha_1, \ldots, \alpha_n), \ldots, H_m(\alpha_1, \ldots, \alpha_n))$. It follows immediately that $F \sim f$.

(c) *Induction.* Let

$$\begin{cases} f(r_1, \ldots, r_n, 0) = g(r_1, \ldots, r_n) \\ f(r_1, \ldots, r_n, r') = h(r_1, \ldots, r_n, r, f(r_1, \ldots, r_n, r)) . \end{cases}$$

Let $G \sim g$ and $H \sim h$. Let

$$H_1(\alpha_1, \ldots, \alpha_n, \alpha, \beta, \gamma, \delta) \equiv H(\alpha_1, \ldots, \alpha_n, \alpha, \gamma) .$$

---

[1] We leave the details to the reader. It is here useful to apply the same techniques as in § 10.

Define F by the following quasi-induction:

$$\begin{cases} F(\alpha_1, \ldots, \alpha_n, *) \equiv G(\alpha_1, \ldots, \alpha_n) \\ F(\alpha_1, \ldots, \alpha_n, (\beta_1\beta_2)) \equiv H_1(\alpha_1, \ldots, \alpha_n, \beta_1, \beta_2, F(\alpha_1, \ldots, \alpha_n, \beta_1), \\ \qquad\qquad\qquad\qquad\qquad\qquad F(\alpha_1, \ldots, \alpha_n, \beta_2)) . \end{cases}$$

Then $F \sim f$, as is easily seen by induction on the last argument of f.

10. A Gödelization of the set of expressions. To each expression α we associate a natural number $\bar{\alpha}$ which is recursively defined by

$$(14) \qquad \begin{cases} \bar{*} = 0 \\ \overline{(\alpha\beta)} = \sigma_2(\bar{\alpha}, \bar{\beta}) + 1 \end{cases}$$

(for σ_2 see § 12.4). $\bar{\alpha}$ is called *the Gödel number* of α. It is not difficult to see that $^-$ is a one-one mapping of the set of expressions onto the set of natural numbers. The inverse of $^-$ is rather complicated. In place of it we use the simpler function $_-$ which has been introduced in no. 9.

Lemma 3. The function $\underline{(\bar{\alpha})}$ is primitive recursive.

This can be shown by giving a quasi-induction for this function. $\underline{(\bar{*})} \equiv \underline{0} \equiv * \equiv C_0^*$. Furthermore

$$\begin{aligned} \underline{\overline{(\alpha\beta)}} &\equiv \underline{\sigma_2(\bar{\alpha}, \bar{\beta}) + 1} && \text{(by (14))} \\ &\equiv \underline{\sigma_2(\bar{\alpha}, \bar{\beta})} * && \text{(by (12))} \\ &\equiv \Sigma_2(\underline{(\bar{\alpha})}, \underline{(\bar{\beta})}) *, \end{aligned}$$

where $\Sigma_2 \sim \sigma_2$ (cf. Lemma 2).

11. The representation of enumerable relations between expressions. We finally want to show, that for each n-ary enumerable relation R between expressions there is an expression ϱ such that for all expressions $\alpha_1, \ldots, \alpha_n$ we have (0).

Let such a relation R be given. First we associate with R an n-ary predicate $\tilde{R}$ by postulating that $\tilde{R} r_1 \ldots r_n$ if and only if $r_1, \ldots, r_n$ are Gödel numbers of expressions $\alpha_1, \ldots, \alpha_n$ such that $R\alpha_1 \ldots \alpha_n$. Conversely we have for arbitrary expressions $\alpha_1, \ldots, \alpha_n$ that

$$(15) \qquad R\alpha_1 \ldots \alpha_n \textit{ if and only if } \tilde{R}\overline{\alpha_1} \ldots \overline{\alpha_n} .$$

$\tilde{R}$ is enumerable (cf. § 2.5). According to Kleene's enumeration theorem (§ 28, Theorem 2), we have for $\tilde{R}$ the biimplication

$$(16) \qquad \tilde{R} r_1 \ldots r_n \textit{ if and only if } \text{there is an } r \text{ such that } T_n r_0 r_1 \ldots r_n r$$

for all $r_1, \ldots, r_n$, where r_0 is a suitable number depending on $\tilde{R}$. The $(n+1)$-ary predicate $T_n r_0 r_1 \ldots r_n r$ (r_0 fixed) is primitive recursive. Hence there exists a primitive recursive function f such that for all $r_1, \ldots, r_n, r$

$$(17) \qquad T_n r_0 r_1 \ldots r_n r \textit{ if and only if } f(r_1, \ldots, r_n, r) = 0\,.$$

Let be F a primitive recursive function such that $F \sim f$ (cf. Lemma 2). Now we have for arbitrary $\alpha_1, \ldots, \alpha_n$:

$R\alpha_1 \ldots \alpha_n$

iff there is an r s.t. $f(\overline{\alpha_1}, \ldots, \overline{\alpha_n}, r) = 0$ (by (15), (16), (17))

iff there is an α s.t. $f(\overline{\alpha_1}, \ldots, \overline{\alpha_n}, \bar{\alpha}) = 0$ ($^-$ maps onto)

iff there is an α s.t. $\underline{f(\overline{\alpha_1}, \ldots, \overline{\alpha_n}, \bar{\alpha})} \equiv \underline{0}$ ($\_$ is one-one)

iff there is an α s.t. $F(\underline{(\overline{\alpha_1})}, \ldots, \underline{(\overline{\alpha_n})}, \underline{(\bar{\alpha})}) \equiv *$.

This last equivalence follows from the fact, that $F \sim f$, and that $\underline{0} \equiv *$. Using Lemma 3, we see that the function G defined by

$$G(\alpha_1, \ldots, \alpha_n, \alpha) \equiv F(\underline{(\overline{\alpha_1})}, \ldots, \underline{(\overline{\alpha_n})}, \underline{(\bar{\alpha})})$$

is primitive recursive. According to Lemma 1 we have a representation ψ of G. Hence

$$G(\alpha_1, \ldots, \alpha_n, \alpha) \equiv * \textit{ if and only if } \vdash \psi\alpha_1 \ldots \alpha_n \alpha *\,.$$

Now we get for all $\alpha_1, \ldots, \alpha_n$

$R\alpha_1 \ldots \alpha_n$

iff there is an α s.t. $\vdash \psi\alpha_1 \ldots \alpha_n \alpha *$

iff there is an α s.t. $\vdash \phi_1 \alpha_1 \ldots \alpha_n \alpha$ (for a suitable ϕ_1 by (ST))

iff $\vdash \vee(\phi_1 \alpha_1 \ldots \alpha_n)$ (by (4))

iff $\vdash \phi_2 \alpha_1 \ldots \alpha_n$ (for a suitable ϕ_2 by (ST)).

Hence ϕ_2 represents the relation R. Thus every enumerable relation is representable by an expression.

References

Among the numerous works by Fitch we draw attention to the following

Fitch, F. B.: A Simplification of Basic Logic. J. symbolic Logic **18**, 317–325 (1953); esp. p. 324, where the expressions H_j occur, for which we have used the notation λ_j which reminds us of Church's λ.

FITCH, F. B.: Recursive Functions in Basic Logic. J. symbolic Logic 21, 337—346 (1956).

On the inversion principle, compare

LORENZEN, P.: Einführung in die operative Logik und Mathematik. Berlin-Göttingen-Heidelberg: Springer 1955.

HERMES, H.: Zum Inversionsprinzip der operativen Logik. Constructivity in Mathematics, ed. by A. Heyting. pp. 62—68. Amsterdam: North-Holland Publishing Company 1959.

§ 33. Further Precise Mathematical Replacements of the Concept of Algorithm

Among the precise replacements which have been given for the concept of algorithm the concept of *calculus in canonical form* originating from POST and the concept of *normal algorithm* originating from MARKOV are worthy of mentioning here. We shall give the definitions of these concepts. For more exact relations between these and the concepts discussed in fuller detail in this book the reader should consult the references given.

1. Calculus in canonical form (POST). Let $\mathfrak{A}$, a finite alphabet, be given. We consider arbitrary words W over $\mathfrak{A}$. Furthermore, we have arbitrary many variables v, which must not be confused with the words over $\mathfrak{A}$. We shall understand by a *rule of inference* a schema of the following kind.

$$\text{Premisses:}\quad \begin{cases} W_{11}v_{11}\,W_{12}\,v_{12}\,\ldots\,W_{1n_1}\,v_{1n_1}\,W_{1n_1+1} \\ W_{21}v_{21}\,W_{22}\,v_{22}\,\ldots\,W_{2n_2}\,v_{2n_2}\,W_{2n_2+1} \\ \ldots\ldots\ldots\ldots\ldots\ldots\ldots\ldots \\ W_{m1}v_{m1}\,W_{m2}v_{m2}\,\ldots\,W_{mn_m}\,v_{mn_m}\,W_{mn_m+1} \end{cases}$$

$$\text{Conclusion:}\quad W_1v_1W_2v_2\ldots W_nv_nW_{n+1}.$$

In this schema the W_{ik} and the W_i are special words (possibly empty, not necessarily different) and the v_{ik} and the v_i are special variables (not necessarily different). We require that

(1) $n_1 \geqq 1, \ldots, n_m \geqq 1,\ n \geqq 1$;

(2) every variable v_i of the conclusion occurs among the variables v_{ik} of the premisses.

If we replace every variable in a rule of inference by a word over $\mathfrak{A}$ (if a variable occurs more than once, then it will naturally have to be replaced by the same word each time), then the premisses become words $W_1^*, \ldots, W_m^*$ over $\mathfrak{A}$ and the conclusion a word W^* over $\mathfrak{A}$. Then we shall say that *W^* is obtained from $W_1^*, \ldots, W_m^*$ by application of the rule in question*. However we make the restrictive proviso that *W^* is not the empty word.*

A *calculus in canonical form* is given by finitely many words over $\mathfrak{A}$ (called *axioms*) and finitely many rules of inference. A word over $\mathfrak{A}$ is called *derivable* in a calculus in canonical form if it is an axiom or if it can be obtained from the axioms by applications of the rules of inference.

Post's calculi in canonical form are of a great generality. However, we can easily give an account of algorithms which are of a similar form but do not however fall under the calculi in canonical form. Let us take as example the logical rule of inference *modus ponens* (cf. (T 5) in § 24.3).

Premisses: $\begin{cases} (p \to q) \\ p \end{cases}$

Conclusion: $q.$

If we have an underlying alphabet $\mathfrak{A} = \{(,), \to, \ldots\}$, then we see that the modus ponens, which can also be written, by the use of the empty word, in the form

Premisses: $\begin{cases} (p \to q) \\ \Box\, p\, \Box \end{cases}$

Conclusion: $\Box\, q\, \Box$

is to all appearences a rule of inference in the sense of Post. However, this schema does not correspond to such a rule of inference: In case of a calculus in canonical form we would in an application of this rule be allowed to substitute for p and q arbitrary words over the alphabet $\mathfrak{A}$ (with the restriction, that q may not be the empty word), thus, e.g. ((for p. However, this is not allowed in the modus ponens, in which only formulae (assertions or something similar) may be substituted for p and q and not the word ((.

Thus, the variables p and q refer to another calculus, namely to a calculus by which we can obtain the formulae (assertions or something similar). LORENZEN based his considerations (cf. references to § 32) on rules of inference in which "alien variables" which refer to another calculus are allowed. Cf. also the paper by CURRY (References).

We say that a calculus in canonical form is *normal* if it has only one axiom and every rule of inference is of the following kind:

Premiss: $W_1 v \Box$

Conclusion: $\Box v W_2.$

One can give effectively for every calculus K in canonical form over an alphabet $\mathfrak{A}$: (1) an alphabet $\mathfrak{A}^*$ containing $\mathfrak{A}$, (2) a calculus K^* in canonical form over $\mathfrak{A}^*$ such that K^* is normal and a word over $\mathfrak{A}$ is derivable in K^* if and only if it is derivable in K.

2. *Normal algorithms* (MARKOV). Let a finite alphabet $\mathfrak{A}$ be given. Let all words occurring in this Section (including the empty word) be words over $\mathfrak{A}$. The symbols $\rightarrow$ and $\cdot$ are not allowed to occur in $\mathfrak{A}$. Let a word over the alphabet $\mathfrak{A} \cup \{\rightarrow, \cdot\}$ be called a *substitution formula* if it is one of the following two kinds:

(1) $$W \rightarrow W'$$

(2) $$W \rightarrow \cdot\, W',$$

where W and W' are words over $\mathfrak{A}$.

A *normal algorithm* is given by a finite *sequence* (not set) of substitution formulae:

$$W_1 \rightarrow (\cdot)\, W_1'$$
$$W_2 \rightarrow (\cdot)\, W_2'$$
$$\cdot\;\cdot\;\cdot\;\cdot\;\cdot\;\cdot$$
$$W_m \rightarrow (\cdot)\, W_m'.$$

The parentheses around the dots mean that there may or may not be a dot in the substitution formula.

A normal algorithm determines unambiguously for every word U over $\mathfrak{A}$ a possibly terminating sequence of words $U = U_0, U_1, U_2, \ldots$ over $\mathfrak{A}$ and for every word U_k with $k \neq 0$ a *rule* by which U_k is obtained from U_{k-1}. $U_0 = U$. Let U_k be already defined. Then there are two cases to be distinguished:

(a) The rule by which U_k was obtained from U_{k-1} contains a dot. (This is not the case for U_0, since U_0, the initial word, is not obtained by a rule at all.) Then the sequence terminates with the word U_k.

(b) The rule by which U_k was obtained from U_{k-1} does not contain a dot, or $k = 0$. We distinguish between two sub-cases.

(b_1) None of the words $W_1, \ldots, W_m$ is a sub-word of U_k.[1] In this case also the sequence terminates with the word U_k.

(b_2) U_k contains one of the words $W_1, \ldots, W_m$ as sub-word. Let i be the smallest number such that W_i is a sub-word of U_k. Let $U_k \equiv C W_i D$ be the standard factorization of U_k with respect to W_i. Then we put $U_{k+1} \equiv C W_i' D$ and say that U_{k+1} is obtained from U_k by means of the rule $W_i \rightarrow (\cdot)\, W_i'$.

If the sequence $U = U_0, U_1, U_2, \ldots$ terminates with U_k as last member, then we write $U_k = \varphi(U)$. We have thus a function, which, in

[1] A is called a *sub-word* of B if there exist words C and D such that $B \equiv C A D$. If A is a sub-word of B, then in general C and D are not unambiguously determined (e.g. for $A \equiv ||$ and $B \equiv |||$ we have $B \equiv \square A| \equiv |A\square$). However, there exists a unique representation $B \equiv C A D$ with a *shortest* C. The latter is called *the standard decomposition of B with respect to A*.

general, is not defined for all words (the process need not terminate at all) and which can be computed by means of the given normal algorithm for the arguments for which it is defined. Thus, φ is a partial recursive function (§19.6). MARKOV put forward a thesis, which corresponds to CHURCH's thesis (§4.3), that every partial recursive function in the domain of words over the alphabet $\mathfrak{A}$ can be obtained by a normal algorithm.

Markov's normal algorithms provide unambiguously determined derivation sequences $U_0, U_1, U_2, \ldots$. An essential difference between Markov's algorithms and Post's calculi in canonical form is that the latter do not determine an unambiguous sequence of derivations and so correspond to many traditional algorithms. On the other hand, Markov's normal algorithms are closely related to the automatically operating Turing machines.

References

POST, E. L.: Formal Reductions of the General Combinatorial Decision Problem. Amer. J. Math. **65**, 197—215 (1943). (Cf. the review by CHURCH in: J. symbolic Logic **8**, 50—52 (1943).)

D'ETLOVS, V. K.: The normal algorithms and the recursive functions [Russ.]. Dokl. Akad. Nauk SSSR. **90**, 723—725 (1953).

MARKOV, A. A.: Theory of algorithms [Russ.]. Akad. Nauk SSSR., Matém. Inst. Trudy **42**, Moscow-Leningrad 1954.

CURRY, H. B.: Calculuses and Formal Systems. Dialectica **12**, 249—273 (1958).

ASSER, G.: Normierte Postsche Algorithmen. Z. math. Logik **5**, 323—333 (1959).

§ 34. Recursive Analysis

We can apply algorithms only to words over a finite alphabet or to things which can be denoted effectively by such words. Such things are the natural numbers, and also the rational numbers, but not the real numbers, or in any case not if we follow the classical point of view (which we do in this book) according to which there are non-denumerably many real numbers. Thus, only certain real numbers α can be defined constructively. We shall require that such a number α can be obtained as the limit of a computable sequence of rational numbers and, indeed, so that it is possible to undertake the estimates necessary for the convergence in a constructive way. In this way we arrive at the concept of computable real number. All algebraic numbers are computable in this sense and so are for instance the numbers e and π.

A great part of analysis has been studied from this point of view. In this paragraph we must confine ourselves to the consideration of some of the basic concepts of the so-called *recursive analysis* and refer the reader for further developments to the references.

1. Computable sequences of natural numbers. In order to be able to apply algorithmic concepts to the rational numbers we must be able to represent these numbers by words, in the simplest case by natural numbers. Such a Gödel numbering is especially simple here. For every rational number ϱ there exist natural numbers p, q and r such that

$$(*) \qquad \varrho = \frac{p-q}{1+r}$$

and, conversely, every triple p, q and r of natural numbers provides us with a rational number according to $(*)$. Thus, we can use the triples (p, q, r) of natural numbers, or also the numbers $\sigma_3(p, q, r)$, as representations of rational numbers. These representations are not unambiguous[1]. (p, q, r) and $(\bar{p}, \bar{q}, \bar{r})$ represent the same rational number if and only if $(1+\bar{r})\,p + (1+r)\,\bar{q} = (1+r)\,\bar{p} + (1+\bar{r})\,q$. This is a decidable relation.

A sequence φ of rational numbers (i.e. a function whose arguments are natural numbers and whose values are rational numbers) can be represented in a similar way by three sequences f, g and h of natural numbers. Thus,

$$(**) \qquad \varphi(n) = \frac{f(n) - g(n)}{1 + h(n)}$$

for every natural number n. Thus we arrive at the

Definition. A *sequence φ of rational numbers* is called *computable*[2] if there exist computable functions f, g and h such that $(**)$ is valid for every n.

In agreement with mathematical custom we shall denote sequences by $f_n, \varrho_n, \ldots$.

An example of a computable sequence of rational numbers is the sequence $\varepsilon_n = \frac{1}{2^n}$. We obtain a representation of the form $(**)$ by putting $f_n = 1$, $g_n = 0$, $h_n = 2^n \dot{-} 1$ (§ 10.4 (10)).

2. Computable convergence. It is well known that a sequence ϱ_n of rational numbers is convergent if and only if for every positive rational (or real) ε there exists an n_0 such that $|\varrho_n - \varrho_l| < \varepsilon$ for all $n, l \geqq n_0$. We need not demand this requirement for every ε. It is sufficient if it is satisfied for the members of a nullsequence, e.g. for the sequence ε_n introduced just above. This means that for every m there must be an n_0

[1] In this respect we are dealing with a more general concept of Gödel numbering than that which was introduced in § 1.3.

[2] In this paragraph the word "computable" can be replaced by "Turing computable", "recursive", or by the name of any other precise replacement of computability.

such that $|\varrho_n - \varrho_l| < \varepsilon_m$ for all $n, l \geqq n_0$. Naturally, the possible choice of n_0 depends on the choice of m. In the convergence criterion it is only required that for every m there exists an n_0 with the given property; it is *not* required that it should be possible to find such an n_0 effectively. We speak of *computable* convergence if there exists a method by which we can find an n_0 for every m. This means that there must be a computable function k_m such that for $n_0 = k_m$ the above mentioned estimate requirement is fulfilled. This way we arrive at the

Definition. A *sequence ϱ_n of rational numbers* is called *computably convergent* if there exists a computable sequence k_m such that

(***) $\qquad |\varrho_n - \varrho_l| < \varepsilon_m$ *for all* $n, l \geqq k_m$.[1]

Obviously, from the point of view of computation, a real number α is well defined if and only if there exists a computable sequence of rational numbers which converges computably and whose limit is α. (We shall also express this in short by saying that the sequence *converges computably to α*.) We shall call such real numbers computable. Thus we have the

Definition. A *real number* α is called *computable* if there exists a computable sequence ϱ_n of rational numbers which converges computably and whose limit is α.

Obviously, there exist only enumerably many computable real numbers (cf. § 2.2). Thus, not every real number is computable. Every rational number ϱ is computable since we have the computable sequence $\varrho_n = \varrho$, which converges computably to ϱ. We give a less trivial

Example. The number e is computable. It is sufficient to show that the sequence $\varrho_n = \sum_{\nu=0}^{n} \frac{1}{\nu!}$ is computable and computably convergent.

If we put $f_0 = 1$ and $g_0 = h_0 = 0$, then $\varrho_0 = \frac{f_0 - g_0}{1 + h_0}$. Let f_n, g_n and h_n be already given with $\varrho_n = \frac{f_n - g_n}{1 + h_n}$. We have that

$$\varrho_{n+1} = \varrho_n + \frac{1}{(n+1)!} = \frac{f_n(n+1)! + 1 + h_n - g_n(n+1)!}{(1 + h_n)(n+1)!}.$$

This coincides with $\frac{f_{n+1} - g_{n+1}}{1 + h_{n+1}}$ if we put

$$h_{n+1} = (1 + h_n)(n+1)! \dot{-} 1$$

$$f_{n+1} = f_n(n+1)! + 1 + h_n$$

$$g_{n+1} = g_n(n+1)!.$$

---

[1] We note the use of the *classical* existence operator in this definition, where the existence of a computable sequence with the property (***) is required. Cf. § 2.1.

The functions f_n, g_n and h_n defined in this way are primitive recursive. This proves the computability of ϱ_n.

Further, for $n < l$

$$\begin{aligned} |\varrho_n - \varrho_l| &= \frac{1}{(n+1)!} + \cdots + \frac{1}{l!} \\ &\leq \frac{1}{(n+1)!}\left(1 + \frac{1}{n+1} + \frac{1}{(n+1)^2} + \cdots\right) \\ &= \frac{1}{n!\,n} \\ &< \frac{1}{2^m} \text{ for } n \geqq 2^m + 1. \end{aligned}$$

We put $k_m = 2^m + 1$. Then the estimate shows the computable convergence of ϱ_n.

Theorem. If α and β are computable real numbers, then so are also $\alpha + \beta$, $\alpha - \beta$, $\alpha\beta$ and (if $\beta \neq 0$) $\frac{\alpha}{\beta}$.

Proof. Let ϱ_n and τ_n be computable sequences of rational numbers which converge computably to α and β respectively. It is known then that the sequences $\varrho_n + \tau_n$, $\varrho_n - \tau_n$, $\varrho_n \tau_n$ and $\frac{\varrho_n}{\tau_n}$ converge to $\alpha + \beta$, $\alpha - \beta$, $\alpha\beta$ and $\frac{\alpha}{\beta}$ respectively. The usual proofs for these facts can be completed without difficulty to show that these sequences are computable and computably convergent.

We shall only consider the case of the quotient in fuller detail. There we must observe that $\frac{\varrho_n}{\tau_n}$ is not necessarily defined for small n, since it is possible that $\tau_n = 0$. First we have to show that there exists a computable sequence η_n, such that $\eta_n \neq 0$ for all n, which converges computably to β. Because $\beta \neq 0$, there exists an m with $\varepsilon_m < |\beta|$ (for the null-sequence ε_m considered above). Let $\varepsilon_{m_0} < |\beta|$. According to (***) there exists a computable function k_m such that $|\tau_n - \tau_l| < \varepsilon_m$ for all $n, l \geqq k_m$. It follows that $\tau_n \neq 0$ for $n \geqq k_{m_0}$. We form the sequence[1]

$$\eta_n = \begin{cases} \tau_n & \text{for} \quad n \geqq k_{m_0} \\ 1 & \text{otherwise.} \end{cases}$$

It is easy to see that η_n is computable.

---

[1] The reader should note that by the following definition only the *existence* of such a sequence is proved (and more is not necessary for the proof in question). For an effective construction of η_n from given τ_n and k_m a knowledge of m_0 would be necessary. Thus, our *method of proving* is not constructive.

Let us put

$$\bar{k}_m = \begin{cases} \max(k_m, k_{m_0}) & \text{for } m \geqq m_0 \\ k_{m_0} & \text{for } m < m_0. \end{cases}$$

$\bar{k}_m$ is recursive.

Let $n, l \geqq \bar{k}_m$. Then a fortiori $n, l \geqq k_{m_0}$ and so $\eta_n = \tau_n$ and $\eta_l = \tau_l$. It follows that $|\eta_n - \eta_l| = |\tau_n - \tau_l| < \varepsilon_m$ (this latter estimate follows for $m \geqq m_0$ from $n, l \geqq \bar{k}_m \geqq k_m$, and for $m < m_0$ from the fact that $n, l, \geqq k_{m_0}$ and so $|\tau_n - \tau_l| \leqq \varepsilon_{m_0} \leqq \varepsilon_m$). Thus,

$$|\eta_n - \eta_l| < \varepsilon_m \quad \text{for} \quad n, l \geqq \bar{k}_m,$$

by which the computable convergence of η_n is shown. We have $\eta_n \neq 0$ for all n. Since η_n converges to $\beta \neq 0$, there exists an m_1 such that $|\eta_n| > \varepsilon_{m_1}$ for all n.

If we presume that the theorem is already proved for the product, then we only need to show in the case of the quotient that if $\beta \neq 0$ is a computable number, then so is $\frac{1}{\beta}$. It is sufficient to show that the sequence $\frac{1}{\eta_n}$ is computable and converges computably. We confine ourselves to proving the last part of the assertion. We use the recursive sequence $\bar{\bar{k}}_m = \bar{k}_{m+2m_1}$. Then, for $n, l \geqq \bar{\bar{k}}_m$

$$\left|\frac{1}{\eta_n} - \frac{1}{\eta_l}\right| = \left|\frac{\eta_n - \eta_l}{\eta_n \cdot \eta_l}\right| < \frac{\varepsilon_{m+2m_1}}{\varepsilon_{m_1}^2} = \varepsilon_m.$$

This completes the proof.

3. Computable decimals. A computable decimal shall mean an integer $g \gtreqless 0$ and a series of the form $\sum_{n=1}^{\infty} f_n \cdot 10^{-n}$, where f_n is a computable function which only assumes the values 0, 1, ..., 9. We shall say that a real number α *can be expanded into a computable decimal* if there exists a computable decimal such that $\alpha = g + \sum_{n=1}^{\infty} f_n \cdot 10^{-n}$. Obviously, every real number which can be expanded into a computable decimal is a computable real number. We also show the converse, i.e. the

Theorem. Every computable real number α can be expanded into a computable decimal[1].

---

[1] SPECKER (see references) discusses primitive recursive real numbers and primitive recursive decimals. These are defined similary to the way we defined computable real numbers, with the only difference that instead of computable functions he uses primitive recursive functions. Specker shows that not every primitive recursive real number can be expanded into a primitive recursive decimal.

Proof. We distinguish between two cases, depending on whether α is rational or irrational[1].

If α is rational, then there exists a representation $\alpha = \frac{a-b}{1+c}$. Now the usual division algorithm leads to a computable decimal representation $\alpha = g + \sum_{n=1}^{\infty} f_n \cdot 10^{-n}$.

If α is irrational, then we can obtain g and the numbers $f_1, \ldots, f_n$ as follows[2]. Let the computable sequence ϱ_n of rational numbers converge computably to α. Let m be any number which is greater than n. Then we can find an $l = l_m$ such that $|\varrho_l - \alpha| < 10^{-m}$. We produce the decimal expansion[3] of ϱ_l until the m-th place behind the decimal point:

$$\varrho_{l_m} = \ldots, a_1 a_2 a_3 \ldots a_n a_{n+1} \ldots a_m \ldots, \quad |\varrho_l - \alpha| < 10^{-m}.$$

We especially consider the places $a_{n+1} \ldots a_m$. If we do not have either of the critical cases that $a_{n+1} \ldots a_m \equiv 0 \ldots 0$ ("zero-case") or $a_{n+1} \ldots a_m \equiv 9 \ldots 9$ ("nine-case"), then the component of ϱ_{l_m} before the point represents the number g and $a_1 = f_1, \ldots, a_n = f_n$. However, in one of the critical cases we cannot make any such assertion.

Now, if for every $m > n$ we would have such a critical case, then it would be either the 0-case each time or the 9-case each time. Furthermore, every ϱ_{l_m} would start with the same beginning $a_1 a_2 a_3 \ldots a_n$. Thus, we would have in the 0-case

$$\alpha = \lim_{m\to\infty} \varrho_{l_m} = \lim \ldots, a_1 a_2 a_3 \ldots a_n 0 \ldots 0 a_{m+1} a_{m+2} \ldots$$
$$= \ldots, a_1 a_2 a_3 \ldots a_n$$

and in the 9-case

$$\alpha = \lim_{m\to\infty} \varrho_{l_m} = \lim \ldots, a_1 a_2 a_3 \ldots a_n 9 \ldots 9 a_{m+1} a_{m+2} \ldots$$
$$= \ldots, a_1 a_2 a_3 \ldots a_n + 10^{-n}.$$

In any case α would be rational, contrary to our hypothesis.

---

[1] Here again we use a non-constructive method of proof. This can be seen from the fact that in general we cannot tell (at least no procedure for it is known yet) whether a computably convergent sequence ϱ_n of rational numbers is rational or irrational. Otherwise we could for instance determine whether *Euler's constant* $\gamma = \lim_{n\to\infty} \left(\sum_{\nu=1}^{n} \frac{1}{\nu} - \log n \right)$ is rational or irrational. (It is easy to see that γ is a computable real number.)

[2] For the sake of simplicity we can further assume that $\alpha > 0$. This is so, since for $\alpha < 0$ there exists a natural number n such that $\beta = n + \alpha > 0$. β is also irrational. From the decimal expansion of β we can immediately obtain the decimal expansion of α.

[3] or *one* decimal expansion, if ϱ_1 has two expansions like for instance $\frac{1}{2} = 0.5000 \ldots = 0.4999 \ldots$.

Thus, there exists an $m > n$ for which we have neither the 0-case nor the 9-case. Such an m can be found by trying systematically $m = n+1, n+2, n+3, \ldots$. So we have found g and $f_1, \ldots, f_n$.

References

SPECKER, E.: Nicht konstruktiv beweisbare Sätze der Analysis. J. symbolic Logic **14**, 145–158 (1949).

MYHILL, J.: Criteria of Constructibility for Real Numbers. J. symbolic Logic **18**, 7–10 (1953).

GRZEGORCZYK, A.: On the Definition of Computable Functionals. Fundam. Math. **42**, 232–239 (1955).

KLAUA, D.: Konstruktive Analysis. Berlin: VEB Deutscher Verlag der Wissenschaften 1961.

AUTHOR AND SUBJECT INDEX

abacus 3
absolute difference 64
ACKERMANN 82, 84, 88, 171
Ackermann's function *84*, 88ff
AL CHWARIZMI 26
algebra of logic V, 28
algorithm Vf, *1*ff, 18ff, 26, 141, 175ff, 201, 231ff
—, Euclidean V, 2
—, idealization of 5
—, realization of 3
—, terminating 2
alphabet 3, 31, 38, 165
alternative 68
antinomy of the liar 29, 176
argument strip 95
ARISTOTLE 28
arithmetic, incompleteness of VI, 175
—, Peano's 30
—, undecidability of VI, 175
arithmetical formula *177*f
— predicate 177f, *179*, *192*ff
— predicates, hierarchy of 192
— sentence *177*f, 201
— term 177
arithmetization 4
ars inveniendi 27
ars iudicandi 27
ars magna 26ff
ASSER 234

basic logic 219
BERNAYS 66
BOOLE 28
BOONE 155
BRITTON 155

calculation, actual 5
calculus 2f, 117, 222
—, propositional 8
— in canonical form (Post) 231f
CANTOR 27
CARDANO 27
characteristic function *14*, *67*, 112
CHURCH 17, 29ff, 156, 163, 171, 176, 186, 206, 209, 219, 234
Church's thesis V, 17, 26, 30, 234
cleaning up machine 51, 52
complement 67, 189, 193
completeness theorem of Gödel 155
computable convergence 235
— decimal 238
— function VI, *9*ff, 17, 59, 61, 75, 82, 212
— real number 236
computing instruction 23
— machine 6
— procedure, non-periodic 43
— step 20, 32
— tape 19, 32, 103
configuration 20, *32*, 108, 149
—, complete 32
—, consecutive 32
—, initial 32
—, internal 25
—, starting 20
—, terminal 32
— word 149, 151
conjunction 67, 71
—, generalized 68
consequence 148
constant 156
— C_0^0 *60*, 122, 181, 212, 225
— C_0^W 37
— C_n^k 63
constructive proof VI, 104, 129
constructivity, theory of 9
copying machine 51, 53
correctness of rules 115, 120
corresponding classes in the Kleene-Mostowski hierarchy 192
— configurations 204
course of values recursion 81
COUTURAT 28
CURRY 9, 219, 234

DAVIS VII, 118, 202
decidability, absolute 12
—, relative 12
decidable VI, 12f
— relation 11
— relative to 12
— set 11, 17, 156
decimal notation 9
decision problem of the predicate calculus 29, 171
— procedure 12
deductable 8, 14
deduction 6ff
—, length of the 7
define 179
defining relation 147
definition by cases 72
δ-function 65
derivability in the λ-K-calculus 209
DESCARTES 27
D'ETLOV 234
diagonal procedure 85, 145, 198
diagrams of Turing machines 44
difference function 41
— index 115
digital computer 20
diophantine equation V, 200
DIRICHLET 207
divide 72
domain of individuals 157

elementary machine 39
empty predicate 69
Entscheidungsproblem of the predicate calculus 171
enumerable predicate 11, *187*ff
— relation *11*
— set *11*
enumeration theorem, Kleene's *112*, 189
ε-function 65
equal 72
equality relation 58
equation 115, 208
— calculus 117
erasure 20
essential undecidability 117
even 72
existential quantifier 10
exponent function 77

factorial function 64
false sentence 178
Fermat's conjecture 12
— predicate 191
FEYS 219
FINSLER 29f
FITCH 219, 230f
Fitch's calculus 222
flow diagram 42
follow 158
formula
— of the predicate calculus 156
— — of the second order 172
— of the propositional calculus 8
free occurrence 156, 159, 172, 177, 208
FREGE 28
F-term 116
function *207*
—, characteristic 14, *67*
—, computable VI, *9*ff, 18, 59, 61, 75, 82, 89, 212
—, λ-K-definable 211ff
—, μ-recursive 59ff, *88*ff
—, non-computable 141
—, non-μ-recursive 139
—, partially computable 10
—, primitive recursive *59*ff
— of zero arguments 37, 96

generable relation 14
— set VI, 14, 15
generalization 68
—, bounded 68f
general procedure 1ff
— recursive function 117
— word problem for groups 146
— word problem for semi-Thue systems 148
— word problem for Thue systems 149
generated vid. generable
GÖDEL 9, 29, 30, 65, 113, 118, 155, 163, 171, 175, 176, 186
Gödel number 4
— numbering *4*, 9, 17, 91, 130f
— numbering of Turing machines 103
Gödel's completeness theorem 28
— predicate 185
GOODSTEIN 187
group system 148f
GRZEGORCZYK 187, 240

halt 20, 33
HASENJAEGER 163, 171
HERBRAND 113, 118

HERMES 163, 231
hierarchy (Kleene-Mostowski) 192ff
HILBERT 66, 82, 202
Hilbert's tenth problem V, 30, 141, 200ff

identification 68
— of variables 61
identity functions U_n^i *60*
immediate consequence 147
implication 68
incompleteness of arithmetic 175
— of the predicate calculus of the second order 29, 171ff
individual 157
— variable 156
induction axiom 176
— procedure 62
— schema *60*, 176
initial function 60
integer 9
integraph 20
internal configuration 25
interpretation 120, 178
interprete 155
inverse relation 148

jump 22

KAHR 171
KALMÁR 17, 26, 140, 155, 171, 187
KEMENY 26
KLAUA 240
KLEENE VII, 29, 30, 66, 89, 93, 97, 112, 113, 118, 129, 130, 140, 192, 202, 219, 240
Kleene-Mostowski hierarchy 192ff
Kleene's enumeration theorem *112*, 189
— normal form theorem 97, 111

λ-calculus 206, 208
λ-K-calculus 206ff
λ-K-convertibility 209
λ-K-definability 206ff
λ-K-definable function *211*ff
λ-operator 207
large left machine 49, 50
— right machine 49, 50
left 20
— end machine 49, 50
— machine 39
— parentheses 208
left search machine 49, 50
— translation machine 51, 52
LEIBNIZ V, 27
letter 3
liar, antinomy of the 29, 176
logic, algebra of 28
—, basic 219
—, modern 28
LORENZEN 9, 223, 231, 232
Löwenheim and Skolem, theorem of 173
LULLUS 26

machine 18f
— a_j 39f
— **C** 51, 52
— **K** 51, 53
— K_n 51, 54
— l 39
— **L** 49, 50
— $\mathfrak{L}$ 49, 50
— **λ** 49, 50
— **r** 39
— **R** 49, 50
— $\mathfrak{R}$ 49, 50
— **ρ** 49, 50
— **S** 49, 50
— **σ** 51, 52
— **T** 51, 52
— table 25
— word 143
MAHN VII
MARKOV 147, 155, 231, 233f
max 74
MENNINGER 9
min 74
minimal basic logic 219
minimal logic of Fitch 219f
modified difference 64
modus ponens 8, 232
MOORE 171
MOSTOWSKI 177, 187, 192, 202
μ-operator 89
—, bounded 74f
—, unbounded 74
μ-recursive function 59ff, *88*ff, 118, 130
— predicate *88*ff
μ-recursiveness VI, 59, 89, 93ff
MYHILL 240

n-copying machine 51, 54
negation 67
normal algorithm (Markov) 233
— calculus in canonical form (Post) 232

normal form theorem 97, 111
— word 153
notation, decimal 9
NOVIKOV 155
number, natural 9
—, rational 9
—, real 9f, 236
— representation 5
numeral 115

odd 72

partially computable function 10
— recursive function 118
particularization 68
—, bounded 69
PEANO 28
Peano axioms 176
Peano's arithmetic 30
permutation of variables 61
PÉTER 18, 26, 66, 78
π-permutation 67
place index 115
POST VI, 26, 30, 35, 147, 155, 192, 231, 234
power function 64
predecessor function 64
predicate, arithmetical *192*ff
—, decidable 11
—, enumerable 11, *187*ff
—, μ-recursive *88*ff
—, primitive recursive *66*ff
—, regular 75
—, undecidable 141ff
— calculus VI, *155*
— calculus, decision problem 29
— calculus, undecidability VI, 163ff
— calculus of the first order 172
— calculus of the second order 172
— variable 156
prime function 77
— number 72
primitive recursion 66
— recursive function 59, *60*f, 65
— recursive predicate *66*ff
print 21, 32
procedure, general 1ff
process Π 65
— Σ 65
product function 57, 64
PUTNAM 202

quantification, bounded 71
—, unbounded 71
quotient 76

RAYMUNDUS LULLUS vid. LULLUS
recursion schema 60
recursive analysis 234ff
— decidability 113
— enumerability 113
— function VI, 65, 113, *117*
— predicate 113
regular function *89*, 101
— predicate *75*, 88
— system of equations 122
relation, decidable 11
—, empty 11
—, enumerable 11
—, finite 13
—, generable 14
replacement rule (RR) 117, 120
right 20
— end machine 49, 50
— machine 39
— search machine 49, 50
ROBINSON 177, 187, 192
ROGERS JR. VII
ROSSER 176f, 186, 189, 192
rule of inference (Post) 231
rule (RR) 114, 116, 120
— (SR) 114, 116, 120
— system VI, 7f
RUSSELL 28

SCHOLZ 163
SCHÖNFINKEL 207
search machine 49, 50
semantics 129, 155, 157, 176
semi-group 147
semi-Thue system 30, *147*, 187
sentence 48
sequence, computably convergent 236
—, convergent 235
set, decidable 11
—, denumerable 11
—, empty 11
—, enumerable 11
—, finite 13
—, generable 14
sg-function 64
$\overline{sg}$-function 64
shift 34
shifting machine 51, 52
σ-function 77ff

sign 3
situation (Turing machines) 20
SKOLEM 187
SMULLYAN 192
soroban 3
SPECKER 238, 240
square, empty 20, 32
—, initial, choice of 55
—, marked 20, 48
—, observed 22
—, scanned 21
standard decomposition of a word 233
— definable function 119
state 32
—, initial 32
—, terminal 32
state of mind 25
stop 33
substitution 69, 161
—, simultaneous 220
— for a variable 62
— formula 233
— in a parameter 78, 80
— operation 208
— process, general 61
— rule (SR) 116, 120
— schema 60
sub-word 233
successor function 57, 60
sum function 57, 64
superimposed 8
SURÁNYJ 171
symbol 3
—, actual 20
—, empty 20
—, ideal 20
— of a diagram, initial 45
— of a Turing machine 25
syntactic 129

tape 19
—, a half 95
— expression 20, 32
TARSKI 30, 163, 176f, 186f
term 115, 177, 208
theorem of Löwenheim and Skolem 173
THUE 155
Thue system 30, 145, *148*f, 154f, 187
TRACHTÉNBROT 171

true sentence 178
TURING 19, 26, 30, 155, 206, 219
Turing-computability 31, 35f, *36*, 55, 59, *93*ff, 98ff, 108, 235
—, standard 95ff, 98ff
Turing-decidability 31, 35, *37*f, 55, 58
— relative to M_1 37
Turing-enumerability 31, 35f, 37f
Turing-enumerable 37
Turing machine VI, 17ff, 25, *31*ff, 34f, 44, 48, 103ff, 110, 142, 203
—, universal 25, 187, *203*ff
—, placed behind a word 34
—, placed on a tape 33
Turing machines and periodicity 55
Turing machines, combination of 44ff
—, diagrams of 44
—, equivalence of 34
—, Gödel numbering of 103
—, interchangeable 47
Turing table 25

undecidability, essential 177
— of arithmetic VI, 175
undecidable predicates 141ff
universal machine 150
— predicate 69
— Turing machine 25, 187, *203*ff

valid 119, 120, 155, 157, 172, 178
variables of the first order 172
— of the second order 172

WANG 26, 171
WHITEHEAD 28
word 3
—, empty *3*, 5
—, ideal 36
— over $\mathfrak{A}$ 3
— problem for groups Vf, 30, 141
— problem for groups, general 146
— problem for semi-Thue systems 145ff
— problem for semi-Thue systems, general 148
— problem for Thue systems VI, 145ff
— problem for Thue systems, general 149

Brühlsche Universitätsdruckerei Gießen

Die Grundlehren der mathematischen Wissenschaften in Einzeldarstellungen mit besonderer Berücksichtigung der Anwendungsgebiete

Lieferbare Bände:

2. Knopp: Theorie und Anwendung der unendlichen Reihen. DM 48,— ; US $ 12.00
3. Hurwitz: Vorlesungen über allgemeine Funktionentheorie und elliptische Funktionen. DM 49,— ; US $ 12.25
4. Madelung: Die mathematischen Hilfsmittel des Physikers. DM 49,70; US $ 12.45
10. Schouten: Ricci-Calculus. DM 58,60; US $ 14.65
14. Klein: Elementarmathematik vom höheren Standpunkt aus. 1. Band: Arithmetik Algebra. Analysis. DM 24,— ; US $ 6.00
15. Klein: Elementarmathematik vom höheren Standpunkt aus. 2. Band: Geometrie. DM 24,— ; US $ 6.00
16. Klein: Elementarmathematik vom höheren Standpunkt aus. 3. Band: Präzisions- und Approximationsmathematik. DM 19,80; US $ 4.95
19. Pólya/Szegö: Aufgaben und Lehrsätze aus der Analysis I: Reihen, Integralrechnung, Funktionentheorie. DM 34,— ; US $ 8.50
20. Pólya/Szegö: Aufgaben und Lehrsätze aus der Analysis II: Funktionentheorie, Nullstellen, Polynome, Determinanten, Zahlentheorie. DM 38,— ; US $ 9.50
22. Klein: Vorlesungen über höhere Geometrie. DM 28,— ; US $ 7.00
26. Klein: Vorlesungen über nicht-euklidische Geometrie. DM 24,— ; US $ 6.00
27. Hilbert/Ackermann: Grundzüge der theoretischen Logik. DM 38,— ; US $ 9.50
30. Lichtenstein: Grundlagen der Hydromechanik. DM 38,— ; US $ 9.50
31. Kellogg: Foundations of Potential Theory. DM 32,— ; US $ 8.00
32. Reidemeister: Vorlesungen über Grundlagen der Geometrie. DM 18,— ; US $ 4.50
38. Neumann: Mathematische Grundlagen der Quantenmechanik. DM 28,–; US $ 7.00
40. Hilbert/Bernays: Grundlagen der Mathematik I. DM 68,— ; US $ 17.00
50. Hilbert/Bernays: Grundlagen der Mathematik II. DM 68,— ; US $ 17.00
52. Magnus/Oberhettinger/Soni: Formulas and Theorems for the Special Functions of Mathematical Physics. DM 66,— ; US $ 16.50
57. Hamel: Theoretische Mechanik. DM 84,— ; US $ 21.00
58. Blaschke/Reichardt: Einführung in die Differentialgeometrie. DM 24,— ; US $ 6.00
59. Hasse: Vorlesungen über Zahlentheorie. DM 69,— ; US $ 17.25
60. Collatz: The Numerical Treatment of Differential Equations. DM 78,— ; US $ 19.50
61. Maak: Fastperiodische Funktionen. DM 38,— ; US $ 9.50
62. Sauer: Anfangswertprobleme bei partiellen Differentialgleichungen. DM 41,— ; US $ 10.25
64. Nevanlinna: Uniformisierung. DM 49,50; US $ 12.40
66. Bieberbach: Theorie der gewöhnlichen Differentialgleichungen. DM 58,50; US $ 14.65
68. Aumann: Reelle Funktionen. DM 59,60; US $ 14.90
69. Schmidt: Mathematische Gesetze der Logik I. DM 79,— ; US $ 19.75
71. Meixner/Schäfke: Mathieusche Funktionen und Sphäroidfunktionen mit Anwendungen auf physikalische und technische Probleme. DM 52,60; US $ 13.15
73. Hermes: Einführung in die Verbandstheorie. DM 46,— ; US $ 11.50
74. Boerner: Darstellungen von Gruppen. DM 58,— ; US $ 14.50
75. Rado/Reichelderfer: Continuous Transformations in Analysis, with an Introduction to Algebraic Topology. DM 59,60; US $ 14.90

76. Tricomi: Vorlesungen über Orthogonalreihen. DM 37,60; US $ 9.40
77. Behnke/Sommer: Theorie der analytischen Funktionen einer komplexen Veränderlichen. DM 79,—; US $ 19.75
78. Lorenzen: Einführung in die operative Logik u. Mathematik. DM 54,—; US $ 13.50
79. Saxer: Versicherungsmathematik. 1. Teil. DM 39,60; US $ 9.90
80. Pickert: Projektive Ebenen. DM 48,60; US $ 12.15
81. Schneider: Einführung in die transzendenten Zahlen. DM 24,80; US $ 6.20
82. Specht: Gruppentheorie. DM 69,60; US $ 17.40
84. Conforto: Abelsche Funktionen u. algebraische Geometrie. DM 41,80; US $ 10.45
86. Richter: Wahrscheinlichkeitstheorie. DM 68,—; US $ 17.00
87. van der Waerden: Mathematische Statistik. DM 49,60; US $ 12.40
88. Müller: Grundprobleme der mathematischen Theorie elektromagnetischer Schwingungen. DM 52,80; US $ 13.20
89. Pfluger: Theorie der Riemannschen Flächen. DM 39,20; US $ 9.80
90. Oberhettinger: Tabellen zur Fourier Transformation. DM 39,50; US $ 9.90
91. Prachar: Primzahlverteilung. DM 58,—; US $ 14.50
92. Rehbock: Darstellende Geometrie. DM 29,—; US $ 7.25
93. Hadwiger: Vorlesungen über Inhalt, Oberfläche und Isoperimetrie. DM 49,80; US $ 12.45
94. Funk: Variationsrechnung und ihre Anwendung in Physik und Technik. DM 98,—; US $ 24.50
95. Maeda: Kontinuierliche Geometrien. DM 39,—; US $ 9.75
97. Greub: Lineare Algebra. DM 39,20; US $ 9.80
98. Saxer: Versicherungsmathematik. 2. Teil. DM 48,60; US $ 12.15
99. Cassels: An Introduction to the Geometry of Numbers. DM 69,—; US $ 17.25
100. Koppenfels/Stallmann: Praxis der konformen Abbildung. DM 69,—; US $ 17.25
101. Rund: The Differential Geometry of Finsler Spaces. DM 59,60; US $ 14.90
103. Schütte: Beweistheorie. DM 48,—; US $ 12.00
104. Chung: Markov Chains with Stationary Transition Probabilities. DM 56,—; US $ 14.00
105. Rinow: Die innere Geometrie der metrischen Räume. DM 83,—; US $ 20.75
106. Scholz/Hasenjaeger: Grundzüge der mathematischen Logik. DM 98,—; US $ 24.50
107. Köthe: Topologische Lineare Räume I. DM 78,—; US $ 19.50
108. Dynkin: Die Grundlagen der Theorie der Markoffschen Prozesse. DM 33,80; US $ 8.45
109. Hermes: Aufzählbarkeit, Entscheidbarkeit, Berechenbarkeit. DM 49,80; US $ 12.45
110. Dinghas: Vorlesungen über Funktionentheorie. DM 69,—; US $ 17.25
111. Lions: Equations différentielles opérationnelles et problèmes aux limites. DM 64,—; US $ 16.00
112. Morgenstern/Szabó: Vorlesungen über theoretische Mechanik. DM 69,—; US $ 17.25
113. Meschkowski: Hilbertsche Räume mit Kernfunktion. DM 58,—; US $ 14.50
114. MacLane: Homology. DM 62,—; US $ 15.50
115. Hewitt/Ross: Abstract Harmonic Analysis. Vol. 1: Structure of Topological Groups. Integration Theory. Group Representations. DM 76,—; US $ 19.00
116. Hörmander: Linear Partial Differential Operators. DM 42,—; US $ 10.50
117. O'Meara: Introduction to Quadratic Forms. DM 48,—; US $ 12.00
118. Schäfke: Einführung in die Theorie der speziellen Funktionen der mathematischen Physik. DM 49,40; US $ 12.35
119. Harris: The Theory of Branching Processes. DM 36,—; US $ 9.00
120. Collatz: Funktionalanalysis und numerische Mathematik. DM 58,—; US $ 14.50
121.
122. Dynkin: Markov Processes. DM 96,—; US $ 24.00
123. Yosida: Functional Analysis. DM 66,—; US $ 16.50

124. Morgenstern: Einführung in die Wahrscheinlichkeitsrechnung und mathematische Statistik. DM 38,— ; US $ 9.50
125. Itô/McKean: Diffusion Processes and Their Sample Paths. DM 58,— ; US $ 14.50
126. Letho/Virtanen: Quasikonforme Abbildungen. DM 38,— ; US $ 9.50
127. Hermes: Enumerability, Decidability, Computability. DM 39,— ; US $ 9.75
128. Braun/Koecher: Jordan-Algebren. DM 48,— ; US $ 12.00
129. Nikodým: The Mathematical Apparatus for Quantum-Theories. DM 144,— ; US $ 36.00
130. Morrey: Multiple Integrals in the Calculus of Variations. DM 78,— ; US $ 19.50
131. Hirzebruch: Topological Methods in Algebraic Geometry. DM 38,— ; US $ 9.50
132. Kato: Perturbation Theory for Linear Operators. DM 79,20; US $ 19.80
133. Haupt/Künneth: Geometrische Ordnungen. DM 68,— ; US $ 17.00
134. Huppert: Endliche Gruppen I. DM 156,— ; US $ 39.00
135. Handbook for Automatic Computation .Vol. 1/Part a: Rutishauser: Description of ALGOL 60. DM 58,— ; US $ 14.50
136. Greub: Multilinear Algebra. DM 32,— ; US $ 8.00
137. Handbook for Automatic Computation. Vol. 1/Part b: Grau/Hill/Langmaack: Translation of ALGOL 60. DM 64,— ; US $ 16.00
138. Hahn: Stability of Motion. DM 72,— ; US $ 18.00
139. Mathematische Hilfsmittel des Ingenieurs. Herausgeber: Sauer/Szabó. 1. Teil. DM 88,— ; US $ 22.00
140. Mathematische Hilfsmittel des Ingenieurs. Herausgeber: Sauer/Szabó. 2. Teil. DM 136,— ; US $ 34.00
141. Mathematische Hilfsmittel des Ingenieurs. Herausgeber: Sauer/Szabó. 3. Teil. DM 98,— ; US $ 24.50
142. Mathematische Hilfsmittel des Ingenieurs. Herausgeber: Sauer/Szabó. 4. Teil. In Vorbereitung
143. Schur/Grunsky: Vorlesungen über Invariantentheorie. DM 32,— ; US $ 8.00
144. Weil: Basic Number Theory. DM 48,— ; US $ 12.00
145. Butzer/Berens: Semi- Groups of Operators and Approximation. DM 56,— ; US $ 14.00
146. Treves: Locally Convex Spaces and Linear Partial Differential Equations. DM 36,— ; US $ 9.00
147. Lamotke: Semisimpliziale algebraische Topologie. DM 48,— ; US $ 12.00
148. Chandrasekharan: Introduction to Analytic Number Theory. DM 28,–; US $ 7.00
149. Sario/Oikawa: Capacity Functions. DM 96,— ; US $ 24.00
150. Iosifescu/Theodorescu: Random Processes and Learning. DM 68,— ; US $ 17.00
151. Mandl: Analytical Treatment of One-dimensional Markov Processes. DM 36,— ; US $ 9.00
152. Hewitt/Ross: Abstract Harmonic Analysis. Vol. II. In preparation
153. Federer: Geometric Measure Theory. DM 118,— ; US $ 29.50
154. Singer: Bases in Banach Spaces. In preparation
155. Müller: Foundations of the Mathematical Theory of Electromagnetic Waves. DM 58,— ; US $ 14.50
156. van der Waerden: Mathematical Statistics. In preparation
157. Prohorov/Rozanov: Probability Theory. DM 68,— ; US $ 17.00
158. Constantinescu/Cornea: Potential Theory on Harmonic Spaces. In preparation
159. Köthe: Topological Vector Spaces I. In preparation
160. Agrest/Maksimov: Theory of Incomplete Cylindrial Functions and their Applications. In preparation
161. Bhatia/Szegö: Stability Theory of Dynamical Systems. In preparation
162. Nevanlinna: Analytic Functions. In preparation
163. Stoer/Witzgall: Convexity and Optimization in Finite Dimensions I. In preparation